FATS IN FOOD PRODUCTS

Profile: David P. J. Moran BSc PhD
Dr David Moran, a consultant to the industry, has spent over 30 years working in the area of edible fats, oils and emulsions, and has cooperated in many related patents, particularly on the topic of yellow spreads and reduced fat varieties.

Profile: Kanes K. Rajah BSc MSc PhD
Dr Kanes Rajah has extensive experience of the edible oils industry, having held various positions with Kraft Foods (UK) Limited (now Kraft General Foods), Kraft R & D Europe Inc., Dairy Crest Foods (Milk Marketing Board of England & Wales) and Ernest George Limited. He is now a consultant to the edible oils and dairy industries on technology and business development.

FATS IN FOOD PRODUCTS

Edited by

D. P. J. MORAN
Jubilee Lodge, Keyston Road, Covington, Cambridgeshire, UK

and

K. K. RAJAH
74 Chadwick Road, Westcliff-on-Sea, Essex, UK

BLACKIE ACADEMIC & PROFESSIONAL
An Imprint of Chapman & Hall
London · Glasgow · New York · Tokyo · Melbourne · Madras

Published by
Blackie Academic and Professional, an imprint of Chapman & Hall,
Wester Cleddens Road, Bishopbriggs, Glasgow G64 2NZ

Chapman & Hall, 2–6 Boundary Row, London SE1 8HN, UK

Blackie Academic & Professional, Wester Cleddens Road, Bishopbriggs, Glasgow G64 2NZ, UK

Chapman & Hall Inc., One Penn Plaza, 41st Floor, New York NY 10119, USA

Chapman & Hall Japan, Thomson Publishing Japan, Hirakawacho Nemoto Building, 6F, 1-7-11 Hirakawa-cho, Chiyoda-ku, Tokyo 102, Japan

DA Book (Aust.) Pty Ltd, 648 Whitehorse Road, Mitcham 3132, Victoria, Australia

Chapman & Hall India, R. Seshadri, 32 Second Main Road, CIT East, Madras 600 035, India

First edition 1994

© 1994 Chapman & Hall

Typeset in 10/12 pt Times by Alden Multimedia, Northampton
Printed in Great Britain by The Alden Press, Oxford

ISBN 07514 0177 3

Apart from any fair dealing for the purposes of research or private study, or criticism or review, as permitted under the UK Copyright Designs and Patents Act, 1988, this publication may not be reproduced, stored, or transmitted, in any form or by any means, without the prior permission in writing of the publishers, or in the case of reprographic reproduction only in accordance with the terms of the licences issued by the Copyright Licensing Agency in the UK, or in accordance with the terms of licences issued by the appropriate Reproduction Rights Organization outside the UK. Enquiries concerning reproduction outside the terms stated here should be sent to the publishers at the Glasgow address printed on this page.

The publisher makes no representation, express or implied, with regard to the accuracy of the information contained in this book and cannot accept any legal responsibility or liability for any errors or omissions that may be made.

A catalogue record for this book is available from the British Library

Library of Congress Cataloging-in-Publication data

Fats in food products / edited by D. P. J. Moran and K. K. Rajah.
    p.   cm.
  Includes bibliographical references and index.
  ISBN 1-85861-006-0
  1. Food—Composition. 2. Food—Fat content. I. Moran, D. P. J.
II. Rajah, Kanes K.
TP372.5.F38 1994
664'.3—dc20   93-18951
                                                                             CIP

∞ Printed on acid-free text paper, manufactured in accordance with ANSI/NISO Z39.48-1992 (Permanence of Paper)

Contents

Preface xiii

List of Contributors xv

1. **Physical Chemistry of Fats**
   R. E. TIMMS 1
   Introduction 1
   Basic Principles 4
      Polymorphism 4
      Phase Behaviour 5
      Melting Point and Solid Fat Content 9
   Physical Properties 11
      Milkfat and Fractions 11
      Mixtures of Milkfat with Other Fats 14
      Solubility of Milkfat in Liquid Oils 16
   Crystallisation 17
      Supersaturation 17
      Nucleation 19
      Crystal Growth 20
      Fractional Crystallisation 21
   Interfacial Properties 22
      Interaction of Lipids with Water 22
      Emulsion Stability 23
   References 24

2. Fats in Cream and Ice Cream
M. ANDERSON, E. C. NEEDS and J. K. MADDEN   29
- Introduction 30
- Interfaces 30
  - Biological Membrane 31
  - Milk Protein Interface 32
  - The Composite Interface 35
- Fat Crystallisation 39
  - Emulsion Stability and Functionality 40
- The Continuous Phase 44
  - Proteins. 44
  - Protein/Polysaccharide Interactions 46
  - Polysaccharide Interactions 47
  - Ice Crystals 48
- Processing Variables 50
  - Homogenisation 50
  - Separation 52
- Product Considerations 53
  - Low-fat Creams 54
  - Whipping Cream. 57
  - Ice Cream 60
- References. 63

3. Butter and Allied Products
R. S. JEBSON 69
- Cream Handling 69
- Cream Treatment 70
  - Neutralization 70
  - Pasteurization. 70
  - Taint Removal 71
  - Cream Cooling and Holding. 76
- Buttermaking 77
  - Theories of Churning 77
  - Fritz Process Continuous Buttermaking Equipment . . 78
  - Recombined Buttermaking 81
- Continuous Buttermaking Variables 82
  - Machine Variables 83
  - Cream Variables 86
- Moisture Measurement and Control 88
  - Methods Based on Heat 88

Salting	89
Storage	90
Salt Slurries	90
Particle Size	90
Incorporation into Butter	91
Packing and Handling	91
Transport	91
Buffer Storage	91
Bulk Packing	92
Patting	96
Palletization	96
Setting and Reworking	96
Setting	96
Reworking	97
Flavour of Butter	98
Feed of the Cows	98
Processing Faults	98
Flavour Defects Caused by Micro-Organisms	99
Oxidative Flavours	100
Colour of Butter	101
Streaky Colour	101
Spotted Colour	101
Primrose Colour	102
Body and Texture	102
Crumbly or Brittle Body	102
Sticky Body	102
Leaky Body	102
Porous or Ice-Cream Texture	102
Types of Butter	103
Lactic Butter	103
Softer Butters	104
References	106

4. Anhydrous Milkfat Products and Applications in Recombination

D. Illingworth and T. G. Bissell	111
Anhydrous Milkfat—Definition, Main Uses and World Production	112
Manufacturing Processes for AMF	114
Direct-from-Cream Process	114
The Principles of Phase Inversion	117

Manufacture from Butter 123
Production of Ghee and Samna Baladay 125
Plant and Equipment used for AMF Manufacture . . . 125
    Phase Inversion Devices 125
    Separation Devices 127
Dehydration of Milkfat 131
Monitoring the Performance of an AMF Plant . . . 134
Quality Assessment of Anhydrous Milkfat 135
Packaging and Handling of Milkfat 137
    Post-dehydration Handling 137
    Measurement of Dissolved Oxygen in Milkfat . . . 137
    Packaging of AMF for Industrial Use. 138
    Consumer Packaging of AMF 139
Use of AMF for Recombining 141
    Recombined Butter and Other Spreads 142
    Other Blends of Milkfat with Vegetable Oils . . . 146
    Recombined Ice Cream 147
    Recombined Milks and Creams 147
    Milkfat in Other Recombined Products 149
References. 149

5. Fats in Spreadable Products
D. P. J. MORAN 155
Introduction 156
Rheology of Fats and Spreads 157
    Fat Crystallisation 157
    Phase Behaviour 159
    Fat Crystal Size 160
    The Crystal Network 161
    The Aqueous Phase 162
    The Structure of Spreads 163
    The Emulsion 165
    Functions of Surface Active Lipids 167
    Defining Spreads Structure 168
Spreads Processing 169
    General Aspects 169
    Churning of Butter Type Products 171
    Use of Scraped Wall Heat Exchangers and Stirred Crystallisers 173
    Relationship of Processing and Product Structure . . . 175

	Alterations in the Properties of Spreads	179
	Process Control	181
	Performance of Spreads	181
	Measurement of Solids	181
	Tests for Spreads Structure	183
	Appearance of Spreads	185
	Effect of Fats on Product Texture	186
	Palate Behaviour of Spreads	187
	Other Functions of Spreads	189
	Higher Fat Spreads (Patent Review)	189
	Structure Control	189
	Emulsifiers	191
	Double Emulsion Products	191
	Miscellaneous Developments	191
	Health Aspects	191
	Fat Substitutes	192
	Low Fat Spreads (Patent Review)	194
	Fat Continuous Products	194
	Water Continuous Products	200
	Very Low Fat Spreads	202
	Products with Less than 40% Fat	202
	The Future	204
	Nutritionally Adapted Spreads	204
	Butter and Milk/Vegetable Fat Products	205
	Multifunctional Spreads	205
	Natural Spreads	206
	Novel Textured Spreads	206
	Processing	206
	Research and Development	206
	References	207
6.	**Fats in Bakery and Kitchen Products**	
	J. PODMORE	213
	Introduction	214
	Production of Margarine and Shortening	215
	Crystallisation Behaviour	216
	Processing	219
	Bakery Fats	220
	Short Pastry	221
	Cake	223

x   Contents

Puff Pastry 227
The Influence of Emulsifiers in Baking 230
Control of Quality in Margarine and Shortening Manufacture 232
Liquid Shortenings 234
Fluid Shortenings 235
Powdered Fats, Flaked Fats and Fat Powders 236
    Methods of Manufacture 236
    Applications of Fat Powders and Powdered Fats 240
Ghee 242
    Ghee Quality 243
    Uses of Ghee 245
Vanaspati 245
Salad and Cooking Oils 246
Frying Fats and Oils 247
Concluding Remarks 251
References 252

7. Milkfat in Sugar and Chocolate Confectionery
V. K. S. SHUKLA 255
Introduction 255
Composition of Milk 256
Milkfat 257
    Fatty Acid Composition 257
    Seasonal and Geographical Variation 258
    Trans Fatty Acids in Milkfat 259
    Flavour Contribution 259
Applications of Milkfat in Confectionery 259
Milk Chocolate 272
Plain Chocolate 273
Toffee 274
Conclusion 275
References 275

8. Fat Products Using Fractionation and Hydrogenation
K. K. RAJAH 277
Part 1: Fractionation 278
Introduction 278
Fractionation Processes 278
    Detergent Fractionation 279
    Solvent Fractionation 279
    Dry Fractionation 279

Main Functional Properties	281
Flavour and Taste	281
Colour	284
Application for Fractionated Fats	284
Frying Oils/Salad Oils/Salad Dressing	284
Cream	285
Ice-Cream	285
Bakery Products	286
Infant Feeding	290
Butter Powder	290
Sauce Improver	290
Yellow Fat Spreads	291
Chocolate	292
Part 2: Hydrogenation	293
Introduction	293
The Hydrogenation Process	294
Catalyst Activity	294
Selectivity	295
Mass Transfer	296
Re-use of Catalyst	296
Effect of Catalyst Type on Selectivity and Product Characteristics	297
Hydrogenation for Food Applications	299
Salad Oil	299
Cooking and Frying Products	304
Bakery Coatings	305
Emulsions and Spreads	305
Chocolate and Confectionary Products	307
Hydrogenated Milkfat Products	309
References	313

9. **Fat Products Using Chemical and Enzymatic Interesterification**
   A. HUYGHEBAERT, D. VERHAEGHE and H. DE MOOR . . . 319

Introduction	320
Chemical Interesterification	323
Non-Milkfat Chemical Interesterification	325
Milkfat Chemical Interesterification	327
Enzymatic Interesterification	333
Enzyme Catalysed Interesterification of Non-Milkfats	333
Enzyme Catalysed Interesterification of Milkfat	339
References	340

10. Flavours Derived from Fats
G. URBACH and M. H. GORDON 347
Part 1: Milkfat as a Source of Flavour (G. Urbach) . . . 347
Introduction 347
Cultured Butter 348
Precursor Flavours 349
    Hydroxyacids 349
    β-Ketoacids 357
    Lower Fatty Acids 360
    Unsaturated Fatty Acids 362
Off-flavours from Irradiated Fats 368
Effect of Anti-oxidants 369
Flavours Not Formed Directly from the Fat 370
Cooked (H_2S) Flavour 373
Neutral Volatiles 373
Flavours of Butterfat Fractions 375
Supercritical Extraction of Butter Flavour 375
Butter Powder 376
Ghee 376
Vologda Butter 377
Heated Butter 377
Effect of Storage on Butter Flavour 377
Part 2: Flavour of Fats Other than Milkfat (M. H. Gordon) 378
Introduction 378
Cocoa Butter 378
Olive Oil 380
Other Fats 381
Oil Degradation Products and Flavour 384
Autoxidation Mechanism 384
    Formation of Volatile Products by Autoxidation . . . 385
Factors Affecting the Rate of Off-flavour Development . . 388
    Reversion Flavours 390
    Ketonic Rancidity 390
    Hydrolytic Rancidity 390
Conclusions 390
References 391
Bibliography 401

Index 407

Preface

The properties of fats and the characteristics of some food products based on fats have been documented in several books. Individual fats such as milkfat, however, have received less attention despite many successful initiatives to increase their utilization in food products. Moreover, the availability of data on the function of fats in the context of major manufactured food products has often been constrained by the general reluctance of manufacturers to disclose details of working practices.

In some areas, such as yellow fat spreads, the market has changed dramatically over the last decade or so by the introduction of a broad class of new products resulting from a trend among consumers in the developed world towards reduced fat consumption. A review of this general area therefore now seems very timely.

In the preparation of this book, we have been fortunate to have had the support of internationally recognised specialists with much relevant experience and achievement in their subject areas. We believe that their contributions not only subscribe to the main aim of this book, by providing useful insight into the functional properties of the major fats in foods, but also offer information concerning recent and novel methods of processing these fats. Opportunities for possible future developments are indicated throughout.

The book begins with a discussion of the physical chemistry of fats, covering aspects such as phase behaviour, polymorphism, crystallisation and interfacial properties, all of which are fundamental to an understanding of the performance of fats in a variety of food applications.

Chapter 2 discusses properties of cream and ice cream, and also the influence of interfaces on these. It includes key subject areas such as the contribution of major ingredients and processing.

In recent years, there have been considerable advances in and optimisation of buttermaking technology. Chapters 3 and 4 give an up-to-date account of this, and the related topic of the preparation and use of anhydrous milkfat.

An up-to-date account of the rapidly evolving area of yellow fat spreads, including low fat spreads, possible future developments and a patent review, is the subject of chapter 5, while chapter 6 covers the application of fats in baked products.

Most chocolate produced worldwide is of the milk chocolate variety, and hence chapter 7 is devoted to the influence of milkfat in that context.

In product usage, many fats have to be modified to achieve the desired product characteristics. Chapter 8 deals with the fractionation and hydrogenation of fats, including methods, resultant properties and key applications. The patent literature is reviewed and milkfat is also described within the context of fat modification.

The subject of enzymatic and chemical interesterification is extremely relevant (chapter 9), especially as enzymatic techniques are of increasing significance as an option for processing edible fats with specific physical properties.

Finally, fats not only provide structure to products but can be dominant in their contribution to taste and flavour, including off-flavour. Chapter 10 discusses the flavour impact of milkfat and other fats under two separate headings.

Each chapter begins with a brief summary to allow the reader rapid access to the scope of the subject matter.

The book is directed at those involved with the technology of the subject—especially scientists and technologists in product development, production managers, quality assurance staff and managers with appropriate responsibilities in procurement, marketing and business development. Lecturers and senior students of food science and technology will find the book a useful source of reference as well as a means of providing general insight into the manipulation of fats in modern food manufacturing.

D.D.J.M.
K.K.R.

List of Contributors

M. ANDERSON
AFRC, Polaris House, North Star Avenue, Swindon SN2 1UH, UK.

H. DE MOOR
Laboratory of Food Technology, Chemistry and Microbiology, Faculty of Agricultural Sciences, State University of Ghent, Coupure 653, B-9000 Gent, Belgium.

M. H. GORDON
Department of Food Science & Technology, University of Reading, Whiteknights, PO Box 226, Reading RG6 2AP, UK.

A. HUYGHEBAERT
Laboratory of Food Technology, Chemistry and Microbiology, Faculty of Agricultural Sciences, State University of Ghent, Coupure 653, B-9000 Gent, Belgium.

D. ILLINGWORTH
Milkfat & Butter Section, New Zealand Dairy Research Institute, Private Bag 11029, Palmerston North, New Zealand.

R. S. JEBSON
Department of Food Technology, Massey University, Faculty of Technology, Palmerston North, New Zealand.

J. K. MADDEN
United Biscuits (UK) Limited, Group R&D Centre, Lane End Road, Sands, High Wycombe, Buckinghamshire HP12 4JX, UK.

D. P. J. MORAN
(Consultant—Yellow Spreads and Fats Technology) Jubilee Lodge, Keyston Road, Covington, Cambridgeshire PE18 0RU, UK.

E. C. NEEDS
AFRC, Institute of Food Research, Reading Laboratory, Earley Gate, Whiteknights Road, Reading RG6 2EF, UK.

J. PODMORE
Research & Development Centre, Pura Food Products Ltd, Dunnings Bridge Road, Bootle, Merseyside L30 6TJ, UK.

K. K. RAJAH
(Consultant—(Oils, Fats, Dairy and Business Development) 74 Chadwick Road, Westcliff-on-Sea, Essex SS0 8LD, UK.

V. K. S. SHUKLA
International Food Science Centre A/S, PO Box 44, Sønderskovvej 7, DK-8520 Lystrup, Denmark.

R. E. TIMMS
(Consultant—Oils and Fats), The Cottages, Halfway Lane, Swinderby, Lincolnshire LN6 9NP, UK.

G. URBACH
CSIRO, Division of Food Science and Technology, Dairy Research Laboratory, PO Box 20, Highett, Victoria 3190, Australia.

D. VERHAEGHE
Laboratory of Food Technology, Chemistry and Microbiology, Faculty of Agricultural Sciences, State University of Ghent, Coupure 653, B-9000 Gent, Belgium.

1

Physical Chemistry of Fats

R. E. Timms

Swinderby, Lincolnshire, UK

SUMMARY

Milkfat has a more variable triglyceride composition than vegetable and other animal fats. Butyric acid is a key fatty acid which influences the physical properties of the fat. Palm kernel and coconut fat are the vegetable fats most similar to milkfat in properties.

The basic principles of polymorphism and phase behaviour of fats are described. Fats occur mainly in alpha, beta prime and beta polymorphic forms and two packing modes for each are possible. Phase behaviour describes how the many triglyceride components in fats and blends are related to physical properties such as melting range and solid content. Solid solutions of triglycerides are important in explaining the performance of the parent fats. Methods for the determination of melting point and solid fat content are discussed, including differential scanning calorimetry and nuclear magnetic resonance. The physical properties are reviewed of milkfat and its fractions, blends of milkfat with other fats and mixtures of milkfat with liquid oils.

Crystallisation is discussed in terms of nucleation and growth stages. The basic principles are described with special reference to the fractionation of fats.

The properties of mixtures of lipids and water can be explained in terms of the hydration of certain lipids and the adsorption of surface-active lipids into the oil–water interface in emulsions. Emulsions are particularly relevant in foods such as cream, butter, spreads, ice cream and cake batters.

INTRODUCTION

Milkfat is an important constituent of most dairy products. Its physical properties have a critical influence on the organoleptic and rheological properties of dairy products and other foods containing milkfat.

A fat is a material that is an intimate mixture of liquid and solid phases whose main constituents are triglycerides. Its physical state may vary from liquid to plastic solid to brittle solid depending on the temperature, which determines the relative amounts of solid and liquid phases. The physical properties of a fat which influence the properties of fatty foods are mainly related to the phase changes involved in these changes of state, i.e. solid to solid, solid to liquid, liquid to solid.

Milkfat is a variable raw material and its composition is strongly influenced by season and diet (Norris *et al.*, 1973; Wood *et al.*, 1975; Banks *et al.*, 1976; Morrison & Hawke, 1979; Banks *et al.*, 1980*a*,*b*; Hawke & Taylor, 1983; MacGibbon & McLennan, 1987). Vegetable fats such as palm oil, palm kernel oil or coconut oil are much less variable (Timms, 1985). Higher- and lower-melting fractions (also called stearin and olein or hard and soft fractions) are now commercially available (Deffense, 1987) and these will be referred to occasionally. Some mention will also be made of interesterified and hardened or hydrogenated milkfats. Although technically interesting, these chemically modified fats have not been used commercially due to the cost of processing and to the concomitant loss of the desirable milkfat flavour (Timms, 1989).

Milkfat originates from the milkfat globule secreted from the mammary gland of the cow. In milk, milkfat is predominantly present as spherical globules, the bulk of which are 1–8 μm in diameter. Each globule is surrounded by a membrane which is composed mainly of protein and phospholipid (Walstra, 1983; Precht, 1988). When the fat is separated or concentrated from milk, the residual membrane influences the properties of products such as butter and cream. In this chapter the properties of 'pure' milkfat only are considered. Commercially, such a product, from which the membrane material has been almost completely removed, is called anhydrous milkfat or AMF. The presence of phospholipids or membrane material is important in interfacial phenomena and this topic will be covered briefly in the section 'Interfacial Properties'.

The chemical composition of milkfat is very different from other fats with similar melting point as shown by the fatty acid compositions given in Table 1. The content of 4:0 (butyric acid) in milkfat is unique. Expressed on a percentage weight basis the figure of 4% underestimates its influence; on a percentage mole basis the content of 4:0 is about 10%. This high level of 4:0 and other shorter-chain acids produces a softer fat than would otherwise be expected from the total saturated fatty acid content of more than 65% (compare the properties of milkfat and cocoa butter with similar saturated fatty acid contents). It also influences the interfacial properties which are important in many dairy products.

The physical properties of milkfat are not very different from other semi-

TABLE 1
Fatty acid composition of milkfat compared with other edible fats

Fat	Fatty acid (% wt as methyl ester)										
	4:0	6:0	8:0	10:0	12:0	14:0	16:0	18:0	18:1	18:2	18:3
Milkfat	4	3	$1\frac{1}{2}$	$2\frac{1}{2}$	3	10	26	13	24	$1\frac{1}{2}$	$1\frac{1}{2}$
MF stearin	$3\frac{1}{2}$	$2\frac{1}{2}$	1	$2\frac{1}{2}$	$3\frac{1}{2}$	12	31	12	20	$1\frac{1}{2}$	1
MF olein	$4\frac{1}{2}$	3	$1\frac{1}{2}$	3	3	10	27	10	24	$1\frac{1}{2}$	$1\frac{1}{2}$
CNO	0	$\frac{1}{2}$	$8\frac{1}{2}$	$6\frac{1}{2}$	48	16	$8\frac{1}{2}$	$2\frac{1}{2}$	7	$2\frac{1}{2}$	0
PKO	0	$\frac{1}{2}$	4	$3\frac{1}{2}$	47	15	9	$4\frac{1}{2}$	14	$2\frac{1}{2}$	tr
HPKO	0	$\frac{1}{2}$	4	$3\frac{1}{2}$	47	15	9	15	$5\frac{1}{2}$	$\frac{1}{2}$	0
HSBO	0	0	0	0	0	tr	11	10	76	$2\frac{1}{2}$	tr
BT	0	0	0	0	$\frac{1}{2}$	$3\frac{1}{2}$	25	23	39	2	1
CB	0	0	0	0	0	tr	26	35	34	3	tr
PO	0	0	0	0	tr	1	45	$4\frac{1}{2}$	39	10	$\frac{1}{2}$

MF = milkfat; CNO = coconut oil; PKO = palm kernel oil; HPKO = hardened PKO (MP = 35°C); HSBO = hardened soyabean oil (MP = 35°C); BT = beef tallow; CB = cocoa butter; PO = palm oil; tr = trace.

solid fats of similar melting point. Data for some basic physical properties, namely, density, specific heat, heat of fusion and viscosity, are given in Table 2, and for solid fat content (SFC) in Table 3. The lauric fats, palm kernel and coconut oils, are most similar to milkfat and it is these fats which are especially useful to replace milkfat in filled-milk products or so-called non-dairy products.

TABLE 2
Density, specific heat, heat of melting and viscosity of milkfat compared with other edible fats (Yoncoskie, 1966, 1967; Walstra & Jenness, 1984; Timms, 1985)

Fat	Density (kg/m^3 at 50°C)	Specific heat (kcal/kg at 50°C)	Heat of melting (kcal/kg)	Viscosity (cP at 50°C)
Milkfat	895	≈0.50	20–24	22
Palm oil	891	0.51	≈23	27
Palm kernel oil	898	≈0.51	≈30	21
Coconut oil	901	≈0.51	≈26	19

TABLE 3
Solid fat content (% by NMR) of milkfat compared with other edible fats

Fat	Temperature (°C)					
	10	*15*	*20*	*25*	*30*	*35*	*40*
Milkfat	50	39	21	13	7	4	0
MF stearin	65	54	37	27	18	9	1
MF olein	42	27	10	2	1	0	0
CNO	80	63	35	2	0	0	0
PKO	67	—	39	15	0	0	0
HPKO	85	—	73	53	21	8	3
HSBO	60	—	40	26	15	4	0
BT	59	55	47	35	25	18	11
CB	82	79	76	69	32	0	0
PO	54	—	26	16	8	5	1

Note: Key as Table 1, data refer to same fats.

BASIC PRINCIPLES

Polymorphism

It has been known for well over a century that triglycerides possess multiple melting points. This multiple melting point behaviour is called polymorphism. Each different form of a triglyceride or fat which has a different melting point is called a polymorph. These polymorphs are different forms of the crystalline solid state. Different solid and liquid forms, e.g. ice and water, are not considered to be different polymorphs, but are different phases.

For fats, the differences between polymorphs are often not very clear and special techniques are often needed to distinguish different polymorphs, especially when they are present together. Especially by X-ray diffractometry can the correct polymorphic description be assigned unequivocally.

Fats and triglycerides occur in any one of three basic polymorphs designated α (alpha), β' (beta prime) and β (beta). Sometimes a form designated γ (gamma) is reported, but this is not a basic form. α is the least stable and lowest melting; β is the most stable and highest melting. Transformations from α to β' to β take place in that order and are irreversible. Most fats and triglycerides possess an α form, although it is often very unstable; some also possess both β' and β forms, others only a stable β' form and no β form or a stable β form and no β' form.

Studies of the crystal structure of triglycerides have shown that triglycerides pack side by side in separate layers and are arranged in pairs, head to tail. Two packing modes are possible resulting in pairs of 2 or 3 fatty acid chain lengths long. In such cases the packing is said to be double or triple chain length spacing and can be indicated by adding -2 or -3 to the basic polymorph symbol, e.g. β'-2 or β-3.

Phase Behaviour

Milkfat consists of a mixture of a large number of triglycerides. For many vegetable oils such as palm oil or soyabean oil, at least 10 triglycerides are present in significant amounts (more than 1%) but the number is much greater for milkfat and for palm kernel and coconut oils, which have a larger number of fatty acids. Each triglyceride has its own polymorphism and melting behaviour. However, in the mixture that we call milkfat the triglycerides do not behave independently. We cannot consider milkfat in terms of its individual component triglycerides, but only in terms of its different phases. We can say that the physical properties of milkfat are determined by its phase behaviour. This was first clearly recognised and explained by Mulder in 1953.

Concept of a phase

A phase is a state of matter (a glyceride mixture in this case) that is homogeneous and is separated from other phases by a definite physical boundary. A phase is completely defined by its composition, temperature and pressure. For the present purpose, pressure can be ignored. Examples of relevant phases are:

(a) Solid–liquid e.g. ice and water; oil and fat.
(b) Two solid phases e.g. coexisting polymorphs in a fat such as β'_1-2 and β_2-3 in cocoa butter; cocoa and sugar in chocolate.
(c) Two liquid phases e.g. oil and water in salad dressing; oil + solid fat + water in butter (a three-phase system).
(d) Membranes The milkfat globule membrane can also be regarded as a separate phase.

Different solid phases in fats are not usually distinguishable except at the microscopic level, although their presence may affect macroscopic properties. For example, the creaming properties of a shortening are affected by the amount of solid and liquid phases and by the presence or absence of β' and β phases. A recent review (Juriaanse & Heertje, 1988) describes the microstructure of shortenings, margarine and butter with clear pictures of the different phases involved.

All natural fats at normal-use temperatures contain at least two phases, solid and liquid. There is always only one liquid phase, but there may be several solid phases.

Concept of a phase diagram

Since a phase is defined by its temperature and its composition, a diagram with temperature along one axis and composition along the other is sufficient to define the phases present in any two-component mixture. Such a diagram is called a phase diagram. The four main types of phase diagram observed with binary mixtures of triglycerides are shown in Fig. 1.

A solid solution (sometimes called a mixed crystal or a compound crystal) is a solid phase in which the two components are randomly mixed. Its properties vary continuously according to its composition just like the more familiar liquid solutions. Only when two triglycerides are very similar will a continuous solid solution be formed as shown in Fig. 1(a). Usually, the solubility of one solid in another is limited (just as with a solid in a liquid) and two solid phases are formed, each of which is usually a solid solution. With mixtures of triglycerides, diagrams of the type shown in Figs 1(b), 1(c) and 1(d) are common.

An interesting study of solid solutions and their relation to the solid fat content and hardness of fat mixtures has been reported by Sherbon & Coulter (1966). They studied mixtures of milkfat stearin and olein and of tributyrin ($C_4C_4C_4$) and tristearin ($C_{18}C_{18}C_{18}$).

Phase behaviour of natural fats

An individual fat can be considered as one composition in an imaginary multicomponent (component = triglyceride) phase diagram. Small compositional changes in the multicomponent diagram will be reflected as natural variations in the physical properties of the fat. Since even the simplest fat such as cocoa butter or palm oil would require consideration of at least five triglycerides to even approximate its phase behaviour, it is clearly not possible to depict the true situation. Instead, consider a eutectic system of two components C and D as shown in Fig. 2. C and D are assumed to have stable β and β' polymorphs respectively.

At composition q_1q_2, we should have a simple binary mixture fat showing a single stable polymorph (β), but melting over a range of temperatures between the solindex and liquidus lines.

At composition r_1r_2, we should have a fat again showing a broad melting

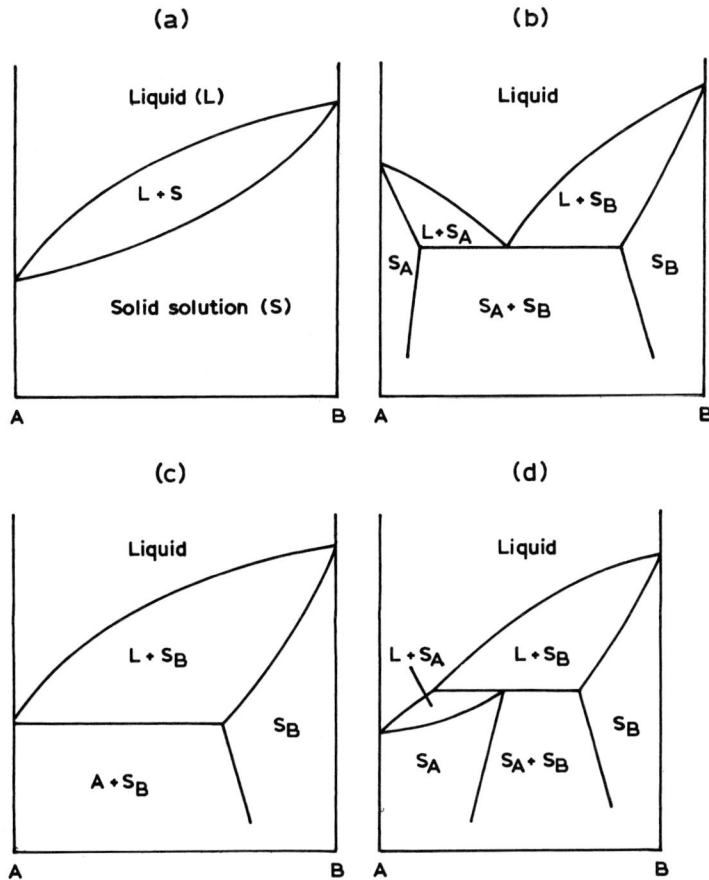

Fig. 1. The four main types of phase diagram observed with binary mixtures of triglycerides: (a) monotectic, continuous solid solution; (b) eutectic; (c) monotectic, partial solid solution; (d) peritectic. (From Timms (1984), reproduced by permission of Pergamon Press PLC.)

range, but now the fat is effectively a mixture of two solid solutions S_C and S_D with both β and β' polymorphs present together. As the temperature is increased, the β' polymorph disappears when the solidus line is crossed, leaving β which finally melts when the liquidus line is crossed. Notice that in this case the 'natural order' of disappearance of polymorphs is observed, i.e. β' before β, but the two are not related as would be the case for a single pure triglyceride. This is the situation which exists in milkfat at normal

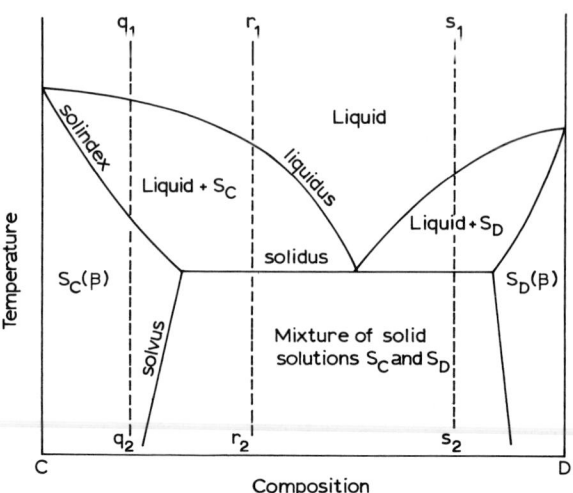

Fig. 2. Schematic phase diagram of two pure triglycerides (C and D, polymorphs β and β') with partial miscibility in the solid state to give solid solutions S_C and S_D, respectively. (From Timms (1984), reproduced by permission of Pergamon Press PLC.)

ambient temperatures. A third phase also exists which is liquid at these temperatures.

At composition s_1s_2, a fat with a $\beta + \beta'$ mixture would again be observed except that β would melt before β', leaving the 'stable' β' to melt to the liquid.

We can suppose the above reasoning to be extended to many components, in which case more than two solid solutions may coexist, their number and composition changing with temperature. When it is remembered that two fatty acids can produce triglycerides showing 15 binary interactions/phase diagrams, the situation in milkfat containing more than 10 major fatty acids can be seen to be very complex. Nevertheless, the simple binary diagram explains many of the features of natural fats including milkfat:

(a) A broad melting range is always observed, not a single clearly defined melting point. Empirically defined melting points are therefore necessary.
(b) Many fats show changing number and type of polymorphs with temperature and mixtures of β and β' are commonly observed even at equilibrium after extensive tempering. Where there is a wide variety of molecular size and type of triglyceride as in milkfat, the β'

form dominates because it is more able to accommodate distortion of the chain packing necessary for solid solutions.
(c) Where both β and β' polymorphs occur in a fat the highest melting form is not necessarily the β form.
(d) Both the composition and the relative proportions of the liquid and solid phases of a fat vary with temperature.

For a more detailed review of the phase behaviour and polymorphism of triglycerides and fats Timms (1984) should be consulted.

Melting Point and Solid Fat Content

As explained above, the melting point of a fat is an empirical property related to the experimental method of determination and not a basic physical property like the melting point of a pure compound. In particular the melting point is directly related to the temperature at which the fat is crystallised or tempered, the higher the temperature the higher the observed melting point. This effect is quite independent of any polymorphic changes.

Consider the schematic phase diagram shown in Fig. 3. A fat shown with composition xx is crystallised and tempered at temperature T_a to produce a

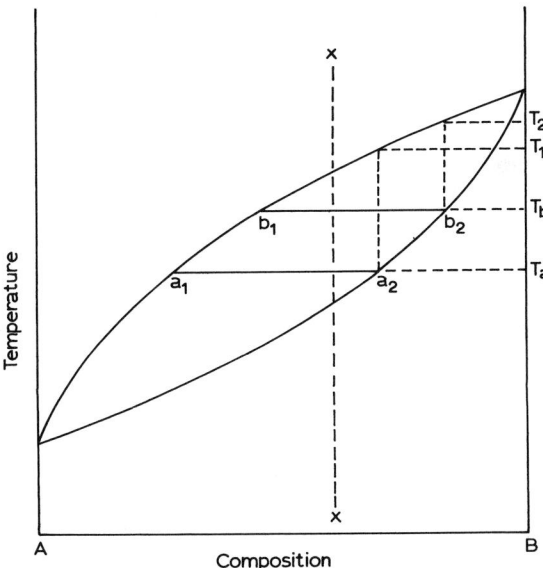

Fig. 3. Schematic phase diagram illustrating the effect of crystallisation/tempering temperature on the observed melting point.

fat comprising a solid phase of composition a_2 and a liquid phase of composition a_1. On heating, the solid solution of composition a_2 melts at T_1, which is reported as the melting point of the fat. If the fat is crystallised and tempered at temperature T_b, a solid phase of composition b_2 and a liquid phase of composition b_1 are produced. On heating, the solid solution of composition b_2 melts at temperature T_2, which is reported as the melting point.

Common methods of melting point determination are:

Slip point	The temperature at which fat in a capillary tube placed in water becomes soft enough to slip or rise up the tube. Also called open-tube melting point, rising point or softening point.
Wiley melting point	The temperature at which a disc of fat becomes spherical when heated in an alcohol–water mixture. Not much used for milkfat.
Barnicoat softening point	The temperature at which lead shot falls through a sample of the fat.
Dropping point	The temperature at which fat drops out of a metal cup with a hole in the bottom.

A more complete description of the melting behaviour of a fat is given by determining the solid fat content at different temperatures. By plotting SFC against temperature a melting curve is obtained as shown in Fig. 4. Like the melting point, the SFC is strongly influenced by the experimental procedure.

Until recently, the SFC of a fat at a given temperature was determined principally by dilatometry (Hannewijk *et al.*, 1964). Nuclear magnetic resonance (NMR), especially pulsed NMR, has now become very widely used and in Europe, but not yet in the USA, has almost entirely replaced dilatometry. Tempering practices vary widely according to different standard procedures and differences in reported SFC can be as much as 20%. Waddington (1980) has reviewed the principles and practice of the NMR technique for the routine determination of SFC in the oils and fats industry and Timms (1988) has reviewed the differences between the various methods.

Although standard methods are available, it should be remembered that they have been mainly developed for application to vegetable fats and are not necessarily appropriate for milkfat or fat blends containing milkfat. Two recent papers (Timms, 1980*a*; Petersson, 1986) have studied this matter for mixtures of milkfat with cocoa butter.

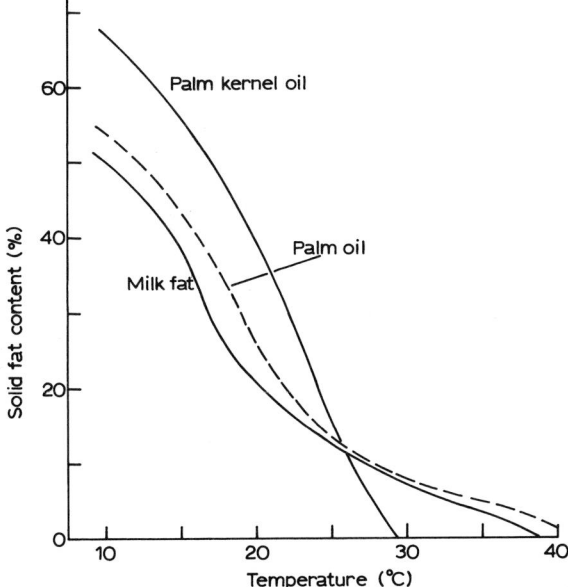

Fig. 4. Solid fat content versus temperature curves for milkfat, palm oil and palm kernel oil.

Differential scanning calorimetry (DSC) has also been used for the determination of SFC of milkfat. This method is of less utility than NMR for routine analyses because of the time involved in measuring one sample, and the sequential sampling makes tempering difficult. The NMR and DSC methods frequently produce substantially different results (van Beresteyn, 1972). This has been attributed to tempering differences and to the use of incorrect heats of melting for the DSC calculation. Lambelet (1983) has studied the problem and has attributed the differences to the presence of an amorphous phase in the rapidly crystallised fat, which is not detected as a solid in the pulsed NMR method. However, Rüegg *et al.* (1984) obtained agreement between NMR and DSC results by accounting for the temperature dependence of the heat of melting.

PHYSICAL PROPERTIES

Milkfat and Fractions

The polymorphism of milkfat was reviewed by DeMan in 1963. α, β' and β forms have been observed and X-ray diffraction patterns, infrared spectra

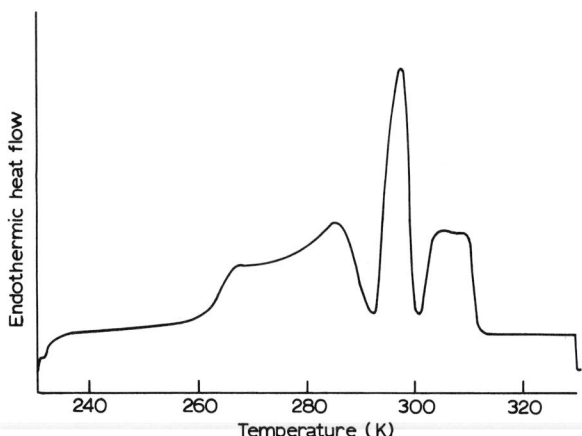

Fig. 5. DSC melting curve of milkfat after tempering for 4 weeks at 13°C (Timms, 1980a).

and methods for obtaining the different polymorphs have been reported (Tverdokhleb et al., 1968; Woodrow & DeMan, 1968).

The phase behaviour of milkfat and the effect of tempering on the hardness have been much studied because of the commercial importance of producing softer butter. Mulder (1953) used phase diagrams and the solid solution concept to explain why less solid fat is produced when milkfat is cooled slowly or stepwise compared with rapid cooling, principles which are applied in practice in the manufacture of softer butter.

Rapid cooling of milkfat to 11°C was shown by van Beresteyn (1972) to produce α crystals which quickly transformed to the β' polymorph. No α polymorph was detected above 22°C. Upon further heating to above 30°C, a β form was detected.

Further understanding of the phase behaviour of milkfat and an extension of the solid solution concept was provided by Timms (1980a,b). He showed that the properties of milkfat can be explained in terms of three fractions associated with three largely independently melting solid solutions. This is reflected in the typical DSC melting curve shown in Fig. 5. The wide range of triglycerides in milkfat results in incomplete miscibility in the solid phase similar to the $r_1 r_2$ theoretical fat discussed in the second section. These three fractions, high melting (HMF), middle melting (MMF) and low melting (LMF), which can be separated by solvent fractionation, are directly related to the triglyceride composition of milkfat as shown in Table 4. At normal ambient temperatures only HMF and MMF are present, the

TABLE 4
Classification of triglycerides in milkfat into three fractions

Fraction	Fatty acid composition of main triglycerides	Typical amount (%)	Melting point (°C)
HMF	Only long-chain saturated acids	5	> 50
MMF	Two long-chain saturated acids + one short-chain or *cis*-unsaturated acid	25	35–40
LMF	One long-chain saturated acid + two short-chain or *cis*-unsaturated acids	70	< 15

HMF = high melting fraction; MMF = middle melting fraction; LMF = low melting fraction.

melted LMF acting primarily as a diluent for the two solid phases. The stable polymorphs of HMF and MMF were found to be β-2 and β'-2 (+ some β'-3) respectively, in agreement with previous data for HMF (Tverdokhleb *et al.*, 1968) and the observed β' + small amount of β observed for whole milkfat.

Both MMF and HMF crystallised in the α-2 form within 10 min of cooling to about 5°C, whereas milkfat itself under the same conditions crystallises in the β' polymorph together with a little β polymorph. Rapid cooling of milk fat to $-10°C$ does produce α polymorphs (Tverdokhleb & Vergelesov, 1974; Timms, 1980b) which are attributed to the LMF which will be substantially crystallised at this temperature. Both MMF and HMF were found to be very stable in the β' polymorph even after extensive tempering at 30°C and 40°C respectively. However, when LMF was added to these fractions HMF transformed to a β polymorph. It was concluded that the presence of LMF in milkfat accelerates polymorphic changes, preventing the stabilisation of the α polymorph at refrigerator temperatures and causing a rapid transformation to the stable $\beta' + \beta$ mixture.

A confirmation of this interpretation was provided by Precht (1980, 1988) who studied the various types of fat globule in butter. He concluded that some globules have an outer shell which tends to be high melting and β whereas the inner part tends to melt between 15 and 30°C and is β'.

When milkfat is hydrogenated, there is a substantial increase in triglycerides of the HMF type and a corresponding increase in the amount of the β polymorph (Timms, 1980b). Fractionation similarly concentrates the HMF in the stearin fraction and increases the amount of β polymorph.

The phase behaviour of interesterified milkfat has also been studied.

TABLE 5
Melting point and solid fat content (% by NMR) of milkfat compared with other edible fats (Timms & Parekh, 1980)

Fat	Melting point (°C)	Temperature (°C)						
		10	15	20	25	30	35	40
Spring MF	32·3	39	22	11	8	5	0	0
Summer MF	34·5	45	31	18	12	8	3	0
PH Spring MF	37·7	69	53	30	22	17	7	0
FH Spring MF	49·9	87	83	73	63	47	33	22
IN Summer MF	39·3	38	28	22	20	14	8	2
MF stearin	39·2	53	42	31	27	20	11	2
MF olein	24·3	42	25	8	2	1	0	0

MF = milkfat; PH = partially hardened; FH = fully hardened; IN = interesterified.

After interesterification, only β' polymorphs were observed whether fast or slow cooling was used (DeMan, 1961), although later work (Timms, 1979) has shown that a small amount of β crystals does develop after storage for several weeks.

Solid fat contents for various milkfats and fractions are given in Table 5. The various modifications show that milkfat can be produced with an increased range of melting, but the range is still limited in comparison to other processed oils and fats.

Mixtures of Milkfat with Other Fats

The phase and isosolid diagrams for mixtures of milkfat with cocoa butter have been reported (Timms, 1980a). The phase diagram is of the monotectic type and cocoa butter was found to form a β-3 solid solution with up to 50% milkfat as shown in Fig. 6. Most of the softening effect of milkfat was shown to be due to the effect of the liquid triglycerides in milkfat.

Isosolid diagrams were also constructed for mixtures of cocoa butter with various modified milkfats (Timms & Parekh, 1980). From the results it was shown that because of eutectic effects no significant improvement in the hardness of chocolate could be expected by hydrogenating or fractionating milkfat.

Isosolid diagrams have also been constructed for mixtures of beef tallow and fractions (Timms, 1979). The 50% isosolid line showed evidence for

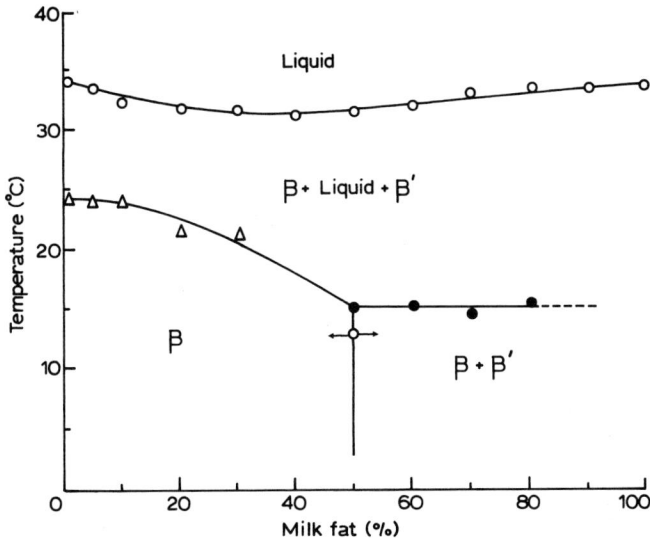

Fig. 6. Cocoa butter/milkfat phase diagram after tempering for 4 weeks at 13°C (Timms, 1980a).

eutectic and compound interactions of the type described by Rossell (1973). Other evidence for compound formation was found with DSC and X-ray studies.

Solid fat contents of mixtures of several oils and fats (palm oil, palm olein, beef tallow, beef tallow olein, beef tallow stearin, cottonseed oil, coconut oil and lard) with milkfat and fractions have been reported (Timms & Black, 1978). There was little evidence of strong eutectic or compound formation. This result is almost certainly because milkfat is such a complicated mixture of triglycerides that it is unlikely that some substantial fraction of the fat would not be miscible in the solid state with the other fat in the mixture. A typical result is shown in Fig. 7 which shows the isosolid diagram for mixtures of milkfat and palm oil. Below 10°C there is clear evidence of a eutectic effect shown by the depression of the 40% and 50% isosolid lines. At lower temperatures triglycerides with short-chain acids will begin to crystallise. It is these triglycerides which are not found in other oils and fats which will tend to show eutectic effects due to immiscibility. Table 6 gives SFCs for mixtures of milkfat with beef tallow, palm oil, coconut oil and cottonseed oil.

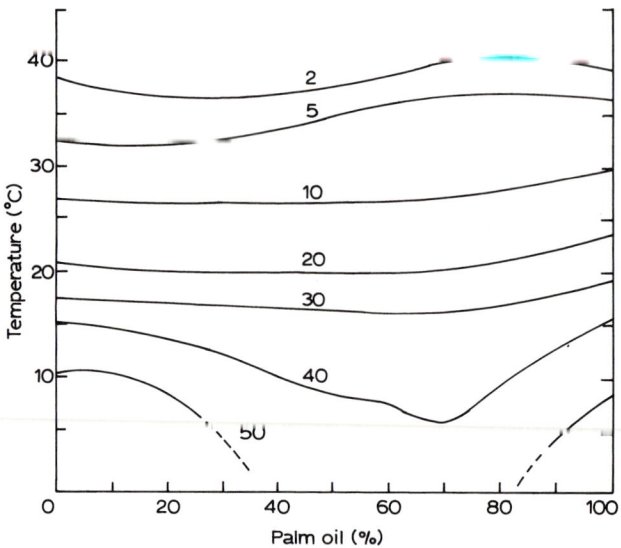

Fig. 7. Isosolid diagram of mixtures of milk fat and palm oil. Each line represents the line of constant SFC given by the number above the line.

Solubility of Milkfat in Liquid Oils

Liquid vegetable oils are used to soften milkfat to produce a softer and more spreadable spread. The softening effect produced is essentially caused by the solution of solid triglycerides in the additional liquid triglycerides, resulting in a reduction of the total solid fat content at a given temperature.

For most edible oils consisting mainly of C_{16} and C_{18} fatty acids, the solubility of triglycerides follows the ideal solubility law:

$$\ln x = \Delta H_m/(1/T_m - 1/T)/R$$

where ΔH_m = heat of melting of the higher-melting component (HMC), T_m = melting point of HMC, x = mole fraction of HMC, T = melting point or solution temperature of the mixture or solution, R = gas constant.

This equation implies that the observed solubility/melting point is independent of the liquid oil used. The data plotted in Fig. 8 show this conclusion to be essentially correct for mixtures of milkfat and vegetable oil, but there are small deviations from the equation at more than 50% milkfat. This deviation, leading to a solubility greater than 'ideal', is due to the differences in molecular size between the triglycerides in milkfat and liquid vegetable oils which increase the entropy of mixing. At 30°C the

TABLE 6
Solid fat contents (% by NMR) of mixtures of milkfat with beef tallow, palm oil, coconut oil and cottonseed oil (Timms & Black, 1978)

Composition of mixture		Temperature (°C)						
Fat	%	10	15	20	25	30	35	40
Milkfat	100	50	39	22	12	6	4	1
BT	100	58	55	47	35	25	18	11
BT	80	46	48	42	31	23	15	10
BT	60	45	45	37	26	18	13	7
BT	40	50	44	31	22	16	10	4
BT	20	52	42	25	16	12	7	3
PO	100	53	41	29	17	10	7	1
PO	80	40	35	23	14	9	6	3
PO	60	38	33	20	12	7	4	1
PO	40	40	33	21	12	7	4	1
PO	20	48	37	20	12	6	4	1
CNO	100	81	63	35	2	1	0	0
CNO	80	70	49	21	2	1	0	0
CNO	60	63	39	17	4	2	1	0
CNO	40	58	34	16	6	3	1	0
CNO	20	55	34	18	10	5	2	0
CSO	100	1	0	0	0	0	0	0
CSO	80	6	5	3	2	1	0	0
CSO	60	15	10	6	4	2	1	0
CSO	40	29	18	10	7	3	1	0
CSO	20	40	28	15	9	5	2	1

BT = beef tallow; PO = palm oil; CNO = coconut oil; CSO = cottonseed oil.

molecular volumes of milkfat and sunflower oil are about 800 ml and 960 ml respectively, which is sufficient to account for the observed deviation from the equation. In a study of this topic, the solubilities of two milkfats, a milkfat stearin and a hardened milkfat in sunflower, soyabean and cottonseed oils were reported (Timms, 1978).

CRYSTALLISATION

Supersaturation

To obtain crystallisation it is necessary to increase the concentration of the solute to be crystallised above the concentration of a saturated solution at

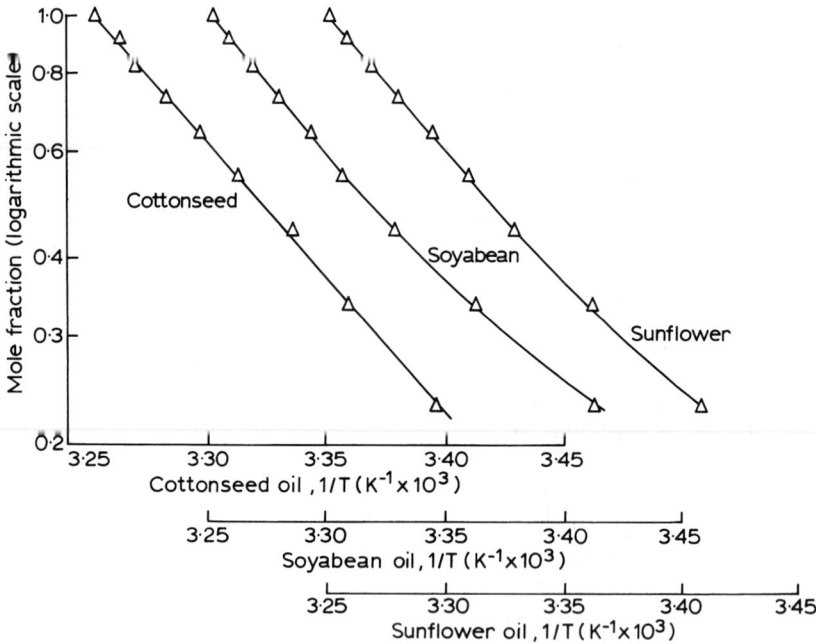

Fig. 8. Solubility of milkfat in cottonseed, soyabean and sunflower oils (Timms, 1978).

the crystallisation temperature. In practice, this is not sufficient to cause crystallisation and solutions can exist indefinitely with concentrations above the saturated level without the formation of any crystals. Such solutions are called supersaturated.

For any system we can draw a saturation–supersaturation diagram as shown in Fig. 9. The continuous line is the normal solubility curve, as for example in Fig. 8, although in a crystallising fat the solute and solvent are not distinguishable as entirely separate entities. Below this line crystallisation is impossible because the solution is not saturated. The dashed line divides the metastable zone from the unstable or crystallisation zone. In the metastable zone crystallisation is possible, but will not occur spontaneously or immediately without assistance such as seeding. Crystallisation will occur spontaneously in the unstable zone.

The position of the dashed-line boundary between the metastable and unstable zones is variable and depends on variables such as cooling rate, agitation and contaminating non-fat particles. In contrast the position of the continuous line depends only on the substance crystallised.

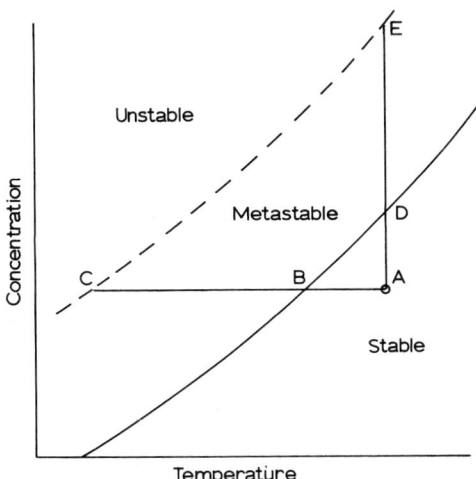

Fig. 9. Saturation–supersaturation solubility diagram.

The reason for the existence of the metastable zone can be understood if crystallisation is considered as a two-step process: nucleation followed by crystal growth.

Nucleation

A crystal nucleus is the smallest crystal that can exist in a solution of a certain concentration and temperature. Crystals smaller than a nucleus would re-dissolve.

When molecules come together to form a crystal there are two opposing forces. First, energy is evolved as the heat of crystallisation, which tends to favour the process. Secondly, the surface of the crystal increases which requires energy to overcome the surface tension. Van den Tempel (1968) has calculated that to form a nucleus of 10 fatty acids in the α polymorph would require a supercooling of 46 K. In practice, supercooling of even 10 K is usually sufficient to induce nucleation in fats and he concluded that most nucleation in crystallising fats is initiated by 'active impurities' such as seed crystals, the walls of the containing vessel or dust.

Once small crystals have formed from the nucleus they can act as active impurities for the nucleation of other triglycerides. Since the various triglycerides in milkfat are closely related, one type of triglyceride can act as a nucleus for other types. As a result, milkfat in bulk shows a small supercooling, often only 1–2 K, and there is little hysteresis between

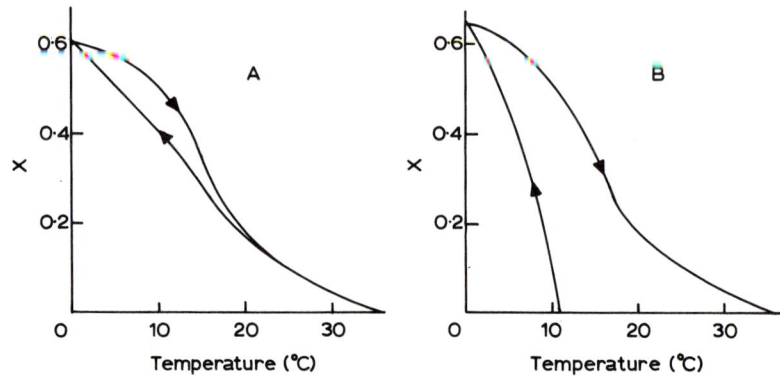

Fig. 10. Proportion of fat (x) which is solid after 24 h holding at temperature T (after previously cooling to 0°C). (A) Milkfat in bulk; (B) the same milkfat in recombined milk, i.e. in small globules. (From Walstra & van Beresteyn (1975); Walstra & Jenness (1984) *Dairy Chemistry & Physics*, pp. 88–97, reproduced by permission of John Wiley & Sons, Inc. © 1984 John Wiley & Sons, Inc.)

crystallisation/solidification and melting curves (Fig. 10(A)). The situation is quite different in dairy products such as milk or cream where the fat is finely divided into separate fat globules (Walstra & van Beresteyn, 1975). In bulk fat, one active impurity in 1 mg would be sufficient to ensure rapid crystallisation. In milk, 1 mg of fat is divided among 10^8 globules, or at least 10^{11} if the milk is homogenised. In each of these fat globules one nucleus must form if crystallisation is to be complete. Consequently supercooling must be deeper than in bulk fat and hysteresis can be considerable (Walstra & Jenness, 1984) as shown in Fig. 10(B).

Nucleation depends not just on temperature and time but also on fat composition. Lipolysis has an important effect on the composition of the fat in dairy products. As more fat is hydrolysed less supercooling is needed, which is attributed to micelles of monoglycerides acting as active impurities in globular milkfat (Walstra & Jenness, 1984). Monoglycerides also influence the crystallisation of other fats but, since most edible fats are refined before use, monoglycerides are of little practical importance except where they are deliberately added as part of an emulsifier system.

Crystal Growth

Once a crystal nucleus has formed it starts to grow by the incorporation of other molecules. These molecules are taken from the adjacent liquid layer which is replenished continuously by diffusion from the surrounding (supersaturated) liquid. The rate-determining step is the incorporation of the new molecule in the correct configuration at the correct place on the growing crystal surface.

During crystallisation/solidification of a fat, the composition of the crystallising solid and the remaining liquid phases changes continuously, effectively following the solidus and liquidus lines on the phase diagram. This makes the crystal growth stage of crystallisation of a fat much more complicated than the crystallisation of a pure solute from a separate solvent.

Crystal growth in fats is relatively slow and if the cooling rate is rapid a larger number of smaller crystals is produced. Rapid cooling also leads to less perfect crystals, since at higher degrees of supercooling molecules are attached to the crystal surface at a faster rate and a new molecule may become attached before its neighbour has become oriented into a perfect position.

Rapid cooling also leads to more solid fat in the final crystallised fat than slow or stepwise cooling (Foley & Brady, 1984), which is simply explained from a consideration of a phase diagram showing solid solution formation (Mulder, 1953; Walstra & Jenness, 1984). However, rapidly cooled fat will slowly change to the equilibrium composition that would have been obtained by slow cooling. The rate of equilibration depends on the temperature and on whether it is constant or fluctuates. Apart from the changing composition of solid and liquid phases in the crystallised fat, the crystal size tends to increase (Ostwald Ripening) and the crystals tend to interlock with time. These effects influence the rheology of the crystallised fat in products such as butter and margarine (Juriaanse & Heertje, 1988).

There have been several studies of the kinetics of milkfat crystallisation and for practical application the process can be considered as a first order reaction with an activation energy in the range 9–11 kcal/mol (Mortensen, 1983).

Fractional Crystallisation

Milkfat, like palm oil, tallow or hardened soyabean oil, is now crystallised on a commercial scale to produce stearin and olein fractions (Amer *et al.*, 1985; Kankare & Antila, 1986; Deffense, 1987). A further crystallisation may be applied to produce a third, middle-melting fraction.

The crystallisation must be performed to yield crystals which can easily be separated from the liquid, uncrystallised oil. For milkfat, the separation method is simple vacuum filtration and large uniform crystals are required if a useful separation is to be achieved. Even when conditions are optimum,

the separated solid phase (stearin) consists of about two thirds entrained liquid phase (olein).
Crystallisation from an organic solvent would improve the efficiency of the separation and the ease of crystallisation. This process has been studied for milkfat, but it has not been economic to operate a commercial process (Jebson & Norris, 1975).

INTERFACIAL PROPERTIES

Interaction of Lipids with Water

Small (1986) has classified lipids into three classes:

Class I Insoluble and non-swelling in water, but spreading at the surface to form a stable monolayer. Triglycerides are the main members of this class.

Class II Insoluble but swelling in water, because water is soluble in the hydrophilic part of the structure to form a liquid crystalline phase. Monoglycerides and phospholipids are important members of this class.

Class III Soluble in water to form micelles. Soaps, lysolecithins and aliphatic detergents are the main members of this class.

Class I and II lipids are important in food products.

Although triglycerides are insoluble in water and are typical Class I lipids, they do have a surface solubility which depends very much on the length of the fatty acid chains. From a consideration of surface areas of long-chain triglycerides it was concluded that to a first approximation the triglyceride state at the surface corresponds to the α or liquid forms, with all three fatty acid chains standing up on the surface and the glycerol and carbonyl groups in the aqueous interface.

However, triglycerides with two long and one short fatty acid chain show more complex behaviour. A study of the $C_{16}C_{16}C_n$ triglyceride series has shown interesting results. When $n \leqslant 6$, the fatty acid chains are pushed into the aqueous interface by surface compression. When $n = 8$ or $n = 10$, the shorter fatty acid chain remains oriented above the aqueous interface, but prevents the surface crystallising so that the surface melting point is below 0°C, compared with melting points of 24°C for $C_{16}C_{16}C_6$ and 38°C for $C_{16}C_{16}C_{12}$ (Small, 1986).

Milkfat contains a substantial proportion (probably about one third) of triglycerides containing a C_4 or C_6 fatty acid. These triglycerides contribute to the whipping and organoleptic properties of dairy products and to the wetting properties of dried dairy products. Their contribution is often overlooked in comparison with the contribution of phospholipids and proteins. Of the other edible oils, only lauric oils contain any of these shorter-chain fatty acids, but even here the amount of triglycerides containing a C_4 or C_6 fatty acid is small—less than 3%.

Phospholipids and monoglycerides are important components influencing the interfacial properties of lipids in food emulsions. Both lipid classes occur naturally, but are almost completely eliminated from edible oils by refining. In dairy products, phospholipids are of special importance (Morisson, 1968; Walstra, 1983). They are major components of the milkfat globule membrane. During the processing of milk to cream and butter, phospholipid migrates (probably as a lipoprotein complex) to the aqueous or serum phase. Buttermilk is rich in membrane material such as phospholipids.

Emulsion Stability

Products consisting of oil and water emulsions can undergo various changes during storage and handling which may lead to breakdown of the products (Mulder & Walstra, 1974). Factors which influence emulsion stability are:

(a) particle size of dispersed phase,
(b) interfacial tension,
(c) elasticity/viscosity of interfacial film,
(d) electrostatic interaction (DLVO theory),
(e) steric hindrance,
(f) liquid crystal formation,
(g) viscosity of continuous phase.

Surfactants adsorb at the oil–water interface reducing the interfacial tension and facilitating a smaller particle size of the dispersed phase. The elasticity and viscosity of the interfacial layer are important factors limiting the coalescence of oil particles. Electrostatic interactions are of limited importance in food emulsions. Adsorption of protein at oil–water interfaces limits coalescence via steric hindrance, an important mode of stability for dairy emulsions. An excellent review of the interactions of surface-active lipids with water, protein and starch components has been given by Krog (1986).

In a study of 30% milkfat-in-water emulsions containing monoglycerides and sorbitan esters, these various factors influencing emulsion stability were investigated (Sogo *et al.*, 1988). The emulsion stability was correlated with the mechanical strength of the milkfat globule membrane.

In whippable emulsions a controlled destabilising of the emulsion during the whipping process is required to obtain the desired volume and stiffness. Such emulsions are usually spray-dried mixtures of fat (usually a hardened lauric oil), protein and surface-active lipid. In the dry powder state the fat phase can be considered to be supercooled due to lipid–protein interactions (Barfod & Krog, 1987). Upon reconstitution in cold water, protein desorbs from the fat globule surface into the water. The protein-induced supercooling of the fat is thus eliminated and spontaneous fat crystallisation occurs. The process can be followed by NMR measurements.

In the final reconstituted whipped emulsion, only a few intact fat globules are present. The foam is stabilised by fat crystals oriented around the air cells. In contrast, in a whipped dairy cream stability is obtained by spherical fat globules adsorbed at the surface of the air cells. Proteins play a dominant role during the formation of the foam (Buchheim *et al.*, 1985).

In water-in-oil emulsions such as margarine, it is necessary to stabilise the liquid emulsion before crystallising the oil in the chilling and working tubes of the processing equipment. A homogeneous product with finely dispersed and stable water distribution is desired. Monoglycerides reduce the interfacial tension between water and oil and are the preferred emulsifiers for this application. Although the monoglycerides are dissolved in the oil phase, they migrate to the oil–water interface during the preparation of the emulsion.

With some manufacturing processes after chilling to produce crystals in the dispersed oil phase, the emulsion is less stable if crystals appear at the interface. Coalescence in such systems could be caused by crystals poking through the interface and piercing the film between the fat globule and the next approaching globule (Van Boekel, 1980). Surface-active compounds like monoglycerides help to wet the fat crystals and to stabilise them in the oil–water interface (Juriaanse & Heertje, 1988), thus modifying or inhibiting coalescence.

REFERENCES

AMER, M. A., KUPRANYCZ, D. B. & BAKER, B. E. (1985) Physical and chemical characteristics of butterfat fractions obtained by crystallization from molten fat. *J. Am. Oil Chem. Soc.*, **62**, 1551–7.

BANKS, W., CLAPPERTON, J. L. & FERRIE, M. E. (1976) The physical properties of milk fats of different chemical compositions. *J. Soc. Dairy Technol.*, **29**, 86.

BANKS, W., CLAPPERTON, J. L. & KELLY, M. E. (1980a) Effect of oil-enriched diets on the milk yield and composition, and on the composition and physical properties of the milk fat, of dairy cows receiving a basal ration of grass silage. *J. Dairy Res.*, **47**, 277–85.

BANKS, W., CLAPPERTON, J. L., KELLY, M. E., WILSON, A. G. & CRAWFORD, R. J. M. (1980b) The yield, fatty acid composition and physical properties of milk fat obtained by feeding soyabean oil to dairy cows. *J. Sci. Food Agric.*, **31**, 368–74.

BARFOD, N. M. & KROG, N. (1987) Destabilization and fat crystallization of whippable emulsions (toppings) studied by pulsed NMR. *J. Am. Oil Chem. Soc.*, **64**, 112–19.

BUCHHEIM, W., BARFOD, N. M. & KROG, N. (1985) Relation between microstructure, destabilization phenomena and rheological properties of whippable emulsions. *Food Microstruct.*, **4**, 221.

DEFFENSE, E. (1987) Multi-step butteroil fractionation and spreadable butter. *Fette Wiss. Technol.*, **89**, 502–7.

DEMAN, J. (1961) Physical properties of milk fat II. Some factors influencing crystallisation. *J. Dairy Res.*, **28**, 117–22.

DEMAN, J. (1963) Polymorphism in milk fat. *Dairy Sci. Abstr.*, **25**, 219–21.

FOLEY, J. & BRADY, J. P. (1984) Temperature-induced effects on crystallisation behaviour, solid fat content and the firmness values of milk fat. *J. Dairy Res.*, **51**, 579–89.

HANNEWIJK, J., HAIGHTON, A. J. & HENDRIKSE, P. W. (1964) Dilatometry of fats. In: *Analysis and Characterization of Oils, Fats and Fat Products, Vol. 1*, Boekenoogen, H. A. (Ed.), Interscience, London, pp. 119–82.

HAWKE, J. C. & TAYLOR, M. W. (1983) Influence of nutritional factors on the yield, composition and physical properties of milk fat. In: *Developments in Dairy Chemistry—2 Lipids*, Fox, P. F. (Ed.), Applied Science Publishers, London, pp. 37–82.

JEBSON, R. S. & NORRIS, R. (1975) *New Zealand Dairy Research Institute Annual Report*, p. 32.

JURIAANSE, A. C. & HEERTJE, I. (1988) Microstructure of shortenings, margarine and butter—a review. *Food Microstruct.*, **7**, 181–8.

KANKARE, V. & ANTILA, V. (1986) Melting characteristics of milk fat and milk fat fractions. *Meijeritieteellinen Aikakauskirja*, **44**, 67–75 (abstracted in *Food Sci. Tech. Abs.*, **19** (1987) 10P219).

KROG, N. (1986) Interactions of surface-active lipids with water, protein and starch components in food systems, *Grindsted Technical Paper*, TP 909-1e., given at AOCS Short Course, Hawaii, May.

LAMBELET, P. (1983) Comparison of NMR and DSC methods for determining solid content of fats—application to milk fat and its fractions. *Lebensm. Wiss. Techn.*, **16**, 90–5.

MACGIBBON, A. K. H. & MCLENNAN, W. D. (1987) Hardness of New Zealand patted butter: Seasonal and regional variations. *NZ J. Dairy Sci. Technol.*, **22**, 143–56.

MORISSON, W. R. (1968) Surface-active lipids in milk and milk products. *Surface-active Lipids in Foods*, SCI Monograph No. 32, London, pp. 75–91.

MORRISON, I. M. & HAWKE, J. C. (1979) Influence of elevated levels of linoleic acid on the thermal properties of bovine milk fat. *Lipids*, **14**, 391.

MORTENSEN, B. K. (1983) Physical properties and modification of milk fat. In: *Developments in Dairy Chemistry—2 Lipids*, Fox, P. F. (Ed.), Applied Science Publishers, London, pp. 159–94.

MULDER, H. (1953) Melting and solidification of milk fat. *Neth. Milk Dairy J.*, **7**, 149–74.

MULDER, H. & WALSTRA, P. (1974) *The Milk Fat Globule*, Commonwealth Agricultural Bureaux, Farnham Royal, UK.

NORRIS, G. E., GRAY, I. K. & DOLBY, R. M. (1973) Seasonal variations in the composition and thermal properties of New Zealand milk fat. II Thermal properties. *J. Dairy Res.*, **40**, 311.

PETERSSON, B. (1986) Pulsed NMR method for solid fat content determination in tempering fats Part II: Cocoa butters and equivalents in blends with milk fat. *Fette Seifen Anstrichm.*, **88**, 128–36.

PRECHT, D. (1980) *Fette Seifen Anstrichm.*, **82**, 142–7.

PRECHT, D. (1988) Fat crystal structure in cream and butter. In: *Crystallization and Polymorphism of Fats and Fatty Acids*, Garti, N. & Sato, K. (Eds), Marcel Dekker, New York, pp. 305–61.

ROSSELL, J. B. (1973) Interactions of triglycerides and of fats containing them. *Chem. Ind. (London)*, **17**, 832–5.

RÜEGG, M., MOOR, U. & BLANC, B. (1984) Eine verbesserte Methode zur Bestimmung des Schmelzdiagramms von Butterfett mit Hilfe registrierender Differential Kalorimeter. *Milchwissenschaft*, **38**, 601–5.

SHERBON, J. W. & COULTER, S. T. (1966) Solid solutions and the hardness of fatty mixtures. *J. Dairy Sci.*, **49**, 1126–31.

SMALL, D. M. (1986) *The Physical Chemistry of Lipids from Alkanes to Phospholipids*, Plenum Press, New York.

SOGO, Y., TANEYA, S. & KAKO, M. (1988) The stability of milk fat/water emulsions containing monoglyceride, sorbitan esters and caseinate. *Develop. Food Sci.*, **17**, 379–91.

TIMMS, R. E. (1978) The solubility of milk fat, fully hardened milk fat and milk fat hard fraction in liquid oils. *Aust. J. Dairy Technol.*, **33**, 130–5.

TIMMS, R. E. (1979) The physical properties of blends of milk fat with beef tallow and beef tallow fractions. *Aust. J. Dairy Technol.*, **34**, 60–5.

TIMMS, R. E. (1980a) The phase behaviour of mixtures of cocoa butter and milk fat. *Lebensm. Wiss. Technol.*, **13**, 61–5.

TIMMS, R. E. (1980b) The phase behaviour and polymorphism of milk fat, milk fat fractions and fully hardened milk fat. *Aust. J. Dairy Technol.*, **35**, 47–53.

TIMMS, R. E. (1984) Phase behaviour of fats and their mixtures. *Prog. Lipid Res.*, **23**, 1–38.

TIMMS, R. E. (1985) Physical properties of oils and mixtures of oils. *J. Am. Oil Chem. Soc.*, **62**, 241–9.

TIMMS, R. E. (1988) Theory and practice of solid fat determination by NMR & dilatometry. Given at AOCS Short Course *Applications of Pulsed NMR Techniques in Food Analysis*, Phoenix, AZ, May.

TIMMS, R. E. (1989) The possibilities for using modified milk fats in the production of confectionery fats, shortenings and spreads. In: *Fats for the Future*, Cambie, R. C. (Ed.), Ellis Horwood, Chichester, pp. 251–61.

TIMMS, R. E. & BLACK, R. G. (1978) Solids contents of mixtures of butter oil and butter fractions with various edible oils and fats. *Dairy Research Report* No. 22, May, CSIRO, Australia.

TIMMS, R. E. & PAREKH, J. V. (1980) The possibilities for using hydrogenated, fractionated or interesterified milk fat in chocolate. *Lebensm. Wiss. Technol.*, **13**, 177–81.

TVERDOKHLEB, G. V. & VERGELESOV, V. M. (1974) *Brief Commun. XIX Int. Dairy Congr.*, **1E**, 211.

TVERDOKHLEB, G. V., GULYAEV-ZAITSEV, S. S., FAL'K, E. YU & GERASIMOVA, ZH. I. (1968) *Izv. vyssh. ucheb. Zaved., Pisch. Tekhnol.*, **39** (abstracted in *Dairy Sci. Abstr.*, **30**, 1036).

VAN BERESTEYN, E. C. H. (1972) Polymorphism in milk fat in relation to the solid/liquid ratio. *Neth. Milk Dairy J.*, **26**, 117–30.

VAN BOEKEL, M. A. J. S. (1980) Influence of fat crystals in the oil phase on stability of oil-in-water emulsions. *Agricultural Research Reports* No. 901, Department of Food Science, Agriculture University, Wagengingen, Netherlands.

VAN DEN TEMPEL, M. (1968) Effects of emulsifiers on the crystallisation of triglycerides. *Surface-active Lipids in Foods*, SCI Monograph No. 32, London, pp. 22–33.

WADDINGTON, D. (1980) Some applications of wide-line NMR in the oils and fats industry. In: *Fats and Oils: Chemistry and Technology*, Hamilton, R. J. & Bhati, A. (Eds), Applied Science, London, pp. 25–45.

WALSTRA, P. (1983) Physical chemistry of milk fat globules. In: *Developments in Dairy Chemistry—2 Lipids*, Fox, P. F. (Ed.), Applied Science Publishers, London, pp. 119–57.

WALSTRA, P. & JENNESS, R. (1984) *Dairy Chemistry & Physics*, John Wiley, New York, pp. 88–97.

WALSTRA, P. & VAN BERESTEYN, E. C. H. (1975) Crystallisation of milk fat in the emulsified state. *Neth. Milk Dairy J.*, **29**, 35–65.

WOOD, F. W., MURPHY, M. F. & DUNKLEY, W. L. (1975) Influence of elevated polyunsaturated acids on processing and physical properties of butter. *J. Dairy Sci.*, **58**, 839.

WOODROW, I. L. & DEMAN, J. M. (1968) Polymorphism in milk fat shown by X-ray diffraction and infrared spectroscopy. *J. Dairy Sci.*, **51**, 996–1000.

YONCOSKIE, R. A. (1966) Determination of heat capacities of milk fat by differential thermal analysis. *J. Am. Oil Chem. Soc.*, **46**, 49.

YONCOSKIE, R. A. (1967) Calorimetric studies of the heat of fusion of milk fat by differential thermal analysis. *J. Am. Oil Chem. Soc.*, **44**, 446.

2

Fats in Cream and Ice Cream

M. Anderson*, E. C. Needs† and J. K. Madden‡
*AFRC, Swindon, UK, †AFRC, Reading, UK and
‡United Biscuits (UK) Ltd, High Wycombe, UK

SUMMARY

Products such as cream and ice cream are emulsions of oil-in-water and can be conveniently discussed in terms of the contribution to their properties or their composition, the interfaces present, the influence of the fat and aqueous phases and the processing applied.

Three types of interface can be recognised, namely, the biological, protein stabilised, and the composite. The biological membrane is subject to damage during processing which may result in product defects such as poor whipping properties and fat plug formation. Most manufactured dairy substitute products are stabilised by a protein membrane, involving caseins and whey proteins. Protein micelles can bridge adjacent fat globules and influence product viscosity. At interfaces proteins, especially casein, are thought to adopt various steric configurations described by the loop–train–tail model. As well as proteins, low-molecular-weight surfactants are present in composite interfaces and a dynamic situation of adsorption/desorption can exist depending on a number of factors. Although such surfactants can be broadly classified using the HLB concept, further considerations have to be taken into account to explain functionality. Among these are direct interactions between surfactants and proteins, polymorphisms and the possible formation of mesophases with water.

The dispersed nature of fats in emulsions has an influence on crystallisation. Other influences may arise from the presence of surfactants and even the structure of the interface. Instability in emulsions can be described as flocculation, clustering, clumping and coalescence. Fat crystals at the interface appear to have a major influence on clumping and the wettability of such crystals can be modified by adsorption of surfactants.

The aqueous phase of dairy type emulsions may contain, in addition to proteins, lactose and inorganic materials, added polysaccharides. These last are used to reduce phase separations and stabilise air bubble distribution. The interaction of certain polysaccharides, such as carrageenans, with proteins can be used, along with their ability to gel, to overcome phase separation in ice cream.

Fat globule size reduction during homogenisation depends not only on the design of the equipment but also on the operating temperature, pressure and ingredient characteristics. The amount of available surface-active material relative to the interfacial area generated during the process can be critical for emulsion stability.

Some of these principles underlying the functionality of oil-in-water dairy emulsions can be illustrated by reference to specific products such as single cream, coffee cream, cream liqueur, whipping cream and ice cream.

INTRODUCTION

Cream and ice cream are well ordered oil-in-water emulsions in which components of the dispersed fat phase, the continuous aqueous phase and the interfacial layer separating them interact in specific ways to determine product functionality. Although statutory regulations rather limit the possibilities for adjusting dairy product composition, formulation of dairy substitute products is much more flexible allowing the processor the advantage of being able to tailor composition for optimum functionality.

Composition, however, is not by itself the sole arbiter of functionality; manufacturing practices impose their own characteristics on the product. Since the objective is to produce a stable, functional emulsion, the effects of processing have to be understood. Of the individual processing stages which can be varied to influence properties, homogenisation, understandably, exerts the largest influence.

In this chapter we describe the principal components involved, how they interact and consider the options available to the processor for adjusting manufacturing practices in relation to the specific functional requirements of cream and ice cream.

INTERFACES

In functional terms the interface can be considered as pivotal since its properties provide the all-important key to how the dispersed fat droplets will behave during processing, storage and use (e.g. whipping). With the range of emulsifiers available, many compositional permutations exist. For our purposes three distinct types of interface can be recognised in cream and ice cream, based on the product and nature of the predominant component. These types are described below under the headings of biological, milk protein and composite interfaces. Interfacial layers should be considered as dynamic surfaces whose nature can change with time as a result of rearrangement, replacement or interaction.

Biological Membrane

Fat globules in milk are stabilised by a naturally occurring membrane thought to be derived from the apical plasma membrane of the mammary gland epithelial cells (Franke et al., 1981). Shennan (1992) has shown that MFGM is not a good model for apical membrane and this raises doubts about the precise origin of the MFGM. Immediately after secretion the outer surface of the membrane has the morphology of a typical biological unit membrane. Subsequently this surface undergoes change during the period between secretion and milking, and after milking as well, involving the loss of a significant proportion of the unit membrane (Wooding, 1971). At the same time irreversible changes to the paracrystalline organisation of the proteinaceous coat of the inner membrane occur (Buchheim, 1986). It follows that much of the membrane material found in skim milk originates from the MFGM (Patton & Jensen, 1976). It is far from clear whether these morphological alterations have any real significance in terms of membrane properties as they relate to cream functionality. One of the primary functions of the membrane in raw milk, however, that of protecting milkfat from the activity of milk lipase, does not appear to be affected by the loss, providing no disruption of the membrane occurs (Anderson, 1985).

The MFGM is composed of polar lipids, predominantly phospholipids and several proteins, lipoproteins and glycoproteins that are distinct from the major milk proteins, casein, β-lactoglobulin and α-lactalbumin. For a compilation of the numerous compositional data on the MFGM the reader is referred to McPherson and Kitchen (1983). Many of these studies indicate that an appreciable amount of neutral lipid is associated with the membrane. In reality this is most unlikely and the bulk of such neutral lipid probably represents an artifact of the method used to isolate the MFGM (Walstra, 1974, 1985).

Heat treatment can cause extensive changes to the composition of MFGM (Dalgleish & Banks, 1991; Houlihan et al., 1992) including the loss of some membrane polypeptides and the adsorption of increasing amounts of serum proteins, especially β-lactoglobulin (Houlihan et al., 1992). Dalgleish and Banks (1991) demonstrated a change in composition at the fat globule surface due to heating by measuring the electrophoretic mobility of fat globules. They went on to discuss possible mechanisms for the interaction between serum proteins and MFGM. These changes do not appear to decrease membrane stability, since, after heating milk and cream (38% fat) to 140°C without agitation for 40 min, Van Boekel and Folkerts (1991) reported no change in the average globule size nor in the width of the globule size distribution curve.

Probably the most important property of the MFGM for the processor is the extent to which it is able to remain undisrupted during such processing stages as separation and heat treatment. Evidence from electron microscopy indicates that deformation of the membrane can result in its rupture (Buchheim, 1986). As a consequence, liquid fat is able to flow out of the globule and is likely to lead to globule clumping. Such damage may lead to loss of whipping properties and fat plug formation in cartons or bottles. However, as we shall see later, this is as likely to be dependent on the composition of the intraglobular fat as on the integrity of the biological membrane.

When milk or cream is homogenised, an approximately three–sixfold increase in fat surface area results accompanied by disruption of the original MFGM. In order to consider how this large increase in fat surface is stabilised we now move on to consider the second type of oil/water interface in which the major milk proteins are the dominant constituents.

Milk Protein Interface

With the exception of unhomogenised milk and cream, the protein interface is a characteristic of most dairy emulsions including dairy substitute products which contain proteins. In this section we consider the nature of the interface formed when the type and quantity of surface-active material is limited to components which can occur naturally in milk supplies since this is the most usual case; the effect of permitted additives of all kinds is described in the next section on composite interfaces. Under these circumstances, the processor has the opportunity to exert some control over the nature of the interface formed during homogenisation, by varying processing parameters.

Surface-active components in milk

Quantitatively, and in terms of emulsifying properties, casein is the most important of the milk proteins. It comprises four individual proteins known as αs_1-casein, αs_2-casein, β-casein, and k-casein which all adsorb at oil/water interfaces; β-casein, the most hydrophobic of these proteins, is the most readily adsorbed (Anderson & Brooker, 1988).

Of the additional surface-active material normally found in milk, the principal whey proteins, β-lactoglobulin and α-lactalbumin predominate. As already mentioned above, some components of the MFGM are surface-active with, in quantitative terms, considerable potential to influence the nature of the interface in homogenised dairy products. When the contribution of MFGM protein towards the total protein content of cream

TABLE 1
Contribution of MFGM protein to total protein in cream

Product	Total protein (%)	MFGM protein (% of total)
Single cream	2·81	4·8
Double cream	2·23	11·8
Whipping cream	1·78	20·1
Milk	3·29	0·9

is estimated (Table 1), the value for double cream could be at least as high as 20% compared with only 1% for milk. This calculation assumes a yield value for MFGM of 1·5 g/100 g fat and that half this material is protein. If the contribution from MFGM phospholipid is taken into account, the potential amount of surface-active material available in double cream as a natural emulsifier is 1·2% of the fat. However, the functional significance of MFGM components in this context has yet to be assessed.

Other sources of naturally occurring emulsifiers in dairy products are the variable amounts of partial glycerides and free fatty acids produced as a result of enzymatic hydrolysis of butterfat by milk lipoprotein lipase. A very small portion of the free fatty acids present arises from the incomplete synthesis of triglyceride in the mammary gland or, under some circumstances, the activity of microbial lipases. The usual concentration of these components is very low in comparison with the amount of protein available and under normal circumstances they can be expected to have little influence on interfacial properties.

Nature of the interface

Morphologically the interface is distinguished from the native MFGM by the absence of any outer unit membrane, and the appearance of casein micelles adsorbed at the surface. Often single micelles are adsorbed to more than one globule and these so-called casein bridges exert an important influence on product viscosity. The individual proteins, in addition to casein, which are present at this interface include α-lactalbumin and β-lactoglobulin (Darling & Butcher, 1978); some native MFGM proteins (Anderson et al., 1977a) are also present.

In terms of the surface properties of individual caseins it would be reasonable to expect preferential adsorption of β-casein, and this does appear to occur at the air/serum interface of milk foam (Brooker et al., 1986). However, in the turbulent flow conditions through the homogenising valve this preference is not observed (Robson & Dalgleish, 1987) and based on the findings of model studies (Oortwijn & Walstra, 1979) the quantitative composition of the interface in normal homogenised dairy products is likely to reflect the composition of the major proteins in skim milk.

Formation of the interface
As we have seen, milk contains a mixture of surface-active materials available to compete for a position at the oil/water interface. The actual composition of the interface under these conditions depends on a number of factors. When a new fat/serum interface is created in the homogenising valve a very rapid adsorption of surface-active species from both the fat phase and the aqueous phase takes place. In general terms, the extent to which an individual species is adsorbed can be said to be related to its composition, concentration, size, and the effect it has on interfacial tension. A discussion of the many contributory factors involved in adsorption is beyond the scope of this chapter, and further details are given in the reviews by Dickinson *et al.* (1988) and Darling and Birkett (1987).

Molecular arrangement at interface
In order to understand the way in which the interface influences product function it is important to consider interfacial structure in molecular terms. When proteins initially adsorb at the interface a variable amount of molecular unfolding takes place (Phillips, 1977). Orientation of protein adsorbed at the interface is described in the familiar model of Graham and Phillips (1979) in relation to the shape of segments of the polypeptide chain. The model proposes three types of segment: trains, in which a segment of amino-acid residues is in direct contact with and spread along the surface; loops, representing coiled segments extending into the aqueous or lipophilic phases dependent on their relative hydrophobicity, and, finally, the C- and N-terminal tails of the protein which locate in a similar manner to the loops.

The position adopted by individual proteins is characteristic of their tertiary and quaternary structure in solution and is modified by such factors as surface concentration, pH, temperature and the nature of the phases involved. At an air/water interface disordered proteins such as β-casein adsorb quickly and at low surface concentrations the interface is likely to comprise primarily trains. In the case of the oil/water interface or at higher surface concentrations in air/water interfaces, the protein is more condensed and this is accommodated by the presence of some loop formation (Graham & Phillips, 1979). β-Lactoglobulin and α-lactalbumin are globular proteins which, in contrast to the caseins, have a highly ordered structure. Although some unfolding occurs when they are adsorbed at the oil/water interface, the process is slow and elements of the native structure are retained (Mitchell et al., 1970).

The applicability of the loop–train–tail model to the orientation of β-casein segments at the interface is discussed in detail by Dickinson et al. (1988) who point to a number of constraints of the concept and the need for additional examination of the model. Nevertheless the model provides a useful framework for visualising interfacial molecular organisation and interaction which in the context of this chapter serves as a base for considering the additional degrees of complexity conferred by the presence of low-molecular-weight emulsifiers and stabilisers.

In products, we must consider the behaviour of a number of proteins coexisting at the interface. It seems likely that the overall properties of the interface are not simply related to those of the component proteins under the same conditions; the presence of one protein may have the effect of modifying the extent to which another unfolds at the interface (Murray, 1987). Such events are likely to take place at some locations of the interface of cream and ice-cream emulsions but their properties are dominated by the aggregated state of the dominant protein, casein. As Darling and Birkett (1987) have pointed out under the dynamic conditions that occur during homogenisation, movement of a surface-active molecule to the interface is by convection rather than by diffusion, and the larger the size of molecule the more rapidly it reaches the interface.

The Composite Interface

Although milk proteins are an integral part of ice cream and dairy substitute creams in which they are included in the form of skim-milk protein or sodium caseinate, these products also contain low-molecular-weight emulsifiers. This section considers how the presence of these molecules may alter the basic protein interfacial model. As already mentioned, the rate at which surface-active molecules arrive at the interface depends on a variety of factors including the extent to which component adsorption reduces interfacial tension. Although low-molecular-weight surfactants lower interfacial tension more than proteins, the time taken for such molecules to arrive at the interface may be considerable when they are present in small quantities (Walstra & Jenness, 1984). Whatever the composition or molecular conformation of the initially formed interface, during the period within which the most thermodynamically stable

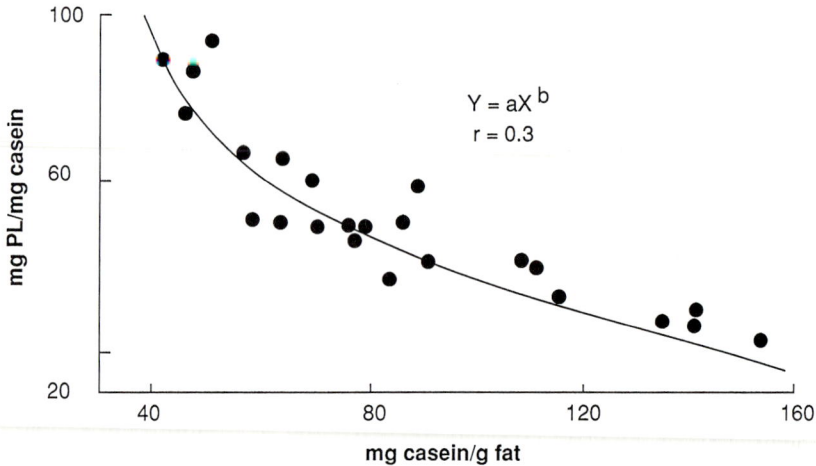

Fig. 1. The amount of casein in the fat phase of homogenised single cream as a function of the relative proportions of phospholipid (PL) and casein in the product.

interfacial state is reached, interchange and rearrangement are likely to occur. Proteins are strongly adsorbed at the interface, having a large number of contact points with the surface. The likelihood of protein desorption is small although in some circumstances one protein can be displaced by another (Musselwhite, 1966); the displacement of gelatin by the more surface-active casein is a well known example of this phenomenon (Dickinson, 1987). However, in the presence of low-molecular-weight emulsifiers, proteins can be partly or completely desorbed and the amount of protein at the interface significantly reduced (Oortwijn & Walstra, 1979). An experimental example of this effect is shown in Fig. 1. Here in a series of single dairy creams containing variable amounts of phosphatidyl choline, the amount of casein associated with the fat phase, interfacial casein, becomes reduced as the phospholipid to casein ratio in the total cream increases. This has important practical significance; McCrae and Muir (1992) investigated the effects on the heat coagulation time of recombined milk of adding egg or soya lecithins. The displacement of protein has also been demonstrated morphologically and quantitatively in ice-cream mixes containing polyoxyethylene sorbitan monooleate, Tween 80 (Goff et al., 1987).

In theory the surfactant which produces the largest reduction in interfacial tension will displace other less surface-active molecules, but in multicomponent food emulsions the situation is complicated by other

TABLE 2
Common emulsifiers with application in cream and ice-cream manufacture

Emulsifier	Reference number	HLB^a value	Use
Monoglycerides	E 471	4	Ice cream, coffee whiteners, imitation creams
Diacetyl tartaric esters of monoglycerides	E 472e	8	Coffee whiteners, imitation creams
Acetic acid esters of monoglycerides	E 472a	2	Whipped products
Lactic acid esters of monoglycerides	E 472b	5–8	Whipped products
Propylene glycerol esters of fatty acids	E 477	3·5	Whipped products
Sodium steroyl-2-lactylate	E 481	10	Imitation creams
Polyoxyethylene sorbitan esters	432–436	10–15	Imitation creams, ice cream
Lecithin	E 322	8–10	
Sucrose esters of fatty acids	E 473	0–18	Ice cream, coffee whiteners

^{a} HLB values are given as a guide only and will vary with fatty acyl chain length and degree of saturation.

factors including component concentration as well as component interactions with each other, and with molecules in the aqueous phase.

Nature of low-molecular-weight emulsifiers

Non-protein emulsifiers used in the food industry are principally esters of medium- and long-chain fatty acids and polyvalent alcohols, sorbitol, sucrose and polypropylene glycol. The properties of the basic molecules may be modified by varying the chain length and degree of unsaturation of the fatty acids or by further derivatisation. Details of the structure and manufacture of food emulsifiers are given in the review of Lauridsen (1976) and their application by Krog (1974); a list of the commonly used emulsifiers in cream and ice-cream products is included for reference in Table 2.

Emulsifiers are characterised by having polar and non-polar moieties and the balance between the hydrophilic and lipophilic parts of the molecule, the so-called HLB value, serves as a general guide to overall polarity. Accordingly the most hydrophilic emulsifiers have high HLB values, the most hydrophobic low values. Functional properties of products can be directly related to HLB values. For example, Needs and

Brooker (1991) have examined the whipping properties of creams containing sucrose ester emulsifiers covering the range of HLB values from 1 to 16. An increase in overrun and a decrease in stiffness were observed as HLB increased. However, in practice the HLB value is not particularly useful in selecting the most appropriate emulsifier in multicomponent systems since, unlike the situation in simple biphasic models, food emulsion properties are likely to be influenced by a number of interactions within the interface and between the interface, dispersed and continuous phases and other factors such as oil phase composition and volume and temperature (see below).

Among such interactions the best documented examples are those involving phospholipids with casein (Griffin et al., 1984), and with β-lactoglobulin (Cornell, 1982; Brown et al., 1983); the effect of such interactions on emulsion functionality remains to be evaluated. However, association between sodium caseinate and commercial monoglyceride (Rahman & Sherman, 1982), which contains a mixture of monoglyceride and diglyceride, improves the viscoelastic properties of the oil/water interface (Doxastakis & Sherman, 1986). In practice emulsifiers are often used in combination to enhance particular emulsion properties. For example, molecular interaction between sorbitan ester and polyoxyethylene sorbitan ester might provide the basis of enhanced stability of oil-in-water emulsions when both emulsifiers are present at the interface (Boyd et al., 1972).

The behaviour of low-molecular-weight emulsifiers is strongly temperature dependent. In the solid state crystal forms occur which are comparable to those already described in Chapter 1, designated as α, β' and β polymorphs in ascending order of stability. In some emulsifiers, above a specific transition temperature in aqueous systems, the hydrocarbon chains become liquid and water is able to penetrate into the polar regions of the molecule to form a liquid crystal lamellar mesophase (Krog & Lauridsen, 1976). Transition temperature is related to the nature of the polar group as well as the chain length and number of double bonds in the fatty acyl chains. At higher temperatures more complex types of mesophase can form including hexagonal and cubic mesophases; in the presence of an oil phase additional transitions may take place. Examples of emulsifiers which form mesophases are monoglycerides, sodium steroyl lactylate, diacetyl tartaric acid esters of monoglycerides, polyoxyethylene sorbitan fatty acid esters, and lecithin. Such emulsifiers are used typically to produce stable emulsions. Further details of mesophase structures and transitions in relation to emulsion functionality are given by Flack and Krog (1974), Krog and Lauridsen (1976) and Fisher and Parker (1988). Emulsifiers which do not form mesophases in water, such as lactic and acetic acid esters of monoglycerides and propylene glycol esters, exist as stable α-form crystals and find application in the controlled destabilisation of whippable emulsions to produce stable foams. This is likely to involve adsorbed crystals at the oil/water interface encouraging globule–globule interaction during whipping.

Molecular arrangement at the interface
Turning to the way in which proteins and low-molecular-weight emulsifiers might coexist at the oil/water interface of a multicomponent emulsion with respect to each other, it is unlikely that they are arranged in separate layers at the surface of the fat droplet (Anderson & Brooker, 1988). The interface is probably more interactive and can be visualised as a composite structure where individual molecules are either adsorbed separately or associate together along discrete segments of the surface. This has been schematically represented in the model of Fisher and Parker (1985).

FAT CRYSTALLISATION

The functionality of oil-in-water emulsions is dependent not only on interfacial properties but also on the characteristics of the fat used in the dispersed phase. The proportion of solid fat and crystal structure influences such factors as mouthfeel and emulsion stability. Here, as with interfacial properties, the situation is complicated by interacting factors which, in addition to triglyceride composition, include fat globule size and temperature treatments during manufacture as well as the nature of the interface.

It has been known for some time that crystallisation in globular fat differs from that in bulk fat. Deeper supercooling is required to initiate the process in globular fat (Phipps, 1957) and the crystallisation rate becomes slower the smaller the globule size (Walstra & van Beresteyn, 1975). The rate of crystallisation affects crystal size; rapid cooling tends to produce small crystals and an increase in the proportion of solid fat. Conversely slow or stepwise cooling tends to produce larger crystals and increase the proportion of liquid fat. In most practical situations nucleation in fat globules is probably heterogeneous since the degree of supercooling required to crystallise the fat in emulsions is less than expected (Mulder & Walstra, 1974).

For dairy creams and ice creams it seems likely that a number of factors which in effect give rise to compositional differences between bulk and globular butterfat account for their respective crystallisation behaviour. Bulk fat has a uniform composition but composition may vary between globules (Mulder & Walstra, 1974). This affects not only fatty acid composition but also the distribution of impurities which can act as nucleation sites for crystallisation. As previously indicated, milkfat contains a number of non-triglyceride constituents. Although it is not clear which of these can catalyse heterogeneous nucleation, the most likely candidates are partial glycerides, since removal of mono- and diglycerides reduces nucleation rate and the addition of monoglycerides increases it (Walstra & van Beresteyn, 1975). Therefore, the more finely dispersed the emulsion becomes, the greater the probability of there being no impurity in some globules which would catalyse heterogeneous nucleation. It has been suggested that the interfacial layer itself might act as a catalyst for crystallisation but there is no evidence for this (Van Boekel & Walstra, 1981). For a comprehensive account of butterfat crystallisation the reader should consult the first chapter of this book and the review by Precht (1988).

In non-dairy creams and ice creams where butterfat is replaced by fats of vegetable origin, typically those fats with high levels of lauric acid, palm kernel oil and coconut oil, the presence of low-molecular-weight emulsifiers not only may affect nucleation but could also, under some circumstances, change the rate at which polymorphic phase transitions take place. The use of emulsifiers to modify crystallisation of some bulk fats is well known. Transition of the α to the β' polymorphic form of tristearin can be retarded in the presence of sorbitan monostearate and glycerol monostearate (Garti, 1988). This phenomenon appears to depend on structure compatibility so that the emulsifier can co-crystallise in the crystal lattice of the triglyceride. Whether intraglobular fat crystallisation is similarly modified when, say, monoglyceride is present at the oil/water interface is a matter of speculation at present.

From a functional point of view, the development of extensive, solid crystal networks during a secondary phase of crystallisation is highly significant. In globular fat the development of networks is confined by the presence of the oil/water interface. However, networks will become more extensive if globules are destabilised, thus changing the rheological properties of the emulsion. Moreover, the extent and location of fat crystals within the globule directly contribute towards destabilisation and functionality. This aspect is discussed in the next section.

Emulsion Stability and Functionality

Any manifestation of instability in an oil-in-water emulsion implies incomplete stabilisation of the dispersed fat droplets by the adsorbed interfacial layer. The way in which emulsifiers influence interactions between fat globules is well documented (Fisher & Parker, 1988). In this section we focus on the role of fat crystallisation in determining stability. Instability which has relevance to dairy products can be broadly described by four phenomena: flocculation, clustering, clumping and coalescence.

Flocculation or agglutination is a loose association between adjacent globules resulting from weak forces. In cold raw milk flocculation is enhanced by the so-called cold agglutinin factor, which is destroyed by pasteurisation. Flocculated globules retain their individual identity and the flocs are easily dispersed by stirring.

Clustering occurs when adjacent globules share part of their interfacial material. This is one consequence of homogenisation in dairy creams where the amount of surface-active material available to stabilise the increase in fat surface area produced may be limited, especially as the fat phase volume increases (Phipps, 1983). In this situation there is widespread linking of globules by the casein micelle bridges (Anderson *et al.*, 1977*a*), referred to in the section 'Nature of the interface', which lead to increased viscosity and can encourage fat separation. Unlike floccules, clusters are difficult to disrupt.

Clumping, or partial coalescence (Boode, 1992), is the aggregation of globules arising from direct fat–fat interaction to produce a continuous fat phase between two or more adjacent globules which partially retain their individual identity. Clumping is possible only when part of the fat is solid and is most likely to occur when the chance of collisions between globules is increased by mechanical effects such as shear, or by high fat concentrations. It is an essential step during the whipping or churning of cream. A good illustration of the undesirable effects of clumping is plug formation in UHT whipping cream. When clumps are warmed to melt the crystalline portion of fat, they either coalesce to form larger globules or oil-off.

Not only is the presence of solid fat required for clumping but the adsorption of fat crystals at the oil/water interface is also important. If intraglobular crystalline fat is examined under polarised light, three crystal forms can be distinguished. These were described by Walstra (1967) as a needle type, with small or large needle-like crystals distributed throughout the globule, a layer type with crystals concentrated around the periphery of the globule and a mixed type which represented a combination of the other two forms.

Subsequently Van Boekel and Walstra (1981) noted that globules containing layer- or mixed-type crystals were unstable to coalescence in a shear field and suggested that this was a direct result of crystal adsorption

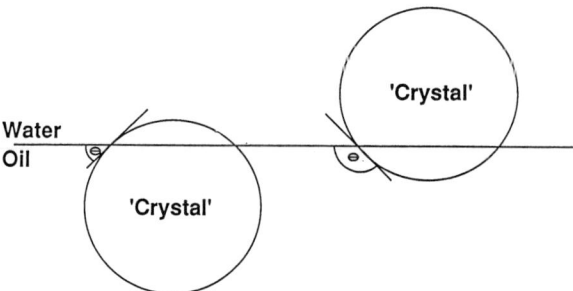

Fig. 2. Influence of the contact angle (θ) on the position of a hypothetical fat crystal adsorbed at the oil/water interface.

at, and protrusion through, the interface. They further proposed that protruding crystals could penetrate the interface of an adjacent globule. Clumping would result because the penetrating crystal would be preferentially wetted by the oil phase of the second globule.

The position of a crystal at the interface can be defined by the contact angle between the oil, water and crystal phases. This is illustrated in Fig. 2. Taking the contact angle through the oil then the smaller the angle the more the crystal is wetted by the oil phase. As the angle increases the crystal becomes wetted more by the water phase and therefore protrudes further. For a given contact angle the extent to which the crystal protrudes is influenced by crystal size and shape. The contact angle varies according to the prevailing value of the interfacial tension. A low interfacial tension, resulting from the adsorption of low-molecular-weight emulsifier, increases the angle.

Experimentally, Darling (1982) found that the churning times of model emulsions, in which different surfactants were used to change the contact angle, were related to the angle. Furthermore, as predicted by the theory and the suggestions of Van Boekel and Walstra (1981), the higher the contact angle the shorter the churning time became. Darling (1982) proposed that the mechanism of destabilisation was more likely to result from a local reduction of the interaction energy between globules at the point of crystal adsorption, than the original theory of interfacial penetration. Boode (1992) has demonstrated that partial coalescence is reversible, at least for a short time after aggregate formation, by changing the wetting properties of the crystals by addition of SDS. Under such conditions the crystals migrate to the aqueous phase and the clumped globules may separate. This supports the view that protruding crystals

TABLE 3
Influence of emulsion characteristics on stability

	Most stable	Least stable
Type of interface	Protein	Protein/emulsifier
Fat globule size	Small	Large
Fat crystal size	Small	Large
Fat crystal distribution	Homogeneous	Peripheral
Solid fat content	Low	High

exert an important influence on emulsion stability and may well account for plugging effects in cream or bottled milk where fat concentration is high. It is probably also important in the more controlled and desirable clumping of globules which occurs during churning (Mulder & Walstra, 1974) and cream whipping (Anderson et al., 1987).

The interactions between fat crystallisation, interfacial composition and functional properties are well illustrated by the phenomena described by Barford and Krog (1987) in spray-dried, non-dairy whippable toppings. Such products typically contain a lauric fat, casein and a lipophilic emulsifier. When spray-dried, the fat phase exists in the usual globular form but part of the fat appears to be in a supercooled state. Barford and Krog suggest that fat crystallisation is inhibited by protein. This effect only appears to be attributable to casein and is confined to lauric fats; no effect was evident with butterfat or hydrogenated soyabean oil (Krog et al., 1987). However, following the rehydration of formulations containing polypropylene glycol monostearate or distilled monoglycerides, the fat undergoes a transition into aggregates of crystalline platelets whose presence is linked to optimum whippability (Krog et al., 1987). Following rehydration, Barford and Krog propose that protein is desorbed from the interface to be replaced by emulsifier; crystallisation of the fat then accelerates.

The way in which variations in the interfacial composition, fat crystallisation and globule size influence stability towards clumping are summarised in Table 3. The influence of these factors, and others such as fat composition, temperature cycling and shear, has been investigated by Boode (1992). She has developed a simulation model and refined the earlier work of Van Boekel and Walstra (1981). When comparing emulsions with different globule sizes it becomes clear that the smaller globules will tend to be the more stable. Similarly, if the emulsions differ in fat crystal size, instability becomes more likely as crystal size increases. It is important to emphasise that these factors are interdependent and that both major and minor constituents play an important role in the functionality of the fat phase in dairy products.

THE CONTINUOUS PHASE

Stability and other functional properties of dairy oil/water emulsions depend not only on the nature of the interface but also on interactions between components of the continuous phase both with each other and with the interface.

The structure of the continuous phase is primarily influenced by its composition. As well as milk proteins, lactose and inorganic constituents, polysaccharides or proteins may be added as stabilisers; in ice cream, ice crystals are also located in the continuous phase. By manipulating the amount of and balance between these constituents, there is scope for controlling the viscosity of the continuous phase so that very viscous or gelled phases may be achieved. Common polysaccharides used in dairy product manufacture are shown in Table 4.

Stabilisers are used to prevent or reduce phase separation, to ensure even distribution of constituents and aid stabilisation of air bubbles in dairy products, especially those with a relatively low fat content such as mousse, ice cream and yoghurt. In ice cream, for example, unlike whipped cream there is insufficient fat to form a network to stabilise the air cells; stability is supplemented by the obstructions posed by the ice crystals and the high viscosity of the unfrozen aqueous phase. Here the aqueous phase exists as a glass if the temperature is sufficiently low.

As we shall see, polysaccharide stabilisers can also influence emulsion stability in other ways, for example, by adsorption at the liquid–liquid interface or through incompatibility with proteins in the aqueous phase resulting in phase separation.

In this section we describe the principal characteristics of stabiliser molecules and how they exert their influence on products through the interactions in which they participate.

Proteins
Casein and whey proteins exert their major influences at the fat/serum and air/serum interfaces of both cream and ice cream as we have described earlier. In particular whey proteins can undergo significant changes during processing of the milk to be used in cream and ice cream (Tamime &

TABLE 4
Common polysaccharide stabilisers of cream and ice cream

Name	Structural characteristics	Charged groups	Industrial source
Guar gum	Linear mannan chain substituted with single galactose residues (40%)	None	Plant (*Cyamopsis tetragonolobus*)
Locust bean gum	Linear mannan chain substituted with single galactose residues (~23%)	None	Plant (*Ceratonia siliqua*)
Carrageenan	Linear chain of partially sulphated galactose and anhydro-galactose residues	—SO_3^-	Red seaweed (*Rhodophyceae*)
Alginate	Linear chain of mannuronic acid and guluronic acid residues	—COO^-	Brown seaweed (*Phaeophyceae*)
Sodium carboxymethyl cellulose	Linear chain of glucose residues, partially carboxymethylated	—CH_2COO^-	Synthetic
Methyl cellulose	Linear chain of glucose residues, partially etherified with methyl groups	None	Synthetic
Xanthan	Linear glucan chain substituted on alternate glucose residues with a trisaccharide unit	—COO^-	Bacterium (*Xanthomonas campestris*)
Pectin	Linear chain of partially methyl esterified galacturonic acid residues with occasional rhamnose residues	—COO^-	Plant (apple, citrus)

Robinson, 1985). These proteins exist in a globular conformation in their native state with their hydrophobic residues folded into the interior of the molecule, protected from the aqueous environment. The degree of protein denaturation that takes place depends on the processing conditions, particularly the heat treatment used. Denaturation exposes functional groups which can then take part in association, and possible aggregation, of protein chains enabling them to bind significant amounts of water. These interactions may involve hydrophobic, hydrogen or electrostatic bonding. The extent of chain association will depend on protein concentration, pH, type and concentraton of ions and competition for water molecules. It is well known that denaturation of β-lactoglobulin by heating above 65°C exposes sulphydryl groups which can interact with k-casein; α-lactalbumin can subsequently interact with this complex (Tamime & Robinson, 1985).

Gelatin is used in some ice-cream formulations because of its ability to form a gel network in the aqueous phase (Ledward, 1986). However, gelatin is also adsorbed at the oil/water interface and as already mentioned can be displaced by casein from a planar oil/water interface. Furthermore, it is known that in casein/gelatin mixtures the two proteins are initially adsorbed at the oil/water interface in similar proportions to their concentrations in the aqueous phase (Dickinson, 1986), followed by the gradual displacement of gelatin by casein from the interface. All of the gelatin will be displaced provided the concentration of casein is sufficiently high. Clearly the situation is much more complex in an ice-cream mix emulsion. However, gelatin is unlikely to have a significant presence at the interface under normal processing conditions, although manipulation of these conditions can change this. In a formulation containing both casein and gelatin a fat globule membrane composed of both proteins can be formed (Musselwhite & Walker, 1969). This is achieved by preparing the mix in two stages and ensuring that the first mix contains a significant excess of gelatin. No added emulsifier is used in this formulation and the ice cream is claimed to have very good stand up properties.

Protein/Polysaccharide Interactions

The ability of proteins adsorbed at the interface to react specifically with polysaccharides has not been widely reported. However, there is one exception and that is the interaction between carrageenans and caseins. k-Carrageenan and k-casein can interact in the presence and, more surprisingly, absence of Ca^{2+} (Snoeren et al., 1975, 1976). This interaction may be responsible for the apparent binding of both k- and λ-carrageenan to casein micelles of milk which was originally observed in electron micrographs

(Hood & Allen, 1977). The interaction between negatively-charged k-casein and carrageenan at neutral pH in the absence of calcium is believed to be an electrostatic one made possible by the presence of positively-charged amino-acid residues on k-casein (Snoeren et al., 1975). α_s- and β-casein also bind to carrageenan, particularly in the presence of Ca^{2+} (Skura & Nakai, 1980, 1981; Ozawa et al., 1984, 1985) suggesting that the negatively-charged polymers are cross-linked by Ca^{2+} ions. Evidence for the binding of carrageenan to casein micelles has been obtained from measurements of electrophoretic mobilities and diffusion coefficients of complexes formed in dilute milk solutions (Dalgleish & Morris, 1988). It appears that at these dilute concentrations (up to 0·02% w/v carrageenan) carrageenans bind to the casein micelles until the surface is covered, followed by aggregation of the complexes. Aggregation of disordered, random coil λ-carrageen and an ordered, helical i-carrageenan was similar. However, k-carrageenan in its random coil conformation did not cause aggregation of casein micelles, but brought about precipitation of the micelles when converted to its helical conformation.

The ability of i- and k-carrageenan to form gel networks (Rees et al., 1982) and to interact with casein micelles is exploited in the stabilisation of some ice-cream mixes which can separate on standing. In particular, ice-cream mix emulsions stabilised by guar gum, locust bean gum or carboxymethyl cellulose can separate during storage into a phase rich in fat and casein, and a serum phase containing polysaccharide. The underlying cause of such behaviour has not been well documented. As both ionic and neutral polysaccharides cause separation, ionic interactions alone cannot be responsible. Instability of the emulsion may be due to a thermodynamic incompatibility of the proteins and polysaccharides in the aqueous phase or to depletion flocculation caused by non-absorbing polymers (Tolstoguzov, 1986; Dickinson 1987).

Geilman and Schmidt (1992) have examined the properties of frozen desserts made from ultrafiltered milk with added sugars. Differences in hardness and melting properties could not be attributed solely to differences between carbohydrates such as glucose and fructose. The observed effects may have resulted from protein–carbohydrate interactions.

Polysaccharide Interactions

The majority of polysaccharides are not surface-active and would not be expected to adsorb at either the air/serum or fat/serum interface. Surface-active polysaccharides such as methyl cellulose, hydroxypropylmethyl cellulose and propylene glycol alginate are as effective as proteins and small molecule surfactants in lowering the interfacial tension of oil-in-water emulsions (Darling & Birkett, 1987). Furthermore, methyl cellulose and ethylhydroxyethyl cellulose can decrease the rate of flocculation of oil droplets in oil-in-water emulsions containing added emulsifiers by their ability to adsorb at the oil droplet surface, and act as steric stabilisers (Bergenstahl, 1987).

The molecular interactions which polysaccharides undergo to control the rheology of aqueous systems has been widely studied and reviewed elsewhere (Rees et al., 1982) and will not be described in detail here. As a consequence of the restricted mobility of polysaccharide chains, they have a large hydrodynamic volume. At relatively low concentrations the chains begin to overlap with one another forming an entangled network capable of binding large volumes of water and leading to high solution viscosities. Such behaviour is entirely non-specific and is shown by disordered polysaccharides such as guar gum and λ-carrageenan as well as sodium alginate in the absence of calcium ions. Other polysaccharides are capable of forming gels through specific chain–chain interactions. Individual chains do not interact completely with one another thus avoiding precipitation. However, conformationally ordered regions of chains associate non-covalently and build up a gel network which can immobilise large quantities of water. Other regions of the chains have features which prevent association in this manner and act to solubilise the polymer. Gels can be formed by polysaccharides such as alginate in the presence of calcium, i- and k-carrageenan, pectin, locust bean gum and modified starches. Xanthan, which does not gel on its own, can interact synergistically with locust bean gum to form a gel. Some polysaccharide systems, e.g. xanthan or sodium carboxymethyl cellulose, which have properties between those of random coils and gels, have been called 'weak gels'. Caldwell et al. (1992) have examined the network structure of a number of polysaccharide stabilisers in chilled solutions by low-temperature scanning microscopy. In the absence of sucrose very different structures were observed; however, the presence of 20% sucrose in the stabiliser solutions dominated the network structures formed which were similar.

Ice Crystals

One major point of difference between cream and ice cream is that ice cream contains ice crystals as an integral part of the product. As the temperature of the ice-cream mix is lowered in the freezer, and subsequently, during the hardening process, water is removed in the form of ice crystals. The average size of the ice crystals is 20–50 μm (Berger, 1990), with a separation between crystals of 6–8 μm. Crystal size can be controlled by the rate of freezing. Both formulation and the temperature of the product determine the amount of ice present. Low-molecular-weight carbohydrates such as glucose, sucrose and lactose are present at quite high concentrations in the ice-cream mix, and will have a major influence on ice formation owing to their ability to depress the freezing point of the mix.

As ice crystals are formed, the residual aqueous phase containing salts, sugars and macromolecules becomes concentrated. This unfrozen matrix exists as an amorphous solid (i.e. a glass) below the sub-zero glass transition temperature and as a viscoelastic liquid (i.e. a rubbery fluid) above this temperature. A typical ice cream at $-29°C$ has a concentration of sucrose and lactose of 86% (w/v) existing as a glass with a sub-zero glass transition temperature of about $-23°C$ (Berger *et al.*, 1972). More recently, glass transition temperatures of -30 to $-43°C$ have been given for typical ice-cream products (Slade & Levine, 1988).

Ice recrystallisation and lactose crystallisation are two events which must be controlled during storage of ice cream. Ice crystallisation is thought to occur by diffusion-controlled maturation where large crystals grow at the expense of smaller ones (Harper & Shoemaker, 1983; Levine & Slade, 1986; Blanshard & Franks, 1987). The glass existing in ice-cream products at sufficiently low temperatures prevents the flow of viscous liquid and stops ice crystallization (Levine & Slade, 1986). Other diffusion-controlled processes such as lactose crystallisation will also be affected. A correlation has been demonstrated between the molecular weight of starch hydrolysis products such as dextrins, maltodextrins and glucose syrups and the sub-zero glass transition temperature (Levine & Slade, 1986). The ability to manipulate this glass transition temperature by incorporating the appropriate starch hydrolysis products enables softer frozen products to be developed (see later).

One of the functions widely attributed to polysaccharide stabilisers is their ability to control ice-crystal formation and growth (e.g. Arbuckle, 1986). Caldwell *et al.* (1992) have measured ice-crystal size by electron microscopy and image analysis. The ice crystals were initially smaller and remained more stable in stabilised ice creams than in control samples. Nucleation of ice in water or in sucrose solution does not appear to be significantly affected by common ice-cream stabilisers such as sodium alginate, carboxymethyl cellulose, guar gum, locust bean gum or xanthan (Muhr *et al.*, 1986). However, reduction in the rates of ice-crystal growth can occur in sucrose solution and this effect is more pronounced for gel-forming polymers such as alginate and agar (Muhr & Blanshard, 1986). The ability of stabilisers to greatly increase the viscosity of the aqueous phase may be important in restricting the mobility of water molecules. Furthermore, the significant increase in the concentration of these polymers in the unfrozen aqueous phase of ice cream undoubtedly leads to molecular entanglement of disordered polymers (Morris *et al.*, 1981). Here, because of the concentration of the polymers and the reduced water activity, specific chain–chain interactions may be promoted, enabling gel formation to take place. Such behaviour might be expected to hinder diffusion-related events such as ice recrystallisation (Levine & Slade, 1986).

PROCESSING VARIABLES

Fundamental to the properties of cream and ice cream is the nature of the emulsion produced during the process of homogenisation. The major effect of other processing variables, and in particular temperature treatments, is on the physical properties of, and interactions between, components present in the formulation. Although such factors are important, homogenisation is perhaps the single processing factor which has the greatest impact on product characteristics.

Homogenisation
Homogenisation is achieved in several different ways. The most common equipment in dairy processing is the valve homogeniser, of which there are a large number of designs. High speed mixers or blenders are also used, but primarily to prepare a prehomogenisation mixture.

The high speed mixer/blender depends upon high hydraulic shear generated by the high speed rotor blades. Material is driven out through the working head by centrifugal forces and as it passes through the perforations in the working head it is subjected to mechanical shear. In a batch mixer a continuous cycle is set up as the material is repeatedly drawn in by the rotating blades. This type of machine has many applications in the food industry (Aarons & Hepner, 1975), but in cream production it is most appropriate in formation of pre-emulsions for imitation products.

The valve homogeniser
The characteristics of the valve homogeniser have been comprehensively reviewed elsewhere (Phipps, 1985). What follows is a general description of the process.

The homogeniser consists essentially of a high-pressure positive displacement pump with between three and seven pistons—to reduce pressure variations—and the homogenising valve itself. There are four principal valve designs in use: simple flat face (e.g. Manton-Gaulin); perforated discs sandwiched between flat faces (e.g. Cherry-Burrell); corrugated faces (e.g. Rannie), and cone-shaped compressed stainless steel wire (e.g. Crepaco). Valves with corrugated surfaces offer increased efficiency and energy utilisation; however, the surfaces are more difficult to machine and maintain.

Hydraulic cavitation, which is similar to the acoustic cavitation described above, has been thought to be the main force responsible for globule break up in the valve homogeniser (Mulder & Walstra, 1974). However, it now appears that cavitation happens under a very limited set of conditions (Phipps, 1985). In particular, at a modest homogenising pressure flow conditions change and cavitation no longer occurs.

Turbulent flow has also been suggested, with some reservation, as the major force responsible for globule size reduction (Mulder & Walstra, 1974). However, it has been shown that effective homogenisation occurs at low turbulence levels (Phipps, 1985).

Phipps (1985) has demonstrated that viscous shear forces play a dominant role in fat globule disruption. Simple shear flow alone will not cause globule break up since the viscosities of the fat and milk serum are too dissimilar. For globule disintegration to occur, the ratio of dispersed and continuous phase viscosities must be less than 4 (Mulder & Walstra, 1974). Fluid flow contracts as material enters the valve slit and there is a sharp increase in velocity which causes a drop in pressure. This velocity falls during passage between the surfaces of the valve. The high convergent flow and the velocity gradient tend to smooth turbulence but produce strong shear forces. Initial extensional flow will elongate globules which break up as a result of deceleration in the radial flow (Dickinson & Stainsby, 1982).

Factors affecting homogenisation

Efficiency of homogenisation is estimated in terms of the extent of globule size reduction. Valve design, as already mentioned, influences homogeniser effectiveness but for a particular piece of equipment there are three main variables controlling efficiency. They are pressure, temperature and characteristics of the starting material. Far from being independent, these variables display a marked interdependence, the most notable of which are between protein content, fat content and pressure. For milk, cream containing 12% fat and other dilute emulsions, there is a linear relationship between log d (the mean diameter of homogenised globules) and log P (pressure) (Goulden & Phipps, 1964). Over a pressure range 5-25 MPa, it has been shown that $d \propto P^{-q}$ where q is the slope of the log plot of d against P. The exponent q is affected by temperature and fat volume fraction.

As the fat content of the emulsion increases the relationship becomes increasingly curvilinear at the higher pressures. For a given homogenisation pressure, the mean fat globule diameter increases with fat content. Varying temperature between 40 and 80°C shows little effect on the homogenisation efficiency in low fat (< 10%) creams. At higher fat levels the temperature increase brings about a progressive improvement. Temperature and fat content both affect viscosity and, as already mentioned, this influences the action of shear within the valve. Another effect of high fat content is an increased probability of rapid coalescence of the newly formed globules.

Other characteristics of the starting material which may affect the extent of homogenisation include globule size distribution, fat to protein ratio, and the presence of non-protein surface-active ingredients in the aqueous phase. The natural variation in globule size distribution in unhomogenised milk and cream is insufficient to affect homogenisation. However, the formation of a blend or pre-emulsion in the manufacture of non-dairy creams and ice cream can influence the structure of the final product. The extent of droplet size reduction during the first stage of a two-stage homogenisation is the usual determinant of the product size distribution.

A typical mean globule size in unhomogenised milk and cream is 3–4 μm. In commercial dairy practice, homogenisation reduces this to between 0·6 and 1·0 μm, according to the conditions used. As the fat volume fraction increases so the concentration of available surface-active material decreases (Table 1) relative to the fat surface area generated during homogenisation. This can have serious implications for both the short- and long-term stability of the emulsion. Phipps (1983) has shown that there is a maximum fat surface area which can be produced that is similar irrespective of homogenisation pressure and the fat volume fraction of the cream. This maximum fat surface area becomes a limiting factor in determining the minimum mean globule diameter for cream with a particular fat volume fraction.

Separation

Separation is the processing factor which directly controls the gross composition of the cream product. Centrifugal separators exploit the differences in density between the milkfat globules and the aqueous skim

TABLE 5
Effect of separating temperature on free fatty acid development (mequiv./100 g fat) in cream

Separating temp. (°C)	Free fatty acid concentration	
	Warmed milk	Cream
30	1·34	2·06
40	1·08	1·47
50	0·84	1·20
60	0·71	0·88

From Needs et al. (1985).

phase. The efficiency of the separator and the fat content of the cream are controlled by the flow rate and also the temperature. Although there are machines designed for cold separation (4°C) and others which pasteurise and separate simultaneously, the optimum temperature is around 50°C. Operating between 20 and 40°C increases problems of globule stability caused by the presence of both liquid and solid fat which can enhance clumping (Te Whaiti & Fryer, 1975). Another problem of separating raw milk below 50°C relates to the presence of milk lipoprotein lipase. This enzyme releases free fatty acids leading to a rancid flavour. Table 5 shows the effect of different temperatures on the level of free fatty acid in milk warmed before separating and in the separated cream (Needs et al., 1985). Below 50°C cream appears to be more susceptible to elevated levels of free fatty acids. The difference between warmed milk and cream values is probably the result of an increase in available substrate resulting from partial fat globule damage during separation.

PRODUCT CONSIDERATIONS

We have already emphasised how in principle the nature of the dispersed oil droplet, its phase volume and the composition of the continuous phase can influence the functionality of oil-in-water dairy emulsions; the fat content of a range of dairy and dairy-substitute products is given in Table 6. Further details on cream legislation are given by Rothwell (1989). Ultimately it is the way in which such compositional variables interact during the stages of the manufacturing process which determines product performance. In this section we consider three types of product which illustrate the factors involved.

TABLE 6
Composition of creams, ice creams and related products

Product	UK legal min. fat (%)	Range fat (%)
1. Creams		
Double	48	48
Single, thickened single	18	18
Whipping, whipped	35	35–40
Spooning		30
Half and coffee	12	12–25
Clotted	48	50–60
Cultured, sour		15–20
Aerosol		32
Plastic		80
Confectionary		40
Sterilised	32	
Powdered, dry		40–70
Butter		40
Synthetic		10–48
2. Frozen desserts		
Dairy/non-dairy		
Ice cream	5	5–11
Mousse		9
3. Miscellaneous		
Powdered coffee whitener		23–35
Liquid coffee whitener		10
Evaporated milk		9
Cream liqueur		12–16

Low-fat Creams

Single cream

There are a number of cream products which have a fat content in the range between 10 and 20%. The shelf life of these creams is limited mainly by the separation of the fat phase during storage to form a cream layer or under certain circumstances a cream plug. Homogenisation reduces this problem. Its use also provides a means of varying viscosity (Rothwell et al., 1989a) but does reduce the heat stability of the cream (Sweetsur & Muir, 1983). To produce creams with the desired properties, appropriate selection of homogenisation conditions and control of the cooling rate are needed.

The creaming rate in low-fat pasteurised creams tends to follow the

Stokes equation, but this is influenced by fat content, globule size and the composition of the globule membrane (Walstra & Jenness, 1984). As a generality, minimum globule size will produce maximum reduction in phase separation. However, there is a lower limit of mean globule size in single dairy creams, around 0·6 μm, resulting essentially from the limited presence of naturally occurring surface-active material, mainly protein. Under severe conditions of pressure and temperature, 'over homogenisation' can take place. This produces globules which are incompletely stabilised allowing clumping to take place (Ogden et al., 1976) and encouraging the formation of large fat aggregates, usually leading to the creation of a fat plug.

As discussed earlier the formation of homogenisation clusters can produce an increase in viscosity thus promoting an apparent creaminess. This effect may be enhanced by reducing the rate at which product is cooled after heat treatment. An initial cooling step down to around 30°C followed by a slow reduction to around 4°C induces a thickening effect. The network of clusters formed is disrupted by stirring. A similar temperature sequence can be used to produce a 'spoonable' cream containing 30% fat or a thickened double cream. However, much less severe homogenisation conditions are used in these creams, say 3.5 MPa compared with 13·8–17·3 MPa for single cream, depending on the plant being used. Increased viscosity can also be achieved by adding polysaccharides or modified starches.

Coffee cream
Coffee cream usually contains between 12 and 18% fat which has been heated to UHT temperature and aseptically packed into individual portion pots. Although sterilisation may also be carried out by an in-bottle process, comments here relate specifically to UHT cream. Coffee whiteners are synthetic emulsions, sometimes liquid but often powdered. Both products must remain stable during storage, but, more especially, must withstand addition to hot coffee at low pH. A physico-chemical defect which sometimes occurs when homogenised cream is added to hot coffee is the formation of an unpleasant curd which rises to the surface. This instability is known as feathering and is usually only associated with homogenised dairy creams. The destabilised material consists of floccular aggregates comprising fat globules linked by a precipitated casein matrix (Anderson *et al.*, 1977*a*). This type of instability should not be confused with a similar problem that sometimes occurs in UHT creams resulting from the growth of *Bacillus cereus* following incomplete product sterilisation.

Susceptibility to feathering is strongly linked to homogenisation conditions and the relative amounts of casein and calcium in the cream. It also tends to increase with the age of the cream. It seems likely that feathering is linked to the formation of clusters in homogenised cream. Increases in viscosity with ageing (Phipps, 1982) or in creams with a pH above 6·5 (Buchheim et al., 1986) have been coupled with a tendency for feathering to occur. Such changes in viscosity accompany an increase in cluster size. Casein loading on the globule surface and the extent to which bridging between globules is present may be crucial for the heat stability of coffee cream. It appears that as casein bridging is reduced so is the risk of feathering. This can be achieved by reducing the calcium content and providing additional protein as sodium or potassium caseinate (Anderson et al., 1977b). Sequestering calcium may also be beneficial, but does not reduce bridging sufficiently to prolong shelf life beyond about 8 weeks.

Optimum conditions to provide sufficient reduction in globule size to avoid phase separation, and at the same time preventing excessive clustering, will differ between plants. However, two-stage homogenisation has been advocated by Abrahamsson et al. (1988) following a comprehensive study of homogenisation conditions. It is generally accepted that the use of the second stage reduces clustering.

Coffee whitener stability is easier to control. Sodium caseinate is usually used as the emulsifying protein which may be protected by interaction with a hydrophilic colloidal stabiliser such as carrageenan and supplemented by the addition of low-molecular-weight lipid emulsifiers.

Cream liqueur

It is not the purpose of this section to examine in detail all aspects of cream liqueur production but to highlight the principal requirements and their underpinning similarities to those of coffee cream. For a comprehensive discussion of cream liqueur production and stabilisation the reader is referred to Banks and Muir (1988).

A cream liqueur combines the flavour of an alcoholic drink with the texture of thickened cream in a product expected to withstand a prolonged shelf life at ambient temperature. These features impose severe requirements on emulsion stability, formulation and processing. The composition of a typical cream liqueur has been described by Banks et al. (1981). Double cream is normally used as the source of fat which is included to a fat concentration of 16% (w/w). Alcohol, usually a blend of neutral spirit and whisky, is included to an ethanol concentration of 14% (w/w). Water makes up 46% (w/w) and the other major component (in terms of weight) is added sugar. To produce an emulsion in which creaming is kept to a minimum and cream plug formation is avoided requires extreme homogenisation conditions. Banks *et al.* (1983) found that two-stage homogenisation, with a pressure of 31 MPa (4500 psi) for both stages, was needed to minimise creaming. A stable emulsion can be achieved if 98% of the fat globules have a diameter of less than 0·8 μm (Muir *et al.*, 1991).

We have already outlined the limitations conferred on homogenisation performance by the naturally occurring levels of protein in cream. Similar problems arise here necessitating protein supplementation. Casein has been found to be the only protein able to provide the required long-term stability, and up to 3·5% (w/w) sodium caseinate is included in most cream liqueur formulations. Moreover, the long-term stability of cream liqueur emulsions, like that of coffee creams, is sensitive to cluster formation and calcium concentration. Commercial sodium caseinate is variable in quality (Muir & Dalgleish, 1987) and performance possibly owing to differences in calcium content. Interaction between sodium caseinate and available calcium may lead to gelation and syneresis during storage. A further improvement to product performance can be achieved by including sodium citrate to complex the calcium ions.

The use of low concentrations (0·5%) of low-molecular-weight lipid emulsifiers such as glycerol monostearate or sodium steryl lactylate is not beneficial. As might be anticipated from earlier discussion they displace casein from the fat surface thereby resulting in a decreased emulsion stability (Dickinson *et al.*, 1989).

Whipping Cream

As well as the normal requirements of any cream product (e.g. emulsion stability during processing and storage) whipping cream must maintain stability as a three-phase system after whipping. A network of fat globules holds the incorporated air bubbles and the continuous liquid phase in a semi-permanent rigid foam structure. Development of the whipped structure and important compositional factors which can affect whipping properties have been discussed in detail elsewhere (Brooker *et al.*, 1986; Anderson & Brooker, 1988; Needs & Huitson, 1991.) Whipping causes partial destabilisation of the cream emulsion. In the initial stages of the process fat globules adsorb to the surface of air bubbles whose interfacial layer is composed of milk serum proteins, dominated by the presence of β-casein. Globule–globule interaction is enhanced under the prevailing shear conditions created by the whipper; the role of fat crystals in promoting such interactions has already been mentioned. Gradually a network of partly coalesced fat globules builds up to surround the air bubbles and form a three-dimensional matrix which provides the foam's rigidity. It is most unlikely that the MFGM plays any role in determining the properties of the foam *per se* but instability such as clumping in unwhipped cream may be linked to premature collapse of the whipped structure (Anderson *et al.*, 1987). Instability is probably attributable to fat globule damage during separation. Examination of whipping creams by freeze fracture transmission electron microscopy reveals that extraglobular fat crystals are rapidly adsorbed at the air/serum interface as predicted (Brooker, 1990). The presence of large numbers of large crystals adsorbed at the air/serum interface may tend to restrict adsorption of fat globules thus preventing the development of a satisfactory fat globule matrix. Fat globule damage resulting in the formation of large extraglobular fat crystals may also lead to release of liquid fat which would have a profoundly destabilising effect on air bubbles. Extensive fat globule damage may produce creams that fail to whip or form unstable foams.

There has been a tendency to associate poor whipping properties with milk fat of low melting point. Banks *et al.* (1989) have investigated effect of melting point and fatty acid composition on whipping properties. Addition of palm, olive, soya or linseed oil to the diet was intended to produce milk fat with an increasing proportion of unsaturated fatty acids. There was a significant increase in the proportion of milk fat that was liquid at 5°C. The whipping properties of cream produced from Friesian milk were unaffected by changes in fat composition. However, it is generally accepted that there is an effect of season on whipping properties, especially late spring, and this has been attributed to an increase in soft fat associated with springtime grass feeding (Rothwell *et al.*, 1989*b*).

Whipping properties are sensitive to processing conditions. Homogenised cream generally performs less well than natural cream. This has been attributed to reduced fat globule size and particularly the size and number of clusters induced by homogenisation (Graf & Muller, 1965). However, it seems more likely that changes in the nature of the oil/water interface are responsible. As we have seen, the adsorption of fat globules at the air/serum interface and the build up of a network of partially destabilized fat globules are key elements in the production of a stable three-dimensional foam. The adsorption of unhomogenised fat globules, stabilised by the MFGM, will result in a greater reduction of interfacial tension than when the primarily casein stabilised globules of homogenised cream are adsorbed; a reduced rate of adsorption may therefore be anticipated with homogenised creams.

Other processing variables have much less effect than homogenisation on

TABLE 7
The main effects of processing factors on the properties of UHT whipping cream

	Cream property			
	Whipping time (s)	Overrun (%)	Viscosity (MPa s)	Fat globule diameter (m)
Separation temp. 40–60°C	***	ns	***	ns
Homogeniser position up–down	***	ns	***	*
Homogeniser pressure 0–4·1 MPa	***	ns	*	***
Heat treatment 135–145°C	**	ns	***	ns
Filling temperature 5–15°C	ns	ns	ns	ns

Statistical significance of changing processing factor from the low to the high setting: ns, non-significant; *, $P<0.05$; **, $P<0.01$; ***, $P<0.001$.

whipping properties but may, nevertheless, exert an influence. Because of the need to preserve the physical appearance of a product during storage while retaining acceptable whipping properties, UHT dairy whipping cream is particularly sensitive to processing conditions. This is illustrated in Table 7 which presents the results of a study into the effects of changing five processing factors from a low setting to a high setting on the properties of UHT whipping cream. The position of the homogeniser induced highly significant changes in whipping time and viscosity, accompanied by a less significant change in fat globule size; overrun was unaffected. Observations by Van Boekel and Folkerts (1991) who examined the effects of UHT heat treatment on the stability of natural and homogenised fat globules suggest that such variations in whipping properties may be linked to changes in fat globule size.

As with other processing conditions, it is not possible to set out an ideal combination of variables, since individual plants have their own characteristics. Nevertheless, patents have been filed (Branciaroli, 1986) which define processing conditions for heat treatment, homogenisation and packing with limits for time, temperature and pressure conditions which claim to produce improved whipping properties. However, these studies illustrate the extent to which processing variables can interact to affect product quality. The way in which separating temperature and homogenisation condition can interact to influence cream properties is shown in Table 8. A combination of homogenising (600 psi) and high separating

TABLE 8
The interactions between homogenisation conditions and separating temperature on the properties of UHT whipping cream

	Separating temp. (°C)	Cream property	
		Whipping time (min)	Overrun (%)
Homogenisation pressure (MPa)			
0	40	81	137
0	60	116	142
4·1	40	177	142
4·1	60	238	120
Level of significance		$P<0.015$	$P<0.001$
Homogenisation position			
Upstream	40	130	167
Upstream	60	140	134
Downstream	40	132	153
Downstream	60	187	148
Level of significance		$P<0.015$	$P<0.003$

temperatures was very effective at reducing free fat. However, whipping time was much increased and there was a significant loss of overrun. High separating temperature in combination with downstream homogenisation resulted in longer whipping times than other combinations of these two factors. Separating temperature had an effect on overrun in creams homogenised upstream while for downstream homogenised cream no effect resulted from a change in separating temperature.

Within the limitations imposed by inherent composition it is difficult to manufacture an entirely satisfactory UHT dairy whipping cream by optimisation of processing conditions alone. Phase separation during storage is a major problem. Because of the high fat content of whipping cream the risk of 'over homogenisation' leading to enhanced cream plug formation is quite high.

Ice Cream

Rather than comprehensively discussing ice-cream formulations and processing in this section, we will focus on some similarities and differences between ice cream and the cream products already described, as well as illustrating how certain compositional and processing changes can lead to different ice-cream products. More detailed information relating to ice-cream formulations and processing can be found in the publications of Arbuckle (1986), Mitten and Neirinckx (1986), Madden (1989) and in the literature reviews of Mann (1992a,b).

The composition of ice cream can vary considerably from country to country and in many cases this is governed by legislation. In the UK, ice cream can be made from either dairy or non-dairy fats. Although a minimum of 5% fat is demanded by law, in practice the amount of fat ranges from 6 to 12% and may be higher for premium dairy ice creams. Similarly, the statutory level of 7·5% milk solids not fat (MSNF) is generally exceeded. However, as little as 2% MSNF is permitted in certain fruit-based ice-cream products (Ice Cream Regulations, 1967).

Ice cream, like whipped cream, also relies on a partial destabilisation of the oil-in-water emulsion for its structure and properties. As many ice-cream products are moulded or shaped after the ice cream has been extruded from the freezer, it is important that the ice cream should have a stiff or plastic consistency. Serious production difficulties can arise if a wet, sloppy ice cream is extruded. Likewise, the consumer expects the ice cream to retain its shape and not to melt down too rapidly. All of these properties can be related to the partial destabilisation of the emulsion in the freezer and the formation of a stable, aerated structure. Thus, the use of Polysorbate 80 to improve the dryness of a wet and sloppy extrusion can now be explained in terms of the action of the emulsifier at the oil/water interface (Goff et al., 1987; Goff & Jordan, 1989), as we have already discussed.

The importance of low-molecular-weight emulsifiers and fat crystals in promoting emulsion instability has already been considered. Both have a role to play in the partial destabilisation of the ice-cream emulsion in the freezer. By quickly cooling the homogenised ice-cream mix to about 5°C before freezing and maintaining this temperature for several hours, crystallisation of the higher melting point triglycerides occurs. Further rearrangement at the fat globule surface and more complete hydration of the milk proteins and stabilisers in the aqueous phase would also be expected to take place. The ice-cream freezer is a scraped-surface heat exchanger which can operate at atmospheric pressure or, more usually for large operations, at reduced pressure. Once inside the barrel of the freezer, the ice-cream emulsion undergoes a further rapid drop in temperature to a final extrusion temperature of $-5°C$ to $-6°C$. The emulsion is subject to powerful shear forces arising from the action of the scraper blades and the presence of fat and ice crystals. Ice crystals are necessary in order for sufficient destabilisation of the emulsion to occur in the freezer barrel (Goff & Jordan, 1989). This probably results from their shearing action at the fat globule surface.

From a study of the effectiveness of whey and casein protein isolates in destabilising ice-cream mixes Goff *et al.* (1989) concluded that whey protein isolate had the potential to replace all or part of the stabiliser and emulsifier in conventional formulations. In experimental formulations which contained no added low-molecular-weight emulsifiers and where the concentration of whey proteins was increased relative to the casein, they observed an increase in the destabilisation of the emulsion in the freezer and a drier ice cream was extruded. This was believed to be caused by the preferential adsorption of whey proteins at the fat globule surface and, as a result of their conformation at the interface, the formation of a less stable interfacial membrane. The use of sodium caseinate gave an emulsion which was too stable while a 95% whey protein isolate was the most effective whey preparation examined.

The air bubbles which are introduced into the cream during freezing make a major contribution to the final texture of the product (Arbuckle, 1986) and can influence the smoothness, lightness and body of the ice cream and whether it is perceived as a cold-eating product or not. Initial stabilisation of the air bubbles in the freezer barrel is likely to occur by a similar mechanism to that discussed already for whipped cream. However, final stabilisation of air bubbles is not by a network of fat globules but depends on the formation of a very viscous or gelled aqueous phase and the presence of ice crystals.

Not all of the water in the ice-cream mix is turned to ice during the residence time of the mix in the freezer barrel. It has been calculated for a typical ice-cream formulation of 8·6% fat, 10·5% milk solids not fat, 16·5% sugar, 0·4% emulsifier and 0·15% stabiliser, that 51% of the water is present as ice when the ice cream is extruded at $-5·6°C$ (Berger *et al.*, 1972). However, the ice crystals which are formed in the freezer should be small in size, less than 50 μm; rapid freezing ensures that this occurs. Large ice crystals give the product an icy texture (Berger, 1990). During subsequent hardening of the product in cold stores operating at -25 to $-30°C$, some 90% of the water is frozen as ice crystals. Ice cream may also be quick-hardened in blast tunnels with an air temperature of about $-40°C$ prior to transfer to cold stores. Control of the amount of ice crystals in ice cream is the basis of soft scoop ice cream which is formulated to be scoopable at temperatures as low as $-18°C$. These products contain increased amounts of low-molecular-weight carbohydrates and polyols in order to depress the freezing point of the ice-cream mix. Dextrose and glycerol have been widely used in the UK (Dea & Finney, 1979; Dea & Pillai, 1980), while the use of fructose has also been described (e.g. Kahn & Eapen, 1981).

The approach has been taken further in formulations which contain polymeric molecules, such as the hydrolysis products of starch, in addition to monosaccharides and polyols (Cole et al., 1984). Such combinations can elevate the glass transition temperature of the mix and enable the products to have an enhanced frozen storage stability with greater heat shock resistance when compared with other soft scoop products (Cole et al., 1984). This technology has been utilised in the launch in the USA of ice-cream cone products which are said to have the texture of soft-serve ice cream. Soft-serve ice cream has significantly less ice than conventional ice cream and has a soft and very smooth texture. However, this texture will deteriorate if it is stored frozen. The patented ice cream cones can be stored for prolonged periods in freezer cabinets and are claimed to be soft enough to squeeze from the packaging immediately on removal from the freezer and to have a soft, smooth texture.

REFERENCES

AARONS, B. L. & HEPNER, L. (1975) *Food Trade Rev.*, **45**(1), 7.
ABRAHAMSSON, K., FRENNBORN, P., DEJMEK, P. & BUCHHEIM, W. (1988) *Milchwissenschaft*, **43**, 762.
ANDERSON, M. (1985) *J. Soc. Dairy Technol.*, **36**, 3.
ANDERSON, M. & BROOKER, B. E. (1988) In: *Advances in Food Foams and Emulsions*, Dickinson, E. & Stainsby, G. (Eds), Elsevier Applied Science, London, p. 221.
ANDERSON, M., CAWSTON, T. E. & CHEESEMAN, G. C. (1977a) *J. Dairy Res.*, **44**, 111.
ANDERSON, M., CHEESEMAN, G. C. & WILES, R. (1977b) *J. Soc. Dairy Technol.*, **30**, 229.
ANDERSON, M., BROOKER, B. E. & NEEDS, E. C. (1987) In: *Food Emulsions and Foams*, Dickinson, E. (Ed.), Royal Society of Chemistry, London, p. 100.
ARBUCKLE, W. S. (1986) *Ice Cream*. AVI Publishing, Westport, CN.
BANKS, W. & MUIR, D. D. (1988) In: *Advances in Food Foams and Emulsions*, Dickinson, E. & Stainsby, G. (Eds), Elsevier Applied Science, London, p. 257.
BANKS, W., MUIR, D. D. & WILSON, A. G. (1981) *Milk Ind.*, **83**, 16.
BANKS, W., MUIR, D. D. & WILSON, A. G. (1983) In: *Physico-chemical Aspects of Dehydrated Protein-rich Milk Products, Proceedings International Dairy Federation Symposium*, Helsingor, Denmark, p. 331.
BANKS, W., CLAPPERTON, J. L., MUIR, D. D. & GIRDLER, A. K. (1989) *J. Dairy Res.*, **56**, 97.

BARFORD, N. M. & KROG, N. (1987) *J. Am. Oil Chem. Soc.*, **64**, 112.
BERGENSTAHL, B. (1987) In: *Gums and Stabilisers for the Food Industry 4*, Phillips, G. O., Williams, P. A. & Wedlock, D. J. (Eds), IRL Press, Oxford, p. 363.
BERGER, K. G. (1990) In: *Food Emulsions*, 2nd edn, Larsson, K. & Frisberg, S. E. (Eds), Marcel Dekker, New York, p. 367.
BERGER, K. G., BULLIMORE, B. K., WHITE, G. W. & WRIGHT, W. B. (1972) *Dairy Ind.*, **37**, 493.
BLANSHARD, J. M. V. & FRANKS, F. (1987) In: *Food Structure and Behaviour*, Blanshard, J. M. V. & Lillford, P. (Eds), Academic Press, London, p. 51.
BOODE, K. (1992) Partial coalescence in oil-in-water emulsions. PhD Thesis, Wageningen Agricultural University, The Netherlands.
BOYD, J., PARKINSON, C. & SHERMAN, P. (1972) *J. Coll. Interf. Sci.*, **41**, 359.
BRANCIAROLI, E. (1986) Heat sterilization of natural cream of milk. UK Patent 2168591A.
BROOKER, B. E. (1990) *Food Structure*, **9**, 223.
BROOKER, B. E., ANDERSON, M. & ANDREWS, A. T. (1986) *Food Microstruct.*, **5**, 277.
BROWN, E. M., CARROLL, R. J., PFEFFER, P. E. & SAMPUGNA, J. (1983) *Lipids*, **18**, 111.
BUCHHEIM, W. (1986) *Keiler Milchwirtschaft. Forschung.*, **38**, 227.
BUCHHEIM, W., FALK, G. & KING, A. (1986) *Food Microstruct.*, **5**, 181.
CALDWELL, K. B., GOFF, H. D. & STANLEY, D. W. (1992) *Food Structure*, **11**, 11.
COLE, B. A., LEVINE, H. I., MCGUIRE, M. T., NELSON, K. J. & SLADE, L. (1984) United States Patent 4,452,824.
CORNELL, D. G. (1982) *J. Coll. Interf. Sci.*, **88**, 536.
DALGLEISH, D. G. & BANKS, J. M. (1991) *Milchwissenschaft*, **46**, 75.
DALGLEISH, D. G. & MORRIS, E. R. (1988) *Food Hydrocoll.*, **2**, 311.
DARLING, D. F. (1982) *J. Dairy Res.*, **49**, 695.
DARLING, D. F. & BIRKETT, R. J. (1987) In: *Food Emulsions and Foams*, Dickinson, E. (Ed.), Royal Society of Chemistry, London, p. 1.
DARLING, D. F. & BUTCHER, D. W. (1978) *J. Dairy Res.*, **45**, 197.
DEA, I. C. M. & FINNEY, D. J. (1979) United States Patent 4,145,454.
DEA, I. C. M. & PILLAI, D. (1980) United States Patent 4,219,581.
DICKINSON, E. (1986) *Food Hydrocolloids*, **1**, 3.
DICKINSON, E. (1987) In: *Gums and Stabilisers for the Food Industry 4*, Phillips, G. O., Williams P. A. & Wedlock, D. J. (Eds), IRL Press, Oxford, p. 249.
DICKINSON, E. & STAINSBY, G. (1982) *Colloids in Foods*, Applied Science Publishers, London, p. 182.
DICKINSON, E., MURRAY, B. S. & STAINSBY, G. (1988) In: *Advances in Food Foams and Emulsions*, Dickinson, E. & Stainsby, G. (Eds), Elsevier Applied Science, London, p. 123.
DICKINSON, E., HARHAN, S. K. & STAINSBY, G. E. (1989) *J. Food Sci.*, **54**, 77.
DOXASTAKIS, G. & SHERMAN, P. (1986) *Coll. Polym. Sci.*, **264**, 254.
FISHER, L. R. & PARKER, N. S. (1985) *CSIRO Food Res. Quart.*, **45**, 33.
FISHER, L. R. & PARKER, N. S. (1988) In: *Advances in Food Foams and Emulsions*, Dickinson, E. & Stainsby, G. (Eds), Elsevier Applied Science, London, p. 45.
FLACK, E. A. & KROG, N. (1974) *Food Trade Rev.*, **40**, 27.
FRANKE, W. W., HEID, H, W., GRUND, C., WINTER, S., FREUDENSTEIN, C., SCHMID, E., JARASCH, E. D. & KEENAN, T. W. (1981) *J. Cell Biol.*, **89**, 485.

GARTI, N. (1988) In: *Crystallization and Polymorphism of Fats and Fatty Acids*, Garti, N. & Sato, K. (Eds), Marcel Dekker, New York, p. 267.
GEILMAN, W. G. & SCHMIDT, D. E. (1992) *J. Dairy Sci.*, **75**, 2670.
GOFF, H. D. & JORDAN, W. K. (1989) *J. Dairy Sci.*, **72**, 18.
GOFF, H. D., LIBOFF, M., JORDAN, W. K. & KINSELLA, J. E. (1987) *Food Microstruct.*, **6**, 193.
GOFF, H. D., KINSELLA, J. E. & JORDAN, W. K. (1989) *J. Dairy Sci.*, **72**, 385.
GOULDEN, J. D. S. & PHIPPS, L. W. (1964) *J. Dairy Res.*, **31**, 195.
GRAF, E. & MULLER, H. R. (1965) *Milchwissenschaft*, **20**, 302.
GRAHAM, D. E. & PHILLIPS, M. C. (1979) In: *Foams*, Ackers, R. J. (Ed.), Academic Press, London, p. 237.
GRIFFIN, M. C. A., INFANTE, R. B. & KLEIN, R. A. (1984) *Chem. Phys. Lipids*, **36**, 91.
HARPER, E. K. & SHOEMAKER, C. F. (1983) *J. Food Sci.*, **48**, 1801.
HOOD, L. F. & ALLEN, J. E. (1977) *J. Food Sci.*, **42**, 1062.
HOULIHAN, A. V., GODDARD, P. A., NOTTINGHAM, S. M., KITCHEN, B. J. & MASTERS, C. J. (1992) *J. Dairy Res.*, **59**, 187.
ICE CREAM REGULATIONS (1967) Statutory Instrument No. 1866, HMSO, London.
KAHN, M. L. & EAPEN, K. E. (1981) United States Patent 4,244,977.
KROG, N. (1974) *J. Am. Oil Chem. Soc.*, **54**, 124.
KROG, N. & LAURIDSEN, (1976) In: *Food Emulsions*, Friberg, S. (Ed.), Marcel Dekker, New York, p. 67.
KROG, N., BARFORD, N. M. & BUCHHEIM, W. (1987) In: *Food Foams and Emulsions*, Dickinson, E. (Ed.), Royal Society of Chemistry, London, p. 144.
LAURIDSEN, J. B. (1976) *J. Am. Oil Chem. Soc.*, **53**, 400.
LEDWARD, D. A. (1986) In: *Functional Properties of Food Macromolecules*, Mitchell, J. R. & Ledward, D. A. (Eds), Elsevier Applied Science, London, p. 171.
LEVINE, H. & SLADE, L. (1986) *Carbohydr. Polym.*, **6**, 213.
MADDEN, J. K. (1989) In: *Foams: Physics, Chemistry and Structure*, Wilson, A. J. (Ed.), Springer-Verlag, London, p. 185.
MANN, E. J. (1992a) Ice cream—part 1. *Dairy Ind. Int.*, **57**, 14.
MANN, E. J. (1992b) Ice cream—part 2. *Dairy Ind. Int.*, **57**, 15.
MCCRAE, C. H. & MUIR, D. D. (1992) *J. Dairy Res.*, **59**, 177.
MCPHERSON, A. V. & KITCHEN, B. J. (1983) *J. Dairy Res.*, **50**, 107.
MITCHELL, J. R., IRONS, L. & PALMER, G. J. (1970) *Biochem. Biophys. Acta*, **200**, 138.
MITTEN, H. L. & NEIRINCKX, J. M. (1986) In: *Modern Dairy Technology, Vol. 2, Advances in Milk Products*, Robinson, R. K. (Ed.), Elsevier Applied Science, London, p. 215.
MORRIS, E. R., CUTLER, A. N., ROSS-MURPHY, S. B. & REES, D. A. (1981) *Carbohydr. Polym.*, **1**, 5.
MUHR, A. H. & BLANSHARD, J. M. V. (1986) *J. Food Technol.*, **21**, 683.
MUHR, A. H., BLANSHARD, J. M. V. & SHEARD, S. J. (1986) *J. Food Technol.*, **21**, 587.
MUIR, D. D. & DALGLEISH, D. G. (1987) *Milchwissenschaft*, **42**, 770.
MUIR, D. D., MCCREA-HOMSMA, C. H. & SWEETSUR, A. W. M. (1991) *Milchwissenschaft*, **46**, 691.
MULDER, H. & WALSTRA, P. (1974) *The Milk Fat Globule*, Commonwealth Agricultural Bureau of Dairy Science and Technology, Farnham Royal, p. 210.

MURRAY, E. K. (1987) In: *Food Emulsions and Foams*, Dickinson, E. (Ed.), Royal Society of Chemistry, London, p. 170.
MUSSELWHITE, P. R. (1966) *J. Coll. Interf. Sci.*, **21**, 99.
MUSSELWHITE, P. R. & WALKER, D. A. (1969) UK Patent Specification 1,158,103.
NEEDS, E. C. & BROOKER, B. E. (1991) In: *Colloids*, Dickinson, E. (Ed.), Royal Society of Chemistry, London, Special Publication no. 82.
NEEDS, E. C. & HUITSON, A. (1991) *Food Struct.*, **10**, 353.
NEEDS, E. C., ANDERSON, M., PAYNE, S. J. & RIDOUT, E. A. (1985) *J. Dairy Res.*, **52**, 255.
OGDEN, L. V., WALSTRA, P. & MORRIS, H. A. (1976) *J. Dairy Sci.*, **59**, 1727.
OORTWIJN, H. & WALSTRA, P. (1979) *Neth. Milk Dairy J.*, **33**, 134.
OZAWA, K., NIKI, R. & ARIMA, S. (1984) *Agric. Biol. Chem.*, **48**, 627.
OZAWA, K., NIKI, R. & ARIMA, S. (1985) *Agric. Biol. Chem.*, **49**, 3123.
PATTON, S. & JENSEN, R. G. (1976) *Biomedical Aspects of Lactation*, Pergamon Press, Oxford.
PHILLIPS, M. C. (1977) *Chem. Ind.*, **5**, 170.
PHIPPS, L. W. (1957) *J. Dairy Res.*, **24**, 51.
PHIPPS, L. W. (1982) *J. Dairy Res.*, **49**, 655.
PHIPPS, L. W. (1983) *J. Dairy Res.*, **50**, 91.
PHIPPS, L. W. (1985) *The High Pressure Dairy Homogeniser*, NIRD Technical Bulletin 6.
PRECHT, D. (1988) In: *Crystallization and Polymorphism in Fats and Fatty Acids*, Garti, N. & Sato, K. (Eds), Marcel Dekker, New York, p. 305.
RAHMAN, A. & SHERMAN, P. (1982) *Colloid Polym Sci.*, **260**, 1035.
REES, D. A., MORRIS, E. R., THOM, D. & MADDEN, J. K. (1982) In: *The Polysaccharides*, Vol. 1, Aspinall, G. O. (Ed.), Academic Press, London, p. 195.
ROBSON, E. W. & DALGLEISH, D. G. (1987) *J. Food Sci.*, **52**, 1694.
ROTHWELL, J. (1989) In: *Cream Processing Manual*, Rothwell, J. (Ed.), Society of Dairy Technology, Huntingdon, p. 110.
ROTHWELL, J., JACKSON, A. C. & FAULKS, B. (1989*a*) In: *Cream Processing Manual*, Rothwell, J. (Ed.), Society of Dairy Technology, Huntingdon, p. 83.
ROTHWELL, J., JACKSON, A. C. & FAULKS, B. (1989*b*) In: *Cream Processing Manual*, Rothwell, J. (Ed.), Society of Dairy Technology, Huntingdon, p. 120.
SHENNAN, D. B. (1992) *Exper. Physiol.*, **77**, 653.
SKURA, M. J. & NAKAI, S. (1980) *J. Food Sci.*, **45**, 582.
SKURA, M. J. & NAKAI, S. (1981) *Can. Inst. Food Sci. Technol. J.*, **14**, 59.
SLADE, L. & LEVINE, H. (1988) In: *Food Structure—Its Creation and Evaluation*, Blanshard, J. M. V. & Mitchell, J. R. (Eds), Butterworths, London, p. 115.
SNOEREN, T. H. M., PAYENS, T. A. J., JEUNINK, J. & BOTH, P. (1975) *Milchwissenschaft*, **30**, 393.
SNOEREN, T. H. M., BOTH, P. & SCHMIDT, D. G. (1976) *Neth. Milk Dairy J.*, **30**, 132.
SWEETSUR, A. W. & MUIR, D. D. (1983) *J. Dairy Res.*, **50**, 291.
TAMIME, A. Y. & ROBINSON, R. K. (1985) *Yoghurt Science and Technology*, Pergamon Press, Oxford.
TE WHAITI, I. E. & FRYER, T. F. (1975) *N.Z. J. Dairy Sci. Technol.*, **10**, 2.
TOLSTOGUZOV, V. B. (1986) In: *Functional Properties of Food Macromolecules*, Mitchell, J. R. & Ledward, D. A. (Eds), Elsevier Applied Science, London, p. 385.

VAN BOEKEL, M. A. J. S. & FOLKERTS, T. (1991) *Milchwissenschaft*, **46**, 758.
VAN BOEKEL, M. A. J. S. & WALSTRA, P. (1981) *Coll. Interf.*, **3**, 109.
WALSTRA, P. (1967) *Neth. Milk Dairy J.*, **21**, 166.
WALSTRA, P. (1974) *Neth. Milk Dairy J.*, **28**, 3.
WALSTRA, P. (1985) *J. Dairy Res.*, **52**, 309.
WALSTRA, P. & JENNESS, R. (1984a) *Dairy Chemistry and Physics*, John Wiley, New York, p. 211.
WALSTRA, P. & VAN BERESTEYN, E. C. H. (1975) *Neth. Milk Dairy J.*, **29**, 35.
WOODING, F. B. P. (1971) *J. Ultrastruct. Res.*, **37**, 388.

3

Butter and Allied Products

R. S. Jebson
Department of Food Technology, Massey University,
Palmerston North, New Zealand

SUMMARY

Churning of cream to butter has been practised for a considerable period of time and current procedures from cream handling to packing and storage are described. Pretreatment of cream is a key operation involving pasteurization and deodorization and the procedure can significantly affect the quality of the finished product. All churning techniques have the objective of partly destabilizing the milk globule interface, so releasing fat which forms the continuous phase of the butter. The introduction of air bubbles assists emulsion destabilization and much modern equipment involves continuous techniques based on the Fritz process. Alternative procedures include the blending of anhydrous milkfat with serum in scraped surface coolers.

Variables in the process include both machine variables (unit configurations and operating conditions), mode of addition of minor ingredients, and cream variables (temperature regimes, holding times and vacreation intensity, etc.). Various monitoring techniques are available for moisture measurement. Methods for packing, transport and handling butter are described and changes which can take place in that product after packing are highlighted.

Problems with finished butter can include off-flavour developments, particularly those caused by the feed of the cows, processing, and oxidative and/or microbiological spoilage. Other product defects may embrace colour body and texture problems. In the production of lactic butter a number of problems can occur if souring of cream is used and the Nizo process has been developed to overcome these. It is claimed that this procedure yields lactic butter of superior keeping quality than that of standard sweet cream salted butter.

Various techniques for softening butter are available including controlled pretreatment of cream and the use of fractionated milkfat.

CREAM HANDLING

Frequently cream is produced by centrifugally separating milk at a separate site, and then transporting it to the butter factory. Such operations necessitate the holding and pumping of cream, which can be deleterious.

The holding of raw cream, even at low temperatures, can result in considerable microbial growth. Although pasteurization will kill most organisms, heat stable proteolytic or lipolytic enzymes, which can catalyse the formation of off-flavours when the butter is stored, may remain (Russell, 1973; Fawcett, 1972, Personal Communication).

Pumping must be carefully done in a manner that minimizes damage to fat globules. Damaged globules may release free fat, which, when cream is held overnight before buttermaking, may cause clumping of the fat globules together forming 'gelled' vats. Free fat can also be attacked by lipases normally present in the cream, giving rise to high free fatty acid levels and off-flavours.

Positive pumps generally cause less damage, but centrifugal pumps can be used, if they are correctly sized for the flow and pressure drop required, and work close to their maximum efficiency. In the piping system, sharp elbows, partly closed valves, or any other feature which causes high pressure drop, and hence excessive turbulence in the cream, should be avoided. Te Whaiti & Fryer (1975) and Mulder & Walstra (1974) have shown that cream at temperatures between 10°C and 35°C is more susceptible to fat globule damage, and hence it is desirable to avoid pumping cream in this temperature range.

To comply with both the microbiological and globule stability requirements, cream which must be held or transported should be pasteurized immediately after separation, and held at temperatures not exceeding 6°C.

CREAM TREATMENT

Cream treatment may comprise the steps: neutralization, pasteurization, taint removal, cooling, and holding.

Neutralization

Neutralization has been practised when conditions of preparation or storage of raw cream have permitted the production of lactic acid in the cream. With increased hygiene on farms and factories, and chilling of milk on farms, acidity does not normally develop in cream, and neutralization is generally no longer necessary, but automatic equipment for dosing alkali for neutralization is available. Details of the methods and benefits of neutralization are given by McDowall (1953a).

Pasteurization

Apart from the destruction of pathogenic and spoilage organisms present in the cream, pasteurization melts the fat in globules giving standard starting conditions for cooling the cream. High temperatures (90–100°C), which will kill more organisms, are usually used.

The equipment used incorporates methods for reducing the level of taint in the cream, either vacuum (flash) pasteurization or vacreation.

Taint Removal

Source of tainting substances

Tainting substances can arise from several sources including: microbial action in the milk, the feed of the cows, or aromas in the milking shed. McDowall (1953b) describes a number of weeds which, if consumed by cows, may taint milk.

Theory

In vacuum pasteurization cream is heated in an exchanger to a high temperature and passes into a vacuum chamber where volatiles present flash off. The relation between the concentration of taint in the cream and that in the vapours may be described approximately by Henry's law:

$$p_t = k x_t$$

where p_t is the partial pressure of taint in the vapours, x_t is the mole fraction of the taint in the liquid, k is a temperature dependent constant specific for the particular taint.

Scott (1954a, b), using Henry's law and mass balance techniques, showed that, for vacuum pasteurization, the reduction in taint is given by:

$$x_o/x_1 = 1/\{(1 - V/L) + mV/L\}$$

where x_o is the concentration of taint in the cream leaving the pasteurizer, x_1 is the concentration of taint in the cream entering the pasteurizer, V is the flow rate of flash vapours (kg/h), L is the flow rate of cream (kg/h), m is the equilibrium ratio of the concentration of taint in the vapour to that in the liquid $((=k/P)$ where P is the total pressure).

Where a greater degree of taint removal is desired, as is important for making sweet cream butter from pasture-fed cows, and desirable even for lactic butter in many circumstances, a steam distillation apparatus, such as the Vacreator (Protech Engineering Ltd, Auckland), is used. Scott (1954a) derives formulae for calculating the reduction in taint concentration for different forms of steam–cream contacting.

For multiple contacting using fresh steam each time, and for cream at the steam saturation temperature, the equation is:

$$x_o/x_1 = (1/\{1 + mV/[LN]\})^N$$

where N is the number of contact stages. Maximum taint removal efficiency is obtained when the steam is evenly divided among the stages.

Greater efficiency is obtained when counter-current contacting of steam and cream is used. For this case Scott(1954b) gives the equation:

$$x_o/x_1 = 1/\{(mV/L)^{N+1} - 1\}$$

When the relation between the concentration of taint in the liquid and that in the vapour is non-linear, a graphical analysis (Scott, 1954a, b) can be used.

In a series of papers (1955, 1956, 1965) McDowall measured the properties of reference substances that could be used for determining the efficiency of steam distillation equipment. He showed (1957) that, for volatile taints, more efficient steam usage is obtained with low temperatures and hence high vacuums. However, with less volatile taints, higher temperatures (higher absolute pressures) should be used. Hence for efficient taint removal, when a variety of tainting substances are present, more than one temperature should be used.

Equipment

Flash pasteurizers are widely used, a flowsheet of one type being shown in Fig. 1. Cream is heated in three regenerative sections and a steam heated section, before passing to holding tubes. It then passes to the first of two vacuum deodorizers. From this point it can be cooled to crystallizing temperature either entirely by regeneration and water cooling, or passed to the second vacuum deodorizer and then further cooled by regeneration and water cooling. Flash pasteurizers have the advantages of low steam consumption, gentle cream treatment, and low noise levels.

Much more intense deodorization is given by the Vacreator (Protech Engineering, Auckland), a flowsheet of which is in Fig. 2. A detailed description is given by Towler (1986). In the Vacreator, there is multiple cream–steam contacting in a manner approximately equivalent to a three-stage counter-current system, in which different stages can be operated at different temperatures, and a final flash stage to a separator operating at high vacuum. Multiple contacting at a range of temperatures ensures efficient removal of tainting substances of different volatilities.

Additional equipment available includes automatic continuous neutralization, 'Ultitem' for direct steam injection pasteurizing at temperatures up to 135°C, 'Economajor' for producing hot water at temperatures up to 75°C by heat exchange from exhaust steam, and the 'Economaster' device which uses the exhaust steam in a reboiler to produce clean vacuum steam which is raised to the required pressure with live steam in a venturi. Steam

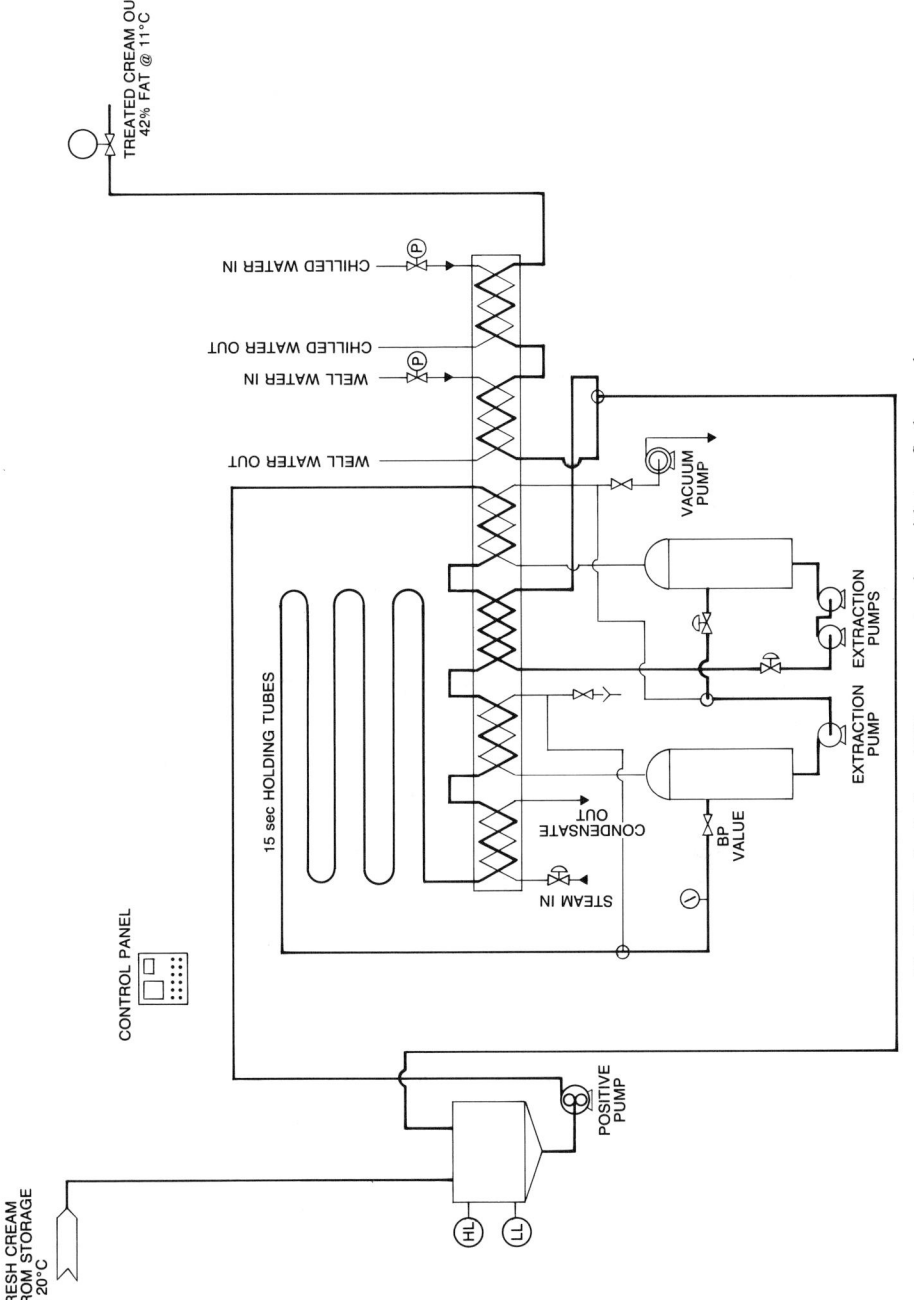

Fig. 1. Flowsheet of an APV cream pasteurizer with two flash vessels.

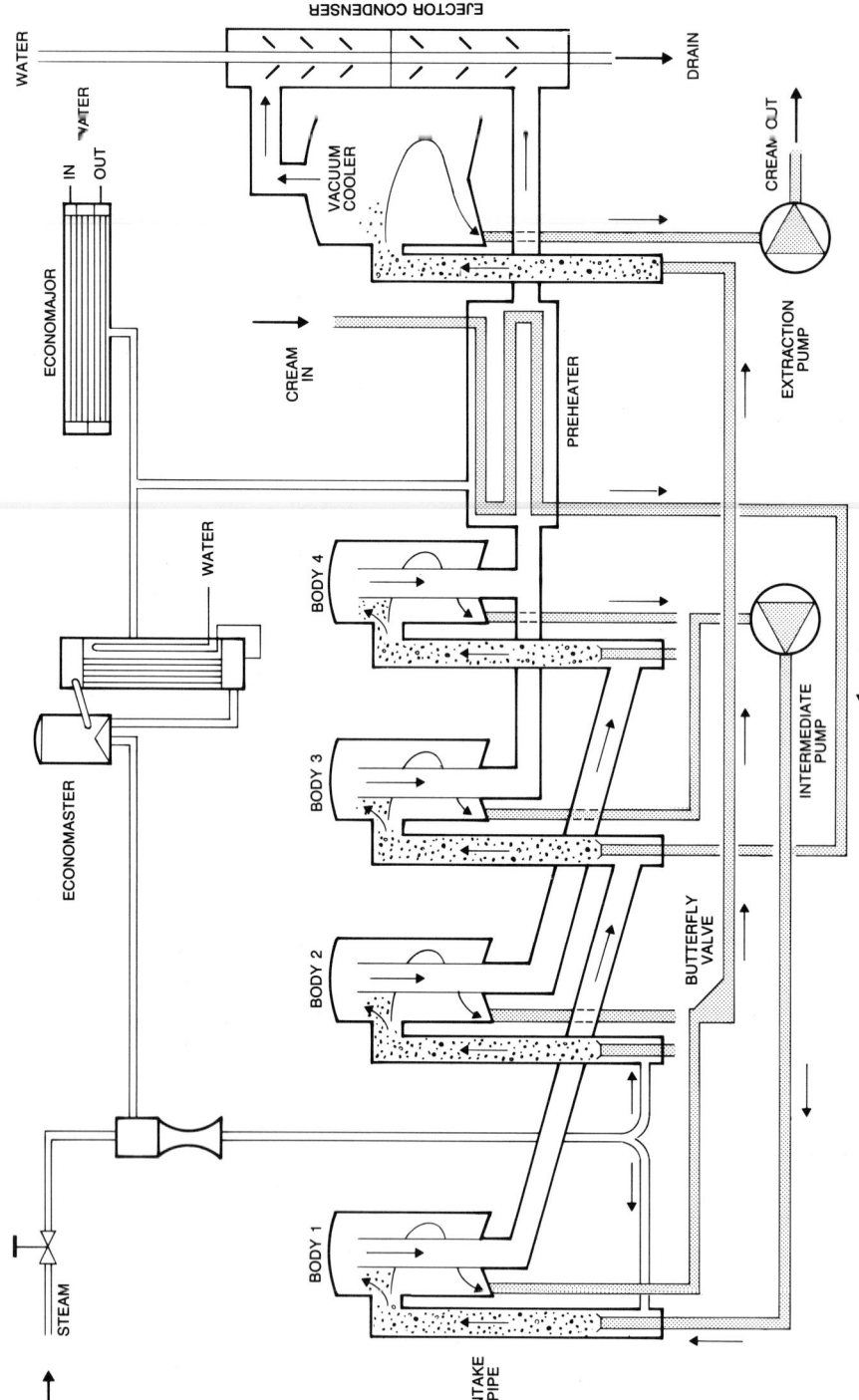

Fig. 2. Flowsheet of a Vacreator with Economaster and Economajor attachments.

TABLE 1
Theoretical comparison of vacuum pasteurization and vacreation concentration of taint in mg/kg after various stages

Taint	Initial concentration	Vacuum pasteurizer
Diacetyl	20	3·6
Acetoin	20	21·6

			Vacreator		
Flash			Body		Final
	1	2	3	4	
8·1	1·67	0·33	0·11	0·01	0·004
20·02	19·92	19·66	17·9	16·3	16·7

savings of 10–15% can be made with this device. Jebson & Lascelles (1977) have proposed a method of reducing energy consumption by using a heat pump on the Vacreator.

Comparison of flash pasteurization and vacreation

Using Scott's formulae given above the theoretical concentration levels of two different taints, diacetyl (volatile, boiling point 88°C) and acetoin (non-volatile, boiling point 143°C), have been calculated for a vacuum pasteurizer and at different stages through the vacreator. The milkfat/serum concentration ratios for the taints given by McDowall (1955) were used in the calculations. The results are in Table 1.

The vacreator is clearly much more effective than the vacuum pasteurizer for the volatile taint, a concentration of 20 mg/kg being reduced to 0·004 mg/kg in the vacreator, but only to 3·6 mg/kg in the vacuum pasteurizer. For the non-volatile taint the concentration actually increases after vacuum pasteurization, which is to be expected, as a greater quantity of the more volatile water would be flashed off. Vacreation does reduce this taint level, although not greatly.

With cream, the efficiency of taint removal depends not only on the taint volatility, but also on the partititon coefficient (ratio of the solubility in fat to that in the aqueous phase). With highly fat-soluble taints, the concentration in the aqueous phase is low, and hence the vapour pressure is also low. Such taints are difficult to remove by vacreation, and virtually impossible to reduce by vacuum pasteurization. The volatility of taints is also affected by the dissolved solids (particularly lactose) in the aqueous phase. These have a boiling point elevation effect (McDowall, 1957).

Butter made from vacreated cream has been shown to have significantly better keeping quality than butter from vacuum pasteurized cream.

Cream Cooling and Holding

The way in which cream is cooled and held will influence the manufacture and properties of the butter made from it. Generally cream is cooled in plate heat exchangers using regeneration, water, and chilled water (Fig. 1).

As the cream cools, the higher melting components in the milkfat crystallize, but, as there is a wide range of different melting point components in milkfat (Taylor & Hawke, 1975), polymorphism and mixed crystal formation occur. The nature and degree of solidification are very dependent on cooling rates, holding times, and the size range of the fat globules (Mulder & Walstra, 1974). Depending on cooling conditions, cream can undergo a degree of supercooling before crystallization commences, and this can influence solidification.

Cream cooling should be organized to obtain the correct ratio of solid to liquid fat for churning, about 45% solid fat being the optimum (Samuelsson & Vikelsoe, 1971a, b; Frede et al., 1982a, b; Mercer, 1985). Cooling techniques which minimize mixed crystal formation and maximize the proportion of liquid fat at any temperature will give a butter with improved spreading properties. Such techniques include the Alnarp process which has been studied by a number of workers, including Samuelsson & Petersson (1937), Dolby (1954) and Dixon (1974). In this process cream is cooled to 4°C for 8 s to allow seeding to occur, reheated to 18°C for 2 h, when slow crystallization of the higher melting fat components takes place, and finally slowly cooled to churning temperature. Butter from cream cooled this way has a hardness approximately 70% of that of butter produced using normal cream cooling methods.

As fluctuations in the performance of a continuous buttermaking machine often occur at silo changes, it is desirable to keep silo numbers to a minimum (Robinson, 1985). Palfreyman (1982, 1988) has investigated the performance of silos for overnight holding of cream. Silos of volume 45–90 m^3, having agitators of vertical pitch-blade, gate, and horizontal propeller types were studied. Mixing of milkfat in the silos was generally found to be consistent, but on occasions cream from most silos had a wider variation than would be ideally acceptable for continuous buttermaking. There was no correlation between the range of milkfat compositions after overnight holding and the other variables (including silo design). It is possible that some factor not measured (for instance fat globule size distribution) may be responsible for the variation in mixing efficiency. Palfreyman also measured fat globule damage/repair in silos, and concluded that most of the damage occurred prior to storage in the silos.

Although running costs were higher, Palfreyman found that the considerable capital advantages of silos with side entry propeller agitators made them the preferred silo design.

BUTTERMAKING

Theories of Churning

Theories of churning are discussed in detail by McDowall (1953c). Under the action of vigorous agitation, cream foams and rapidly increases in volume until it is about 190% of the original. As churning continues the foam bubbles become smaller and more numerous, and the walls thinner as the total surface area of the film increases. There is no further increase in volume. The fat globules tend to concentrate at the surface. A point is reached at which the film is so thin that there is no room for the fat globules to slide past one another, and collisions become frequent. At the correct ratio of solid to liquid fat in the globules, they are disrupted in the collisions, the liquid fat leaks out, cementing the globules together to form butter granules, and coating the foaming agents so that the foam is no longer stable, and collapses. With continued agitation butter granules build up in size until they are suitable for draining off the buttermilk.

The butter is then worked, to break up water droplets to a sufficiently small size (majority less than 1 μm), so that bacterial growth cannot occur. Salt, water and/or flavouring substances are added to meet the specification of the butter being made.

A similar mechanism is thought to take place whether batch churns or Fritz process continuous buttermaking machines are used, but most large modern creameries use the continuous process.

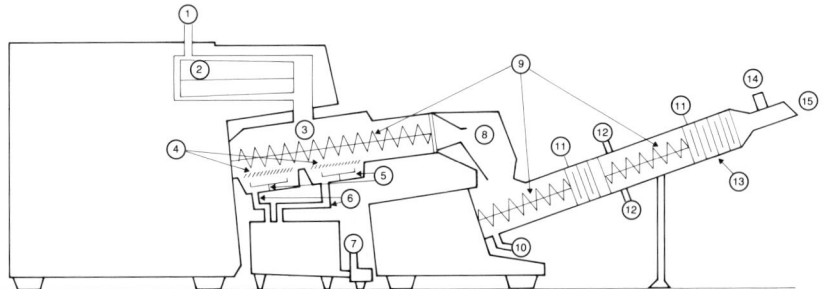

Fig. 3. Diagram of a Contimab MD130 continuous buttermaking machine. 1, cream inlet; 2, churning cylinder; 3, butter granule–buttermilk separation; 4, buttermilk screens; 5, buttermilk screen cleaners; 6, buttermilk level siphons; 7, buttermilk pump; 8, squeeze drying; 9, transportation augers; 10, buttermilk drain; 11, injection points; 12, vacuum section; 13, working section; 14, moisture meter sensor; 15, butter outlet.

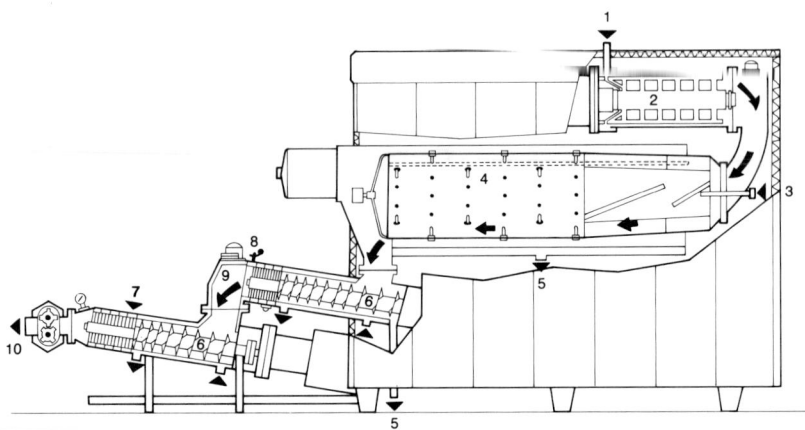

Fig. 4. Diagram of a Pasilac HCT2 continuous buttermaking machine. 1, cream inlet; 2, churning cylinder; 3, chilled buttermilk 'wash'; 4, butter granule–buttermilk separation; 5, buttermilk drains; 6, working section; 7, injection points; 8, throttle gate; 9, vacuum section; 10, butter pump.

Fritz Process Continuous Buttermaking Equipment

All Fritz-type continuous buttermaking machines have a number of elements:

- (i) cream feed pump,
- (ii) high speed churning section,
- (iii) separating section to drain the buttermilk from the butter granules,
- (iv) working system to break up water droplets, and distribute additives (water, salt and/or cultures) evenly,
- (v) vacuum section,
- (vi) salting and dosing,
- (vii) final cone.

The main types of machine available are the Contimab (Fig. 3) (Simon Frères, Cherbourg), the Pasilac (Fig. 4) (Silkeborg) (APV Pasilac AS), Westfalia (Fig. 5) (Westfalen) and Ahlborn (GEA Ahlborn GmbH & CoKG, Sarsdedt).

Cream feed pump

Positive pumps with variable speed drives are used. The Contimab has a mono type, and the others gear pumps.

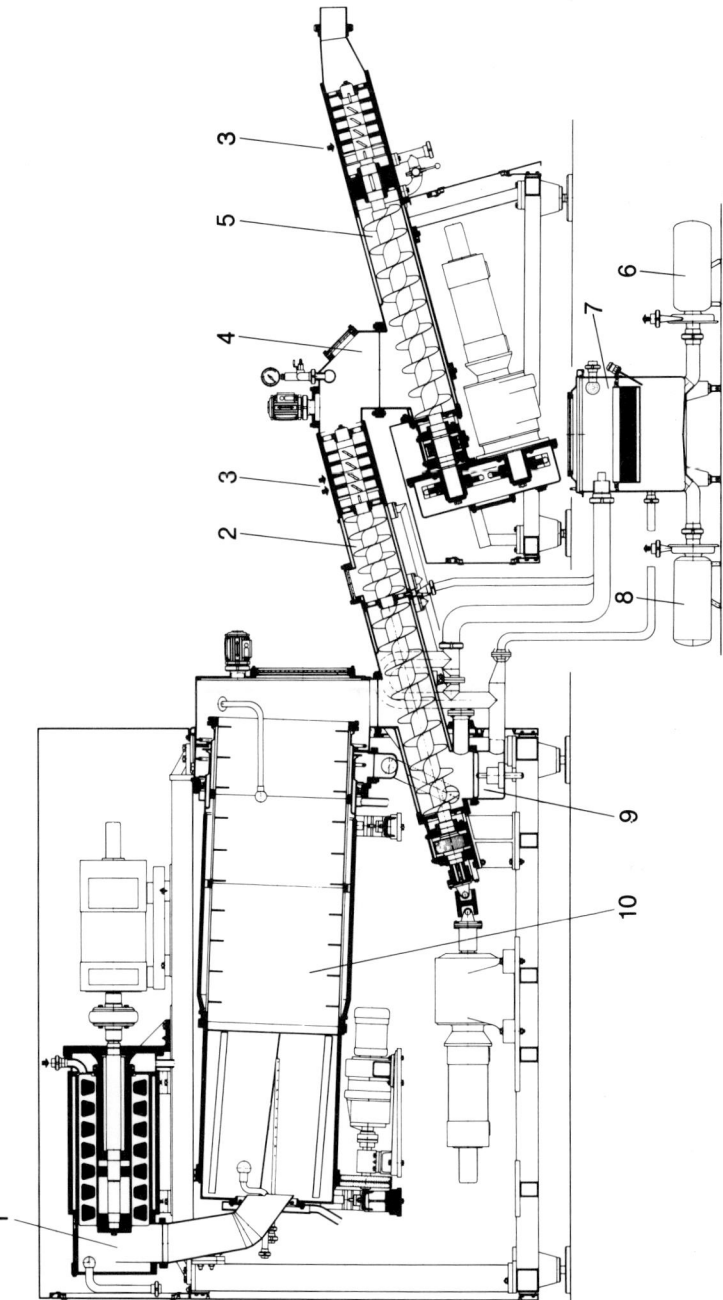

Fig. 5. Diagram of a Westfalia BUD continuous buttermaking machine. 1, churning cylinder; 2, 1st working section; 3, injection points; 4, vacuum section; 5, 2nd working section; 6, buttermilk pump 1; 7, buttermilk balance tank; 8, buttermilk pump 2; 9, buttermilk screen; 10, butter granule–buttermilk separating section.

Churning section

The churning time in continuous machines is a few seconds, compared with 10–20 min in batch churns. All continuous machines employ variable speed gate-type agitators, often with stepped blades, but in the more modern Contimabs the blades are attached to a drum which fills most of the churning space. This prevents build-up of granules in the churning section and lowers power consumption.

The Contimab, Westfalia and Alborn use expanding cone vee-belt drives, but the Pasilac has a hydraulic system. Chilled water circulates in the jackets of the churning sections to minimize the temperature rise in the butter.

Buttermilk draining

In the Contimab the granules float on a pool of buttermilk in the beginning of the working section, the level of which is controlled by a siphon. In the older MC series, butter is forced up out of the augers through a rectangular tunnel which forces the granules together and squeezes out the buttermilk. The size of the tunnel outlet appears to be important to moisture control, but cannot be altered during a run. In the later Contimab MD series of machines, a drying tunnel is not used, but a gooseneck at the rear of this section is used to control the buttermilk level to a point about half way up the screw.

In the Pasilac and Alborn, buttermilk and granules drop into the separating section, which has a fixed screen, and beaters revolving at 35–42 rev/min. Granules and buttermilk spend a few seconds on a solid portion of the draining cylinder (to build up granule size) before moving to the screening section. The granules then drop through a chute into the working section.

The Westfalia separating section has a fixed screen, and beaters revolving at 35–42 rev/min. These granules and buttermilk also spend a few seconds on a solid portion of the draining cylinder before moving to the screening section.

Working

Working is accomplished by moving the butter with parallel contra-rotating augers and forcing it through a series of orifice plates. On the downstream side of the plates cruciform beaters contribute to the working and flow of the butter. Beaters of different angles can be used. Flutes are cut in the auger flights in the first part of the working sections to assist draining of the buttermilk. The degree of working is controlled by the size of the holes in the orifice plates, and by the type of beater; the greater the pitch angle the more transfer and the less the working. Most machines have two consecutive working sections with independent variable speed drives.

Vacuum working
As well as helping to avoid laminations in bulk butter, vacuum reduces the air content of the butter, and hence its volume. In the Contimab, vacuum is drawn half way up the second working section, but in the Pasilac and Alborn vacuum is drawn between the two working sections.

Salting and dosing
Salt is added as a slurry of 40–60% salt, from a well agitated slurry tank, by small variable speed positive pumps through three injection points in the first of the final set of orifice plates on the Contimab, and on the final orifice plate before the vacuum section on the Pasilac. The addition of water to adjust the moisture content, and of flavouring substances for lactic butter, is done at the same place.

There must be very small clearances between the augers and the worker body immediately prior to the salt dosing point, or salt can drain back into the buttermilk.

Outlet cone
This shapes butter to a final ribbon, and can be restricted by a movable plate to give more working.

Arrangement of multiple machines
When more than one buttermaking machine is required to supply a single packing line, it is desirable to only partially work the butter in each machine, and feed the butter to a common blender. These blenders resemble the working section of a continuous buttermaking machine, and are of sufficient capacity to handle the butter from all machines feeding them. If a blender is not used, it is very difficult to control each machine so that the amount of working, and hence the moisture distribution, is exactly the same. Butter colour depends to some extent on the moisture distribution, and, if different coloured butters are fed to a single packer, they will be partially blended and the packed butter will have an undesirable streaky or mottled appearance.

Recombined Buttermaking
There has been considerable development from the original Cherry Burrell, Nu-way, and Creamery Package processes (McDowall, 1953d), processes

in which butter is made by mixing liquid anhydrous milkfat with serum and/or salt, and cooling. Jebson (1982) has described the raw materials needed, and the process of recycling cooled butter around the scraped-surface cooler. This modification overcomes the brittleness problems of the earlier processes. Munro (1982) showed that the use of a pinworker made recycling unnecessary.

At the New Zealand Dairy Research Institute the process has been further developed (Truong & Munro, 1983). Cream is converted to anhydrous milkfat by the direct-from-cream process described in Chapter 5, and mixed with dairy liquids and/or salt slurry. The mixture is then passed through a series of scraped-surface coolers and pinworkers in which it is cooled from 40°C to 10°C. Because it takes a finite time for crystals to form and grow, the scraped-surface heat exchanger supercools the liquid fat, only some of which crystallizes in the heat exchanger. The majority of the fat crystallizes in the pinworkers, where the turbulence and shear encourage the formation of evenly sized crystals.

The Ammix process has the advantages of a greater flexibility in the composition and physical properties of the butter than is normally possible with the Fritz-type processes.

CONTINUOUS BUTTERMAKING VARIABLES

The properties of butter, particularly moisture content and hardness, are affected by a number of processing variables. These include:

Machine variables	*Cream variables*
Beater speed (and beater type)	Fat content
Cream flow rate	Temperature
1st worker speed	Cooling regime
2nd worker speed	Holding time for crystallization
Worker configurations	Vacreator maximum temperature
Draining cylinder speed	Vacreation intensity
Cannon angle (Contimab)	Fat globule size distribution
Salt addition rate	Cream handling
Water addition rate	Acidity
Gooseneck height (Contimab)	Seasonal factors
Vacuum	

The effects of some of these variables are discussed by Dolby *et al.* (1966),

Anderson (1969), Hughes *et al.* (1976), Mogensen & Danmark (1984), Mercer (1985) and Lorwood (1988). A summary of the effects of the above variables on butter moisture, butter temperature and butter hardness, including some unpublished work, is given below.

Machine Variables

Beater speed

This is the basic control. At low speeds there is little churning, and the 'butter' moisture is high. As churning speed is increased, more energy is put into the cream, the butter granule size is increased so that drainage is better, and the butter moisture decreases, until an optimum granule size is obtained. If the beater speed is further increased, large wet granules formed by excessive energy, which do not drain well, are formed, and the butter moisture rises. It can be imagined that at very high beater speeds most of the serum would be re-emulsified into the butter and the 'butter' would have a moisture content close to that of the original cream.

The relation between beater speed and butter moisure is then a curve of basically parabolic shape. Greater stability of control is obtained by setting the beater speed so that the butter moisture is that at the bottom of the curve. A beater speed below that of minimum butter moisture gives 'under-churning', and speeds above that of minimum butter moisture give 'over-churning'. Lorwood (1988) has determined the curves on a commercial buttermaking machine, and an example is shown in Fig. 6.

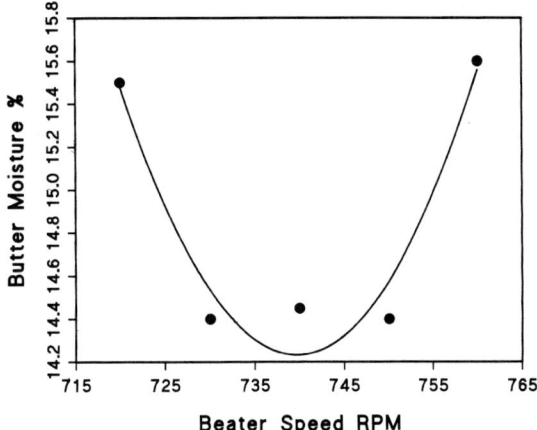

Fig. 6. A typical butter moisture versus beater speed curve.

Mercer (1985) showed that the butter granule temperature at the minimum butter moisture is related to the cream temperature through the equation:

Butter granule temperature = 0·4 (cream temperature) + 9°C

However, a seasonal trial by Lorwood (1988) failed to confirm this relation at all times of the season, and she found that a temperature difference of 2°C occurred between the cream temperature and the butter granule temperature at the point of minimum moisture.

In the older continuous buttermaking machines 'gate'-type agitators were used. The churning space was open, and under conditions of underchurning, with fine granules, the mixture did not flow readily through the churn, and the level of butter in the churning cylinder would build up. The load in the churn was higher, and energy input to the butter and buttermilk greater. When the beater speed is higher than that for minimum butter moisture extra energy is being expended, and this will raise the butter temperature. The curve of beater speed versus butter temperature is similar to that for moisture versus beater speed.

In some modern designs of churning cylinder there is a cylindrical screen attached to the beaters. This screen prevents build-up in the churn, and ensures that motor loads do not rise in underchurning conditions.

As the temperature of the butter rises, the hardness falls, so, at the point of minimum butter moisture, the hardness will be highest and the butter firmest. Softening is greater when beater speed is excessive, as there will be some working of the butter. Increasing the butter moisture also softens the butter, and this is a further reason for the hardness curve to be similar to the moisture curve.

Worker speeds

As the speed of the workers is increased, the time for draining of the buttermilk from the butter is reduced, and the butter moisture rises. There is normally a drain cock below the second worker. If this is closed, the speed of the second worker will not affect the butter moisture.

A certain amount of energy is required to work a given flow rate of butter. This will cause the same rise of temperature whether the work is done quickly or slowly, i.e. the temperature rise is basically independent of the working speed. If the working section is cooled, a higher speed will allow less time (and less butter surface area) to cool the butter, and the temperature rise will be lower.

The hardness of the butter leaving the machine is primarily affected by the amount of working given, but it is also modified by the temperature at which the working is done. The higher the temperature, the greater the proportion of liquid fat, the softer the butter, the less shearing of fat crystals, and the less effective the working. Hardness is also affected by moisture content, higher moisture contents causing reduction. Hence increasing the worker speed will tend to decrease the hardness from this cause.

The worker configuration influences the amount of working given to the butter, and is determined by the number of orifice plates in the workers, the sizes of the holes in the plates, the alignment of the augers, and the type of beaters between the orifice plates.

Adjustable working plates are now supplied with most modern buttermaking machines. They consist of two touching orifice plates. One of these can be moved by turning a knob on the top of the working section, giving immediate control over the effective hole size. It is more common to make adjustments on the second worker. On this worker the alignment of the flights can be adjusted to give either a higher pumping efficiency or more backmixing. Reducing the backmixing is thought to reduce the amount of work done by the orifice plates, as with less backmixing there are fewer of the holes in use. (At normal flow rates the working sections are about 25% full.) Adjustments to the flights are usually made to ensure air is removed from the butter in the vacuum section. Effects on working are of secondary consideration.

More orifice plates and smaller holes will increase the rate of working and reduce drainage. Hence butter moisture and butter temperature will increase.

Cream flow rate

As the cream flow rate is decreased at constant beater speed, the energy input per unit volume of cream is increased. The effect of this depends on whether the butter is being overchurned (in which case the butter moisture will increase) or underchurned, when the butter moisture will decrease.

The minimum butter moisture will be lower at lower cream flow rates, and the beater speed at which the minimum occurs will also be lower. When the cream flow rate is decreased, a given change in beater speed has a greater percentage effect, the slopes of the beater speed versus moisture curves are steeper, and the buttermaking machine is harder to control.

Draining cylinder

The rotational speed of the butter granule draining cylinders has not been found to have a significant effect on butter moistures or other butter properties.

Auger angle
On Contimab machines, the angle of the auger to the horizontal is variable. As the angle is increased, the moisture drains back more easily, and butter moisture is reduced. Other butter properties are not affected.

Salt addition rate
It would be expected that, as salt is added as a slurry of up to 60% moisture, an increase in the salting rate would increase the butter moisture. In general this is true, but, when cream fat is high, churning is easy, and overchurning, which results in soft, high moisture granules, is likely. When such granules are extruded through the orifice plates, channels, through which both salt and moisture may drain, can form. If the salting rate is increased, the slurry may enlarge these channels, actually decreasing the moisure content, and contaminating the buttermilk with salt (Hughes *et al.*, 1976).

Gooseneck height
Raising the height of the gooseneck on Contimab machines raises the level of the buttermilk (on which the granules float), giving more time for draining. The butter moisture is slightly reduced.

Cream Variables

Palfreyman (1982, 1988) has shown that variation in cream properties, such as the fat percentage, can be smoothed out by the action of cream silo agitators during overnight holding. However, a number of buttermakers have reported uneven running of buttermaking machines when the cream supply to the silos the previous day has varied in fat content or temperature. If fluctuations occur during cream preparation, it is more difficult to maintain constant butter properties, particularly butter moisture.

Cream fat content
At low cream fat contents extra energy and hence high beater speeds are necessary to churn the cream. Often the power required exceeds the capacity of the motor, and the butter tends to be underchurned. Frequently the power limitations will necessitate low cream flow rates, and in these cases minimum butter moistures may be lower, but control will be more difficult.

As the cream fat content is increased, butter moisture will decrease, and beater speeds can be lowered. At cream fats above the optimum, it is difficult to reduce power to the level required, the butter tends to be overchurned, and the machine is more sensitive to slight changes in the operating and cream variables. At cream fats below optimum, the machines are much less sensitive to changes in the variables, e.g. the beater speed versus butter moisture curve is flatter, but power consumption is much higher. Mercer (1985) discusses these points in more detail.

At low cream fat contents, the high power consumption causes a large increase in butter temperature, and the butter is soft and difficult to work properly.

Cream temperature

At low temperatures the proportion of solid to liquid fat is too high, churning is difficult, and high beater speeds must be used. Sufficient energy must be put into the cream to raise the temperature to near the optimum. At high churning speeds unnecessary damage is done to the fat globules, and more moisture is beaten into the butter. At higher cream temperatures, there is insufficient solid fat present to assist fat globule disruption, high beater speeds must be used, and high butter moistures are likely, together with high buttermilk fat contents.

If cream temperatures are slightly above optimum, churning is easy and overchurning likely, with consequent high butter moistures.

Cooling regime

Shock cooling will produce a cream of higher solid fat content than slowly cooled cream, and hence is similar in effect to lowering the churning temperature. The cooling rate, which is controlled by the temperature to which the cream is cooled in the plate heat exchanger, relative to the final temperature in the silo, will also affect the fat crystal size and globule stability. Shock cooling will produce small fat crystals, less liquid fat, and a brittle butter body, irrespective of the churning temperature. Shock cooling will also reduce fat losses to the buttermilk, because there is a greater proportion of solid fat present.

Fat globule size distribution

Small fat globules are harder to disrupt than large, and hence more churning power is required. Very small globules, such as occur in homogenized cream, may be impossible to churn.

Vacreation increases the range of globule sizes, and flash pasteurization increases the average globule size. Vacreated cream is the more difficult to churn, and the fat losses in the buttermilk are higher.

Large fat globules occur in the early spring, and with Jersey rather than

Friesian milk. They are more easily damaged under intense vacreation, and in pumping and handling milk and cream.

Pasteurization temperature
The higher the pasteurization temperature, both in a Vacreator (especially if an 'Ultitem' direct steam injection unit is fitted) and in a flash pasteurizer, the greater the damage to the fat globules, and higher churning speeds, with greater fat loss to the buttermilk, will probably be needed.

Vacreation intensity
Increase in vacreation intensity increases the shear forces generated in the steam–cream mixing and is likely to damage further the fat globules, with similar consequences to changes in the pasteurization temperature.

Acidity
Acid cream churns more easily than sweet cream (Dolby et al., 1966).

Seasonal factors
In New Zealand, in the early spring, the milkfat is soft, the proportion of liquid fat at churning temperature is higher, and churning temperatures must be reduced to compensate. It is often necessary to shock-cool the cream. As the season progresses the milkfat firms, and normal conditions can be used. In droughts the milkfat may be very firm, and churning temperatures may have to be raised.

In other countries, particularly where cattle are housed, the milkfat hardness will depend on the feed used, and seasonal patterns may be quite different.

MOISTURE MEASUREMENT AND CONTROL

Methods Based on Heat

The Kohman test (McDowell, 1953e) is the usual factory control test, the main difficulties being the time taken and the subjectivity of the endpoint, although Pilborough (1981, Personal Communication) has shown that reproducibility can be improved if a temperature detection endpoint is used. Rouse (1983) obtained a single tester standard deviation of 0·033% with this test, but the between operators standard deviation was 0·13%.

In the IDF Oven Method (IDF, 1960) weighed dishes containing butter and ground pumice (to prevent spattering) are placed in an oven at 102°C until constant weight is reached. In a modification (Ministry of Agriculture

& Fisheries, 1979) a temperature of 108°C is used and a constant time of 1 h. Control is required over oven type and air flow, pumice preparation and sample cooling. Rouse (1983) studied this method and found the pumice condition very important for reproducibility. As pumice is a form of silicon dioxide, it is likely that water is adsorbed onto the surface to form silicic acid. Not only the pumice particle size, but also the particle size distribution will affect the surface area, and hence the amount of water adsorbed.

A number of different methods based on the measurement of the dielectric constant of butter (which varies with the water content) have been used. These includes the Wohlburn, Brabender and Alkonix (Alfa-Laval, Lund) systems. The major difficulty with this type of measurement is that the dielectric constant depends on a number of factors other than moisture content including: salt content, air content, solids-not-fat content, butter density, butter temperature, and butter pressure. Of these possibly the most important is air content, as many of the changes that change moisture also change air, and changes in air change the amount of fat passing the sensor, and hence the ratio of moisture to fat.

The Alkonix system compensates for a number of these variables separately and is thus less sensitive to changes. Nevertheless it is still necessary to calibrate the system at each vat change.

The moisture content of a butter sample is inferred by measuring the absorbance of various wavelengths of infra-red light, in the InfraAlyzer (Alfa-Laval, Lund), Inframatic (Bran+Luebb, Norderstedt) and Dickiejohn (Dickiejohn, USA) systems. In each case the measurement is rapid (about 20 s) but off-line, and the instrument must be calibrated for each cream vat.

As no automatic moisture control system has proved entirely satisfactory, Rouse (1983) examined the use of statistical methods and in particular the Shainin control chart (Smith, 1971) for butter moisture control. He found that, in general, moisture stayed relatively constant, and, when it did change, it changed quickly and by quite large amounts. An experienced operator on his own, and one using the chart, detected the change equally quickly, and were equally successful in controlling the machine. Gradual changes, which could be more quickly detected using the chart, did not occur often. There was therefore no advantage in using the chart.

SALTING

Salt is incorporated into butter primarily to enhance flavour, but also for microbiological stability. The salt concentration is usually between 1·2 and

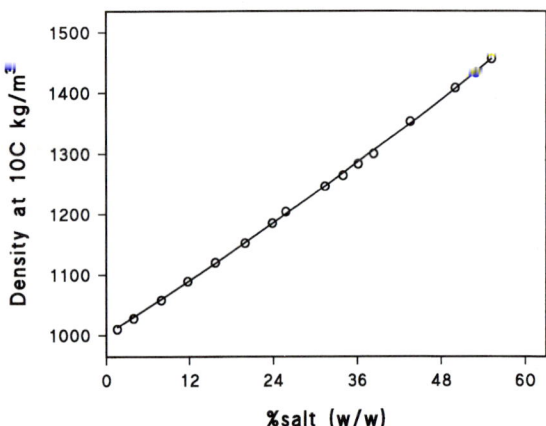

Fig. 7. Curve showing the relationship between the percentage salt in a slurry and its density.

1·6%. Too little gives a flat or insipid flavour, and with too much the flavour is harsh or saline. When the milkfat is soft, it melts more rapidly in the mouth, the salt is more easily detectable, and levels have to be lowered.

Storage

Salt is deliquescent at relative humidities above 75%, and if humidities oscillate around 75% caking can occur. Caking is slowed by modifying the crystal structure for coarse salt, but for the very fine grades necessary for continuous buttermaking an approved anticaking agent is added. For prolonged storage caking can be prevented by keeping the salt room 10°C above ambient.

Salt Slurries

The minimum moisture content of continuous machines is too high for salt to be added in solution, and it is generally incorporated into butter as a slurry of 40–60% strength. Slurries must be kept well agitated, but can be made up in large tanks in a salt room (which should be separate from equipment rooms to minimize corrosion by salt dust) and pumped through a ring main to buttermaking machines. Slurry strength can be quickly checked with a hydrometer in the 1·3–1·5 range, Fig. 7 giving the relation between density and slurry strength.

Particle Size

When solid particles of salt are added to butter during working, the particles slowly dissolve in the moisture in the butter. If the particles have not completely dissolved at the end of working, they will draw water from the butter surrounding them and continue to dissolve. The area of butter around the dissolved particle, being lower in moisture content, is darker in colour than the surrounding butter, and has the appearance of a dark spot. The defect is known as spotted colour, and butter with this defect tends to have a harsh flavour.

To prevent this defect occurring, salt particles must be sufficiently small that they are completely dissolved before working is complete. Jebson (1967) showed that in small continuous buttermaking machines a salt maximum particle size of 100 μm, and in larger machines a particle size of 120 μm, is needed to avoid spotted colour. As 0·03% of particles larger than this will cause spotted colour, the usual specifications call for a particle size of 50 μm.

Incorporation into Butter

The slurry is usually injected at the beginning of the last working section by positive pumps. Stockwell (1972) found that the delivery pressure (and hence flow) from these pumps is highly variable. A likely explanation is that variable numbers of solid particles can sit on the pump valve seats causing variable leakage at each stroke.

A better system is to use a centrifugal pump to produce a constant pressure through a recycle line, with a specially developed control valve (Protech Engineering, Auckland) to adjust the flow, and a mass flow meter to measure the flow.

PACKING AND HANDLING

Transport

Butter is generally moved by being pumped by specially designed gear pumps mounted at the end of the final working section of the buttermaking machine. The pumps force the butter through closed tubes to butter packing, although in some older factories an open ribbon of butter slides down a chute to the hopper of a packing machine. The pressure drop in a pumped butter line is high, and normally butter can only be pumped short distances.

Buffer Storage

Buffer storage is required to allow the buttermaker to continue running during short stoppages of the packer. The buffer should normally be kept

Fig. 8. Photograph of a Pasilac butter silo.

as empty as practicable to allow time to correct faults in the packer. A typical system is the Pasilac silo shown in Fig. 8. Butter is pumped into the top of the silo, and air rams apply a controlled back pressure. The silo is theoretically a 'first-in-first-out' form of storage, but there is some evidence that butter tends to flow near the middle of the silo leaving butter at rest near the walls. Other buffering systems include the Egli trolley, and the Simon trolley (Simon Frères).

Variations in holding time in silos are undesirable as the butter sets up or hardens during holding, and then can partially rework and become leaky in subsequent handling.

In the recombined system the mixing tank can be used for buffer storage with the butter passing through the coolers at a rate slightly greater than the packing rate and the excess passing back to the mixing tank via a remelting system.

Bulk Packing

Bulk butter is usually packed in 25 kg cartons having first been wrapped in a suitable film. There are several types of automatic packing machine available, but the two main ones in current use are the Pasilac FMG packer and the Simon Frères Conticub.

The FMG packer consists of:

(a) a carton former and conveyor,
(b) electronic tare scales,
(c) two wrapping and filling stations,
(d) electronic final weighing with automatic weight adjustment,
(e) parchment folding and top pressing station,
(f) a microprocessor control cabinet which also supervises the semi-automatic CIP (cleaning-in-place) sequences.

It will pack up to 5 t of butter per hour. In Fig. 8, butter is continuously pumped to the packer where it is diverted to each filling station in turn. A single conveyor passes under both filling stations, with sequencing to alternately permit feeding to, and withdrawing from, each station. At each station either one or two layers of wrapping material can be cut to length and wrapped around a metal box which is then lowered to form a snug fit in the base of the waiting carton. Butter is then fed into the carton from the base of the metal box, which rises under an adjustable back pressure. At a preset height butter flow ceases, the metal box lifts away, and the filled carton is conveyed to the final weighing station where the microprocessor controls the addition of further butter until the desired final weight is achieved. In subsequent stations the wrapping material is folded onto the top of the block, which is also pressed flat. The carton is then passed to a separate station for gluing, closing, taping and numbering.

The Conticub comes in single, shown in Fig. 9, and double piston versions. Butter is pumped into a cylinder until its piston has reached a preset position. Two layers of wrapping material are fed from rolls around a rotating nozzle, by a cam. A carton is then formed and pushed over the nozzle. The piston then moves forward discharging the butter into the carton which retreats under back pressure from a ram. The butter is cut from the nozzle, and the carton tipped upright and conveyed to scales where, under control of a microprocessor, more butter is added to achieve the required final weight. The wrapping material is folded onto the top of the block, and the carton conveyed to the gluing and taping machine.

Systems with pumped butter and automatic packing have the advantages:

(a) reduced labour costs,
(b) improved hygiene as butter is less exposed to atmosphere and handling,
(c) softer butter can be packed;

Fig. 9. Photograph of a Conticub butter packer.

but there are the difficulties:

(a) loss of free moisture,
(b) body problems after long holding times,
(c) laminated or layered body.

Free moisture is probably associated with partial working after holding. Dolby (1965) showed that butter that had set up for some time when partially reworked would aggregate droplets and hence could release free moisture. Butter held in a buffering device has the opportunity to set up, especially if it is not 'first-in-first-out'. Dolby also showed that butter from shock-cooled cream sets up more rapidly than that from Alnarp or slowly cooled cream, but butters held for less than 30 min did not show free moisture on partial reworking. Hence free moisture should be minimized by:

(1) holding the butter a minimum time before packing,
(2) producing butter which will set up a slow rate,
(3) working the butter thoroughly.

Body problems at long holding times occur through the older butter setting up, being reworked during packing, and, as a result, having a different moisture distribution and hence different colour from the other butter. The defect is observed as packer streak in the boxes of butter. The methods proposed for minimizing free moisture should also assist in overcoming this problem.

Laminations have been observed in bulk butter packed by the methods described above, but not in the earlier extrusion bulk packing methods used in New Zealand (McDowall, 1953f). Laminated butter is divided into many thin layers which take the form of concentric hemi-ellipsoids. It has been observed by Bissell (1987, Personal Communication) that laminations are more clearly defined in butter having more than 0·2% air content, but that the absence of air is a necessary but not sufficient prerequisite for freedom from laminations.

A possible theory is that air bubbles are not evenly distributed and shear takes place in regions of greater air bubble density. Where shear occurs the butter is worked causing a softening and greater tendency for shear to continue along that surface. Adjacent portions would be less likely to shear and continue setting. Thus a series of layers of greater and less working develops.

Besides removing air, it is likely that laminations would also be avoided with reduced holding times between manufacture and packing, and lower setting rates.

Patting
Consumer sized portions of butter can be patted using automatic patting machines in sizes from the 10-g minipats to 500-g pats. Collating machines which pack the pats in boxes are also available.

Palletization
A batch of 48 cartons can be stacked either manually, or with automatic palletizing machines onto wooden pallets, to produce a container load of 1·2 t of butter. A total of 14 pallets can be loaded into a standard 20 t container.

SETTING AND REWORKING

Setting
When butter first emerges from a churn or continuous buttermaking machine it is soft, but it gradually hardens, more rapidly at first but gradually over a period of 30 days. Figure 10 from Taylor *et al.* (1973) shows a typical setting curve. Mulder (1949) and DeMan & Wood (1958) suggest that thixotropic changes are major contributors to the setting of butter. However as the immediate environment of the crystals does not change as a result of churning, it is difficult to envisage how thixotropic changes, of the magnitude indicated by the setting, can be initiated by churning. Mulder points out that, as there are no heat changes evident

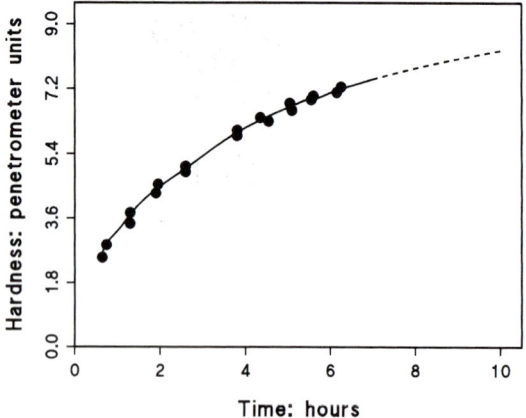

Fig. 10. A typical setting curve for butter.

during setting, crystal growth is unlikely to be a major cause, and again it is difficult to imagine a stimulus for a major change in crystal growth rates as a result of churning.

An alternative theory is proposed. When the milkfat crystallizes inside the milkfat globules, the crystal sizes are limited by the globule size, smaller globules housing smaller crystals, but within a single globule the crystal size is likely to be reasonably uniform. After churning there will be a large range of crystal sizes present in one environment. When there is a range of crystal sizes present in a saturated solution, the larger crystals grow, and the smaller crystals dissolve, with no net heat change. Hence in freshly made butter, there can initially be a rapid growth of larger crystals which will grow into one another creating a rigid structure. The growth rate will exponentially diminish as the numbers of small crystals dissolving decrease and setting curves similar to Fig. 10 could be expected. DeMan & Wood (1958, 1959a) showed that the setting process is arrested by freezing, but continues again when the butter is brought to a higher temperature. When the butter is frozen there would be very little liquid fat available for solution to occur, and the liquid present would have a high viscosity, so diffusion reactions would be very slow. When the butter is back at a higher temperature crystallization and solution would continue.

Taylor *et al.* (1973) showed that the setting rate is higher at higher temperatures. This would also be explained by a lower liquid viscosity at higher temperatures permitting a higher diffusion and hence higher crystallization and solution rates. They also showed that Vacreator-treated cream sets at a faster rate than plate-pasteurized cream. Dolby (1953) showed that Vacreator-treated cream has a much wider fat globule size distribution than plate-pasteurized cream and, in particular, there are many more fine globules. The many small crystals in these would dissolve at a faster rate, and contribute to a higher setting rate.

Reworking

If butter which has set up is worked again, it becomes much softer. It will set again but the final hardness is lower than the original hardness of the butter. The effect of reworking has been studied by Huebner & Thomsen (1957), DeMan & Wood (1959b), Dixon (1966, 1967), and Taylor *et al.* (1971). They found that a maximum decrease in hardness on reworking occurred when butter had been stored at a temperature of or above 5°C for 30 days but there was little difference in the reduction in hardness after 7 days' holding. There was a greater reduction in hardness with initially hard butter from shock-cooled cream than with the softer butter from Alnarp-treated cream. Rework temperatures in the range of 5–18°C had no significant effect on the reduction in hardness on reworking.

FLAVOUR OF BUTTER

The most important selling attribute of butter is its flavour, which is the main reason for its higher selling price than that of other fats. Hence in butter manufacture, flavour is the most important factor. Butter flavour is made up of three components: aroma, taste and texture, the total physiological response being a composite of these constituents. Chemically, many compounds contribute to butter flavour, the most important being acetoin (20 mg/kg) and diacetyl (5 mg/kg). There are also numerous aldehydes, ketones, fatty acids and lactones (Heath, 1983).

In general butter should have a clean 'buttery' flavour with a pleasant mouthfeel, melting completely in the mouth leaving no fatty aftertaste. It must be free from flavour defects, including those caused by: the feed of the cows, faulty processing, micro-organisms, oxidation.

Feed of the Cows

This is basically a farm management problem, and is controlled in the first instance by grading the milk or cream at reception at the factory, but also by the vacreation treatment of the cream.

Processing Faults

Neutralized or soda flavour
This defect is thought to be a consequence of the action of sodium carbonate or bicarbonate on some of the constituents of cream. The defect is caused by over neutralizing, or poor neutralization, e.g. poor mixing of neutralizer and cream. Butter made from neutralized high acid cream will almost inevitably have a neutralized flavour.

Scorched or cooked flavour
This flavour results from incorrect pasteurization of cream, when part or all of the cream reaches too high a temperature, or is held at a high temperature for too long. A particular cause can be direct injection steam, supplied through a pressure reducing valve from a high pressure source. If the steam is dry it superheats on passing through the valve, and may reach a temperature sufficiently high to scorch the cream. The normal solution is to humidify the steam to prevent superheating.

Flat flavour
The butter lacks the characteristic clean buttery flavour, and contains less than normal concentrations of butter flavouring constituents. It may result from:

(a) mature pasture,
(b) cream with a low fat content,
(c) high intensity of steam during vacreation,
(d) excessive washing of butter granules.

Faults (b) and (d) are unlikely to occur with continuous buttermaking, as it is difficult to use low fat cream in a continuous buttermaking machine, and provision for intense washing is not made in these machines.

Harsh flavour
This flavour results from too high a salt content for the softness of the milkfat, or from undissolved salt crystals in the butter.

Flavour Defects Caused by Micro-Organisms

Rancid flavour
This flavour results from the lypolysis of triglycerides to give free fatty acids. The shorter chain length acids (butyric, caproic, caprylic and capric) are primarily responsible for the taint. The enzyme lipase is generally responsible, and may be naturally present in the milk, or be introduced to the cream by microbiological contamination. Excessive or violent handling of milk or cream, resulting in disruption of fat globule membranes will encourage the occurrence of this taint.

Buggy flavour
A term describing the presence of an unclean flavour, produced by contaminant micro-organisms, e.g. *Bacillus coli*.

Putrefactive taint
A flavour resembling the odour of decomposing meat is caused by the organism *Pseudomonas putrefaciens*. This bacterium can contaminate water supplies and, when in butter, has usually come from the water.

Mouldy flavour
Moulds will grow on the surface of butter, if there is free moisture present, and if the temperature is above 0°C, and produce a mouldy flavour.

Oxidative Flavours

Storage flavour

This defect arises from slight oxidation of milkfat which occurs on storage. However, well made butter can be kept at least 2 years if stored at $-10°C$ and retain acceptable flavour.

The presence of this flavour indicates that significant oxidation has taken place. Possible causes of the oxidation include:

(a) presence of copper, iron or other oxidizing agents,
(b) exposure of the butter to ultra-violet light,
(c) relatively high storage temperatures,
(d) very long storage times.

Fishy flavour

An unpleasant flavour (in butter) often described as resembling sardines or herrings. McDowall (1953g) discusses possible causes. These include:

(*a*) *High acidity.* High acidity is a common factor in cases of fishiness, but on its own high acidity does not cause fishy flavours to develop. In fact Pearce *et al.* (1976a) showed that lactic butter made with a defined starter has a very good keeping quality, better than that of salted sweet cream butter.

(*b*) *High salt.* Like high acidity, high salt on its own does not produce fishy flavours, but in combination with acid can cause rapid deterioration of butter. There is evidence that it is not actually salt but contaminants in salt (possibly magnesium) that are the main causative agents in the development of fishiness.

(*c*) *Copper and iron contamination.* These metals are both catalysts for oxidation. Lactic butter with good keeping quality could only be made after tinned copper cream holding vats and tinned brass pumps once used had been replaced with stainless steel.

(*d*) *High churning and storage temperatures.* High temperatures increase reaction rates, and, as oxidation of fats is autocatalytic, higher temperatures will enable initiation reactions to proceed faster, and the butter will take longer to arrive at the storage temperatures. Hence these factors will assist the formation of fishy flavours when other factors are also favourable. Overworking has also been suggested as a factor, but its effect may well be associated with the consequent higher temperatures.

(*e*) *Neutralizer addition.* Excessive or undissolved alkali can react with casein in cream. The products appear to accelerate oxidation.

(*f*) *Stored frozen cream.* This product will have undergone some oxidation during storage, and, if used for buttermaking, will accelerate oxidation.

COLOUR OF BUTTER

The carotenoids, principally β-carotene, are the main colouring agents in milkfat. The level of colour in milkfat depends on the breed of cow (Jerseys give a more highly coloured fat than Friesians), the time of season, and particularly the type of feed, pasture giving much more colour to the fat than dry feed.

The colour of butter is not only determined by the level of carotene in the milkfat, but also by the size distribution of the water droplets; the finer the droplets, the greater the scattering of light, and the lighter the colour of the butter.

Uneven colour in butter is generally regarded as a fault (Dolby, 1956). The more common colour defects are given below.

Streaky Colour
The butter has waves or streaks of different shades of yellow through it. It can arise through insufficient working of the butter during manufacture, as, when underworked, the moisture distribution, and hence the colour, is uneven. It can also arise when butter is held for varying lengths of time (as can happen in some types of butter holding devices). If there is a variation in holding times, some of the butter will partly set up, and, when partially worked by subsequent handling, the moisture distribution will change. A third way in which it may occur is when butter made in different machines is put together in a packing or patting machine. It is inevitable that the different machines will work the butter to different degrees and, unless the mixed butter is thoroughly blended before packing, a streaky colour will result.

Spotted Colour
Butter with this defect has small dark yellow spots throughout. It is due to the presence of undissolved salt particles remaining at the end of the working process. Each salt crystal attracts water from the surrounding butter, and dissolves in it, denuding the surrounding butter of water, and making it darker in colour.

In continuous buttermaking it may be caused by using salt that is too coarse, and does not dissolve in the available working time, or by holding salt slurry for too long a period (e.g. overnight). When crystals agitated in a saturated solution are held, the smaller crystals will dissolve, and the larger crystals grow.

Primrose Colour
If butter is packed in a film which is permeable to water vapour (for instance in parchment) and then stored, water may evaporate from the surface, leaving it darker in colour. This fault is known as primrose colour.

BODY AND TEXTURE
The ideal body of butter is close, smooth and plastic, so that it readily spreads on bread. Correct manufacturing procedures will enable the production of butter of this character, but there are a number of possible body faults.

Crumbly or Brittle Body
The butter is hard and brittle, and a trier plug of this butter has a dull appearance. The plug shows a marked tendency to fracture. The basic cause is insufficient liquid milkfat present, generally caused by too much or too rapid cooling of the cream. Crystal sizes tend to be small, and this contributes to the brittleness.

Sticky Body
The trier plug of butter has a dull appearance, and the plug tends to stick to the trier, with the back of the trier smeared with butter. The stickiness is caused by overworking the butter which raises its temperature, melting a portion of the fat and resulting in too little solid fat in the butter. If the cream is excessively cooled the extra energy required to churn the cream may raise the butter temperature causing sticky body. In this case other faults, such as porous texture or leaky body, may be present as well. The defect does not usually persist on storage.

Leaky Body
Butter with this defect will have droplets of water on a cut surface. The defect is due to insufficient working of the butter, but may also occur if the butter is held so that setting commences, and is then partly reworked (Dolby, 1965). Butter with leaky body may be subject to accelerated microbiological deterioration on storage.

Porous or Ice-Cream Texture
Butter with this defect has an open or porous structure. The defect is normally found only in butter made in continuous buttermaking machines,

and may be accomplished by leaky body. It occurs after excessive cooling of the cream followed by underchurning, and vigorous working of the butter in the presence of air. It can be overcome by the use of correct cream cooling procedures and vacuum working of the butter.

TYPES OF BUTTER

Lactic Butter

Lactic butter is produced when sour cream is churned. Originally cream was soured by adventitious organisms and batch churned. Organisms which produce lactic acid grow well in milk, and are very common, so, when cream is held, it generally sours with lactic acid development. A more consistent flavour is obtained if a defined starter using a mixture of known strains of micro-organisms is used. A blend of *Streptococcus cremoris* 270, *Leuconostoc cremoris* 501 and *Str. diacetylactis* DRC_1 in the ratio 4/1/1 consistently gave butter of very good flavour (Russell *et al.*, 1972; Pearce *et al.*, 1976*a, b*).

The production of lactic butter by the souring of cream has a number of disadvantages:

(a) A sour buttermilk, which may be difficult to dry, is formed.
(b) The souring cream may become contaminated with bacteriophage or other undesirable micro-organisms.
(c) pH and flavour may be variable.
(d) Batches must be the size of the cream holding vats.

A process, developed by the Nizo Dairy Research Institute (Frielink, 1977), in which starter is added to the granules, overcomes these problems. Sweet cream is cooled and churned, and the buttermilk drained from the granules. Starter is added, the churn rotated for a few minutes and drained again. The moisture is made up to 16% with starter. For this process a starter mixture which develops a higher acidity is desirable. Work at the New Zealand Dairy Research Institute showed that the butter acidity came partly from addition of starter to make up the moisture content, and partly from a transfer of acid and flavouring to moisture already present in the granules. The granule size at draining is important for this process. The size must be large enough for good draining, but small enough for good transfer between the added starter and the moisture in the granules.

Provided the butter is unsalted and copper contents are kept very low, lactic butter was found to have a better keeping quality than standard salted sweet cream butter.

When butter is made in continuous machines the above method is not available, although an extra holding and draining section has been proposed. Generally lactic butter is made by adding lactic acid and a starter distillate to sweet cream butter, the addition being made at the point of salt injection.

Softer Butters

Butter as normally prepared is too firm to be spread on bread directly from a refrigerator, and much research has been conducted into methods of softening butter. A comparison of a number of these methods used on New Zealand butters is shown in Fig. 11.

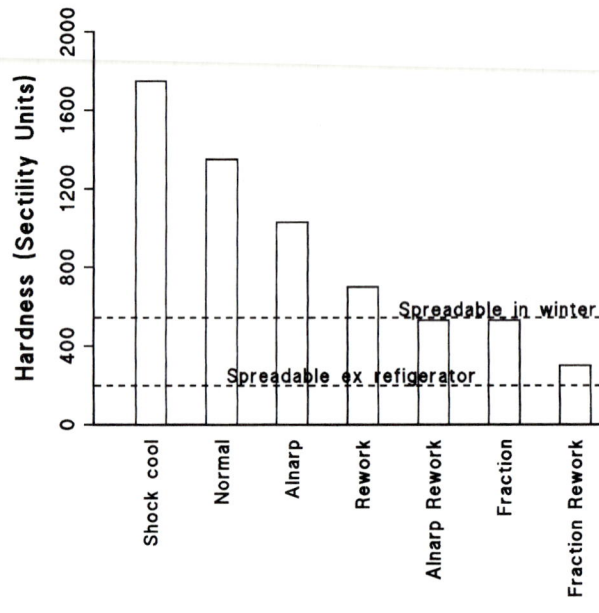

Fig. 11. Comparison of the hardness (in sectility units) of butter made in different ways. Shock cool: butter made from cream which is cooled in the plate heat exchanger to churning temperature. Normal: butter made from cream which is cooled in the plate heat exchanger to within 3–4° of churning temperature, and then cooled slowly in the cream vat to churning temperature. Alnarp: butter made from cream which has been subjected to a modified Alnarp cream cooling procedure. Rework: normal butter which has been held for 30 days at 4°C and then reworked. Fraction: butter made from 50% normal cream and 50% recombined cream which has been made from the liquid fraction of milkfat held at 25°C and the crystalline fraction removed. Fraction rework: fraction type butter which has been held for 30 days at 4°C and then reworked.

The Alnarp treatment, described under cream cooling, if combined with reworking can achieve a 50% reduction in hardness, when compared with normal butter, and 500-g pats of butter made this way become spreadable at normal ambient temperatures about 30 min after they are taken from a refrigerator. However, if this butter temperature is allowed to rise above 18°C, on subsequent cooling, the hardness increases to that of normal butter (Taylor & Jebson, 1975).

If milkfat is fractionated into high and low melting point fractions by one of the methods described in Chapter 6 (the fractionation being carried out at 25°C), and the fraction is mixed with an equal quantity of milkfat and made into butter, that butter after reworking will be spreadable about 15 min after removal from a refrigerator (Dolby *et al.*, 1971). If a softer fraction than this is used the butter becomes too soft, and lacks standup at warm ambient temperatures.

If a butter is required that is spreadable at refrigerator temperatures, stands up well at warm ambient temperatures, yet melts completely in the mouth, then the melting curve of the fat with temperature must be similar to that of the ideal curve in Fig. 12. Good quality polyunsaturated margarines have melting curves similar to the ideal, but the curve for butter

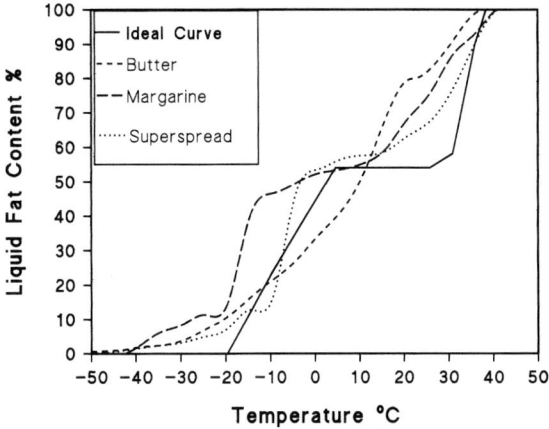

Fig. 12. Melting curves for different spreads. Ideal curve: the theoretical requirement for a spread which is uniformly spreadable from refrigerator temperatures to high ambient temperatures, but which melts completely in the mouth. Butter: the melting curve for butter made by normal methods. Margarine: the curve for a good quality polyunsaturated margarine. Superspread: the curve for a butter made from milkfat from which the intermediate melting fats have been removed.

is rather different. In order to obtain from milkfat a fat close to the ideal, the intermediate melting fraction must be removed, and the fractionation methods used must be efficient and give clean separations of liquid and solid fractions.

These requirements can be met by a two-stage fractionation using acetone as the solvent (Jebson *et al.*, 1974, 1975; Munro *et al.*, 1976; Norris, 1976). Milkfat is dissolved in acetone, and crystallized successfully at 10°C and −10°C. The acetone is removed from the fractions by evaporation and steam distillation. The fraction crystallized at 10°C is combined with the fraction liquid at −10°C, and used to make butter by a version of the Ammix process. Considerable care is necessary to prevent the development of off-flavours. Although technically proven, the process has never become commercial.

REFERENCES

ANDERSON, R. T. (1969) Trials of Silkeborg continuous buttermaking machine. *New Zealand Journal of Dairy Science & Technology*, **4**, 29–33.

DEMAN, J. M. & WOOD, F. W. (1958) Hardness of butter. I. Influence of season and manufacturing method. *Journal of Dairy Science*, **41**, 360–8.

DEMAN, J. M. & WOOD, F. W. (1959*a*) Hardness of butter. II. Influence of setting. *Journal of Dairy Science*, **42**, 56–61.

DEMAN, J. M. & WOOD, F. W. (1959*b*) Influence of temperature treatment and season on the dilatometric behaviour of butterfat. *Journal of Dairy Research*, **26**, 17.

DIXON, B. D. (1966) *Australian Journal of Dairy Technology*, **21**, 143.

DIXON, B. D. (1967) Spreadability of butter: pilot reworking studies. *Australian Journal of Dairy Technology*, **22**, 202–8.

DIXON, B. D. (1974) Spreadability of butter: pilot reworking studies. II. *Australian Journal of Dairy Technology*, **29**, 227–30.

DOLBY, R. M. (1953) The effect of different cream treatments during the pasteurization process on the size distribution of fat globules in cream and butter. *Journal of Dairy Research*, **20**, 201–4.

DOLBY, R. M. (1954) The effect of temperature treatment of cream before churning on the consistency of butter. *Journal of Dairy Research*, **21**, 67–77.

DOLBY, R. M. (1956). Some colour defects in butter. XIV International Dairy Congress, 2, Part 1, p. 133.

DOLBY, R. M. (1965) Changes in moisture distribution caused by partial reworking of butter shortly after churning. *Journal of Dairy Research*, **32**, 263–7.

DOLBY, R. M., JEBSON, R. S. & LE HERON, B. S. (1966) Fritz process buttermaking in New Zealand. XVII International Dairy Congress, pp. 43–8.

DOLBY, R. M., JEBSON, R. S. & RUSSELL, R. W. (1971) Recombined butter from fat fractions. *New Zealand Dairy Research Institute 43rd Annual Report*, pp. 22–3.

FREDE, E., PRECHT, D. & PETERS, K. H. (1982*a*) A new method for control of cream ripening. *Milchwissenschaft*, **37**, 657–60.

FREDE, E., PRECHT, D. & PETERS, K. H. (1982b) Practical tests on the registration of crystallization curves for milkfat. *Milchwissenschaft*, **37**, 733–36.
FRIELINK, J. G. (1977) Method for producing aromatic soured butter without acidifying it. Netherlands Patent No. 7513464.
HEATH, H. B. (1983) *Flavour Technology Profiles. Product Applications*, AVI, Westport, CN.
HUEBNER, V. R. & THOMSEN, L. C. (1957) Spreadability and hardness of butter II. Some factors affecting spreadability and hardness. *Journal of Dairy Science*, **40**, 839–46.
HUGHES, I. R., JEBSON, R. S. & TUTTIETT, P. T. (1976) Moisture control in continuous buttermaking. *New Zealand Journal of Dairy Science & Technology*, **13**, 29–36.
INTERNATIONAL DAIRY FEDERATION (1960) Determination of the moisture content of butter. *FIL: IDF 10*.
JEBSON, R. S. (1967) Preparation of salt for continuous buttermaking machines. *New Zealand Journal of Dairy Science & Technology*, **2**, 5–10.
JEBSON, R. S. (1982) Recombined butter. IDF Document No. 116, pp. 30–2.
JEBSON, R. S. & LASCELLES, D. R. (1977) Application of heat pumps in the dairy industry. *New Zealand Journal of Dairy Science & Technology*, **12**, 116–22.
JEBSON, R. S., TAYLOR, M. W., BISSELL, T. G. & LOCHORE, J. C. (1974) Spreadable butter. *New Zealand Dairy Research Institute 46th Annual Report*, pp. 35–6.
JEBSON, R. S., TAYLOR, M. W., MUNRO, D. S., BISSELL, T. G., BEWICK, B. J. G., NORRIS, R., LOCHORE, J. C., ARCHER, K. M. & MCDOWELL, A. K. R. (1975), Spreadable butter. *New Zealand Dairy Research Institute 47th Annual Report*, pp. 32–5.
LORWOOD, D. (1988) Moisture in continuous buttermaking machines. Diploma in Dairy Science & Technology Thesis, Massey University, Palmerston North, New Zealand.
MCDOWALL, F. H. (1953a) *The Buttermaker's Manual*, New Zealand University Press, Wellington, pp. 280–303.
MCDOWALL, F. H. (1953b) *The Buttermaker's Manual*, New Zealand University Press, Wellington, pp. 143–58.
MCDOWALL, F. H. (1953c) *The Buttermaker's Manual*, New Zealand University Press, Wellington, pp. 474–9.
MCDOWALL, F. H. (1953d) *The Buttermaker's Manual*, New Zealand University Press, Wellington, pp. 511–34.
MCDOWALL, F. H. (1953e) *The Buttermaker's Manual*, New Zealand University Press, Wellington, pp. 1536–45.
MCDOWALL, F. H. (1953f) *The Buttermaker's Manual*, New Zealand University Press, Wellington, pp. 579–81.
MCDOWALL, F. H. (1953g) *The Buttermaker's Manual*, New Zealand University Press, Wellington, pp. 777–80.
MCDOWALL, F. H. (1955) Steam distillation of taints from cream. I. Theoretical considerations and properties of the reference substances, diacetyl and acetoin. *Journal of Dairy Research*, **22**, 311–27.
MCDOWALL, F. H. (1956) Steam distillation of taints from cream. II. Investigations on commercial equipment with the use of diacetyl and acetoin as reference substances. *Journal of Dairy Research*, **23**, 46–65.

McDowall, F. H. (1957) Steam distillation of taints from cream. III. Factors affecting the rate of removal of the reference substances, diacetyl and acetoin. *Journal of Dairy Research*, **24**, 104–90.

McDowall, F. H. (1965) Steam distillation of taints from cream. X. Vapour liquid relationships for benzyl mercaptan. *Journal of Dairy Research*, **32**, 147–149.

Mercer, W. B. (1985) A study of selected variables in continuous buttermaking. Diploma in Dairy Science & Technology Thesis, Massey University, Palmerston North, New Zealand.

Ministry of Agriculture & Fisheries (1979) *Standard Laboratory Methods*. Chemistry Section 3.8.4a. Dairy Division, Ministry of Agriculture & Fisheries, Wellington.

Mogensen, G. & Danmark, H. (1984) Process technological factors in continuous buttermaking affecting energy consumption, fat loss, and butter quality. *256. Bereting Statens Forsogsmejeri Hilerod*.

Mulder, H. (1949) Physical structure of butter. 12th International Dairy Congress, pp. 81–5.

Mulder, H. & Walstra, P. (1974) *The Milkfat Globule*, Commonwealth Agricultural Bureaux, Farnham Royal, pp. 101–28.

Munro, D. S. (1982) Alternative equipment for the manufacture of recombined butter. IDF Document No. 142, p. 33.

Munro, D. S., Taylor, M. W., Bissell, T. G. & Lochore, J. C. (1976) Spreadable butter. *New Zealand Dairy Research Institute 47th Annual Report*, pp. 32–5.

Norris, R. (1976) Fractionating fats: butter spreadable over a wide temperature range. NZ Patent 172101.

Palfreyman, K. R. (1982) Performance evaluation of cream crystallizing silos. Diploma in Dairy Science & Technology Thesis, Massey University, Palmerston North, New Zealand.

Palfreyman, K. R. (1988) An evaluation of the mixing performance and fat globule damage in cream crystallizing silos. *New Zealand Journal of Dairy Science & Technology*, **23**, 373–84.

Pearce, L. E., Limsowtin, G. K. Y., Natarajan, A. M., Jebson, R. S. & Lochore, J. C. (1976a) Manufacture and storage of lactic butter made with a defined starter. *New Zealand Dairy Research Institute 48th Annual Report*, p. 38.

Pearce, L. E., Natarajan, A. M., Limsowtin, G. K. Y., Jebson, R. S. & Lochore, J. C. (1976b) Direct addition of starter to butter granules. *New Zealand Dairy Research Institute 48th Annual Report*, p. 39.

Robinson, R. K. (1985) *Modern Dairy Technology*, Elsevier Science Publishers, London, pp. 93–116.

Rouse, S. T. J. (1983) Butter moisture control using statistical methods. Diploma in Dairy Science & Technology Thesis, Massey University, Palmerston North, New Zealand.

Russell, R. W. (1973) The effect of cream handling conditions on butter quality. *New Zealand Journal of Dairy Science & Technology*, **8**, 124–6.

Russell, R. W., Fryer, T. F. & Jebson, R. S. (1972) Unsalted acid butter. *New Zealand Dairy Research Institute 44th Annual Report*, p. 23.

Samuelsson, E. & Petersson, K. I. (1937) *Svenska Mejeritidningen*, **29**, 65, 73.

Samuelsson, E. & Vikelsoe, J. (1971a). Liquid fat of butter in relation to its consistency. *Maelkeritidende*, **84**, 159–69.

SAMUELSSON, E. & VIKELSOE, J. (1971*b*). Estimation of the amount of liquid fat in cream and butter by low resolution NMR. *Milchwissenschaft*, **26**, 621–5.
SCOTT, J. K. (1954*a*). The steam stripping of taints from liquids I. An analysis of processing methods with particular reference to the Vacreator. *Journal of Dairy Research*, **21**, 354–69.
SCOTT, J. K. (1954*b*) The steam stripping of taints from liquids II. Counterflow stripping with particular reference to possible use of Vacreator equipment. *Journal of Dairy Research*, **21**, 370–82.
SMITH, C. S. (1971) *Quality and Reliability—an Integrated Approach*, Pitman, London.
STOCKWELL, D. T. J. (1972) Continuous buttermaking—a process capability study. M. Tech. Thesis, Massey University, Palmerston North, New Zealand.
TAYLOR, M. W. & HAWKE, J. C. (1975) The triacylglycerol compositions of bovine milkfats. *New Zealand Journal of Dairy Science & Technology*, **10**, 40–8.
TAYLOR, M. W. & JEBSON, R. S. (1975) The efects of temperature changes on the hardness of butter. XIX International Dairy Congress, p. 673.
TAYLOR, M. W., DOLBY, R. M. & RUSSELL, R. W. (1967) *New Zealand Journal of Dairy Science & Technology*, **6**, 172–6.
TAYLOR, M. W., DOLBY, R. M. & RUSSELL, R. W. (1971) The reworking of butter. *New Zealand Journal of Dairy Science & Technology*, **6**, 172–76.
TAYLOR, M. W., DOLBY, R. M. & RUSSELL, R. W. (1973) The effect of manufacturing conditions on the setting rate of butter. *Journal of Dairy Research*, **40**, 393–402.
TE WHAITI, I. E. & FRYER, T. F. (1975) Factors that determine the gelling of cream. *New Zealand Journal of Dairy Science & Technology*, **10**, 2–7.
TOWLER, C. (1986). *New Zealand Journal of Dairy Science & Technology*, **21**, 79–87.
TRUONG, H. T. & MUNRO, D. S. (1983) Ammix buttermaking. XX International Dairy Congress, pp. 341–2.

4

Anhydrous Milkfat Products and Applications in Recombination

D. Illingworth and T. G. Bissell*
New Zealand Dairy Research Institute, Private Bag 11029,
Palmerston North, New Zealand

SUMMARY

AMF in its various forms is defined and the main uses and world production figures are listed. Details are given of the manufacturing processes used for AMF, covering all possible feedstocks—direct from cream and different types of butter—together with a section describing the production of ghee and samna balady. The importance of cream pretreatment in the direct-from-cream process is stressed, and the phase inversion stage is then covered, with emphasis placed on the factors having the most influence on the efficiency of the process. The manufacturing process is described by reference to the equipment used at each stage of the process—cream concentration, phase inversion, separation and dehydration. Equipment manufactured by Westfalia and Alfa Laval is described in detail. Methods for monitoring the performance of the plant are confined to measurement of the fat content of various streams within the plant. The quality of the AMF produced depends mainly on free fatty acid levels, measurement of oxidative deterioration, such as peroxide value or anisidine value, and contamination by other oils and fats. Reference is made to the analytical methods which may be used to measure these directly, together with methods for determining trace metals such as copper and iron which catalyse oxidation. The effect of dissolved oxygen on the storage stability of AMF and methods for measurement of dissolved oxygen are discussed. A description of the various ways of packaging AMF for both industrial and retail use leads into sections covering the use of AMF and its fractions in recombined products. These include recombined butter and other spreads and blends, with descriptions of plant and equipment which can be used, ice cream, recombined milks and cream and cheese. The handling of AMF and the quality of other materials used in recombining is discussed.

* Dr Bissell was tragically killed in a mountaineering accident during the preparation of this chapter.

ANHYDROUS MILKFAT—DEFINITION, MAIN USES AND WORLD PRODUCTION

Anhydrous milkfat, variously called AMF, anhydrous butteroil, anhydrous butterfat, butterfat or butteroil, is the product exclusively obtained from milk, cream or butter by means of processing to remove almost completely all traces of moisture and solids-not-fat material. The International Dairy Federation (IDF) standard (International Dairy Federation, 1977) lays down precise definitions of various grades of milkfat, as shown in Table 1.

AMF is used primarily in recombined dairy products such as whole milk, cream, ice cream, cheeses and recombined butter. The principal users are countries of the Middle East, South and Central America and the Pacific region where milk production is inadequate to meet demand or where there

TABLE 1
IDF specifications for anhydrous milkfat, anhydrous butteroil and butteroil

	Definition		
	Anhydrous milkfat	*Anhydrous butteroil*	*Butteroil*
Composition			
Milkfat (%) minimum	99·8	99·8	99·3
Water (%) maximum	0·1	0·1	0·5
Free fatty acid (FFA) as oleic acid (%) maximum	0·3	0·3	0·3
Copper (mg/kg) maximum	0·05	0·05	0·05
Iron (mg/kg) maximum	0·2	0·2	0·2
Peroxide value (PV) as milli-equivalents O_2/kg	0·2	0·3	0·8
Coliforms	Absent in 1 g	Absent in 1 g	Absent in 1 g
Taste and odour	Clean and bland (at 20–25°C)	No pronounced unclean or other objectionable taste or odour	No pronounced unclean or other objectionable taste or odour

Note: The recommended physical structure should be in the form of a smooth, fine grain structure. Anhydrous milkfat is defined as the product obtained from prime quality milk, cream or butter to which no neutralizing substances have been added. Anhydrous butteroil or butteroil are the products obtained from butter or cream, both of which may be of variable age. Both may contain traces of neutralizing substances.

TABLE 2
Butteroil production (10^3 tonnes)[a]

Country	1986	1987	1988	1989	1990	1991
EEC	218	265	265	135	115	106
New Zealand	33	73	67	53	40	47
Australia	25	18	23	22	26	37
Sweden	6	8	5	6	7	7
Switzerland	4	4	4	5	5	5
Uruguay	1	0·1	0·1	0·2	0·2	0·2

[a] These figures include all specifications listed in Table 1.

are no established dairy industries. Large quantities of AMF are imported into the Middle East and areas of South-East Asia to supply ghee to workers from the Indian sub-continent. A sizeable tonnage of AMF is used within the country or region of manufacture. For example, in EEC countries, notably Belgium, The Netherlands, France, Germany and Denmark, AMF is fractionated into high- and low-melting fractions using a physical process (Tirtiaux, 1976).

Although milkfat has some unique properties, mainly its natural flavour, its complex triglyceride composition results in physical properties that make it inadequate for many uses and a melting range from $-40°C$ to $+40°C$. High-melting fractions have melting points in excess of $38°C$ and are used as speciality bakery fats in products such as puff pastry and croissants (Rodenburg, 1973; Munro & Illingworth, 1986). Low-melting fractions have melting points in the range $21-28°C$ and in general there is some concentration of colour and flavour. Because they are produced in larger quantities than the high-melting fractions, they have tended to present a disposal problem. The perceived advantages of increased milkfat flavour and ease of melting in these fractions have led to manufacturers offering them as alternatives to AMF for the recombining trade. AMF is also used in dairy spreads and other blended products for the home market or export. In another major producing country, New Zealand, commercial fractionation of milkfat has been introduced only recently for the production of tailor-made fats for industrial use (Illingworth et al., 1989). By comparison with the EEC, the volumes produced are quite small. Blended products of AMF with vegetable oils are produced primarily for export, although recent changes to the food regulations have seen the appearance of full-fat dairy spreads on the domestic market.

Table 2 (Anon., 1992) shows production figures for the major producing and exporting countries over a 6-year period. Although the EEC has cut production in recent years, it still dominates world production, with New Zealand and Australia making significant contributions. The small quantity produced in the United States is not exported and figures are not available. Similarly production figures for samna balady and ghee, the traditional forms of AMF of the Middle East and India, are not published.

MANUFACTURING PROCESSES FOR AMF

In Europe AMF has been produced traditionally from lactic or cultured butter but increasingly salted and unsalted sweet cream butter are used as starting materials. In Australia, New Zealand and other producing countries, there is a tradition of manufacture either direct from cream or from sweet cream butter.

The different raw materials require slightly different processing conditions. This is best illustrated by comparing the results of centrifuging samples of each type of raw material as shown in Fig. 1. Both 80% fat cream (after phase inversion) and sweet cream butter form an emulsion layer because some globular fat, containing both milk proteins and phospholipid material from the globule membrane, is still present in the uninverted form. Processing must take into account means to limit the emulsion formation. Sour cream (cultured) butter and sweet cream butter that has the serum pH adjusted to 4·5 (this includes those butters that are artificially soured sweet cream butters as from the NIZO process (NIZO, 1977)) do not show an emulsion layer but contain more precipitated or sedimented solids owing to protein denaturation.

Alfa Laval (Anon., 1970*a*) and Westfalia (Lehmann *et al.*, 1988) are the principal suppliers of plants that can process one or all of these raw materials, and typical plants are described in detail further on. Both types of processes entail certain practices that are essential for the production of a high quality final product.

Direct-from-Cream Process

Pretreatment of cream
Before considering the manufacturing process, some consideration must be given to the prior handling of the cream. Good cream handling practices as used for butter manufacture should be followed to avoid lipolysis and possible gelling of the cream prior to and during processing.

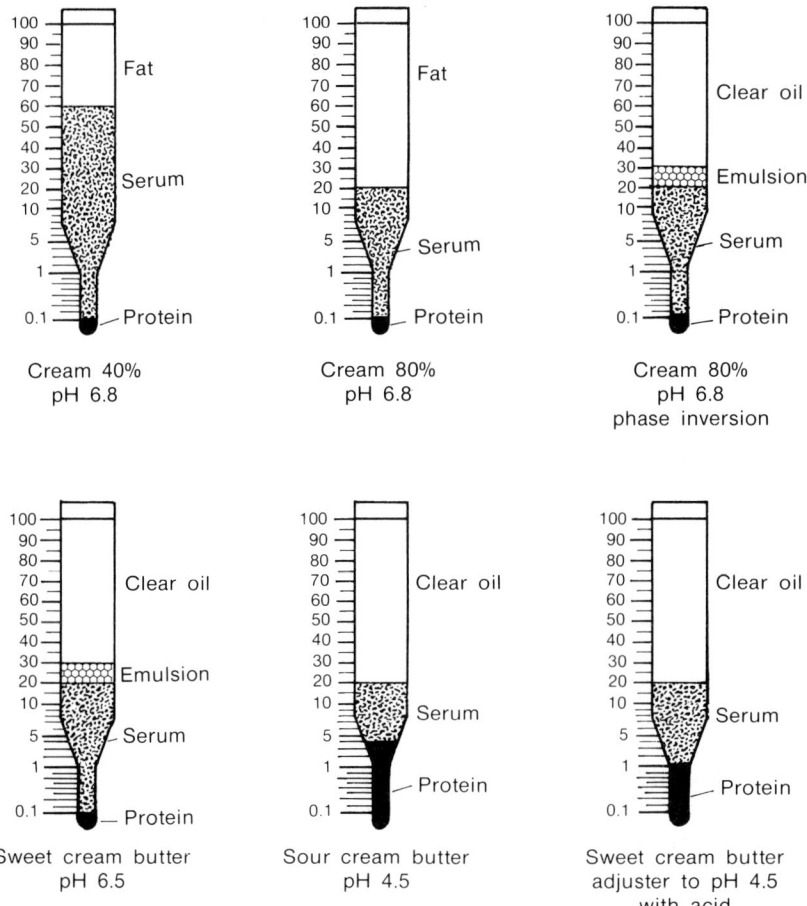

Fig. 1. Centrifuged samples of possible raw materials for AMF manufacture (courtesy of Westfalia).

Lipolysis refers to chemical breakdown of the triglycerides of milkfat by enzymes to produce free fatty acids (FFA). Lipolytic enzymes occur naturally in milk and cream, but are readily de-activated by heating cream at 85°C for 15 s. However, lipolytic enzymes arising from microbial growth can survive even UHT conditions (140°C for 2 s) and the only control is to keep bacterial counts low at all stages of processing. High levels of FFA lead to the development of soapy flavours which can carry through to the

AMF, because of the presence of free decanoic and dodecanoic (medium chain) fatty acids (McDaniel et al., 1969).

Gelling or thickening of cream results from clumping of the fat globules. This occurs when the fat globule membrane becomes damaged and allows free fat to be released.

Both types of deterioration may be minimized by immediate pasteurization after separation followed by chilling to less than 10°C prior to storage or transportation. Excessive agitation and aeration of cream or any practice, such as homogenization, that causes disruption of the fat globule membrane should be avoided.

If cream is held for significant lengths of time at temperatures higher than 10°C, the resulting increased levels of acid-producing bacteria lead to high lactic acid levels in the cream. There are two main reasons why this must be removed: (1) to maintain the quality of the recovered buttermilk or milkfat serum, and (2) to prevent the precipitation of protein during pasteurization (above 0·2% lactic acid).

The usual practice is to treat the cream, using a solution of sodium hydroxide (maximum concentration 4% w/w), immediately prior to pasteurization in a plate pasteurizer, to reduce the lactic acid level to within 0·06–0·12%, but preferably not below 0·09%, as that is considered to be bad practice.

Pasteurization of cream is carried out in a plate pasteurizer prior to AMF manufacture and achieves the following: (1) destruction of bacteria—primarily to give a high quality buttermilk or milkfat serum; (2) inactivation of lipolytic enzymes—the pasteurization temperature must be at least 85°C (higher temperatures may give improved keeping quality); (3) better resistance to oxidation in the final product and a longer shelf life.

Post-pasteurization processing

Having undergone the necessary pretreatments, the cream can then be processed in the AMF plant. A typical layout of a plant for making AMF from cream is shown in Fig. 2. Treated cream at 55–60°C with a fat content of about 40% is fed to the cream concentrator which is a self-desludging centrifugal separator. Serum leaving the heavy phase outlet may contain up to 1–2% fat and can be fed directly to a buttermilk separator for recovery. Recovered fat is returned as a cream to the cream concentrator feed tank, whilst the milkfat serum (< 1% fat) may be spray dried.

Cream leaving the concentrator (70–80% fat) is fed to a phase inversion device, where fat globule membranes are ruptured and the oil-in-water emulsion inverts.

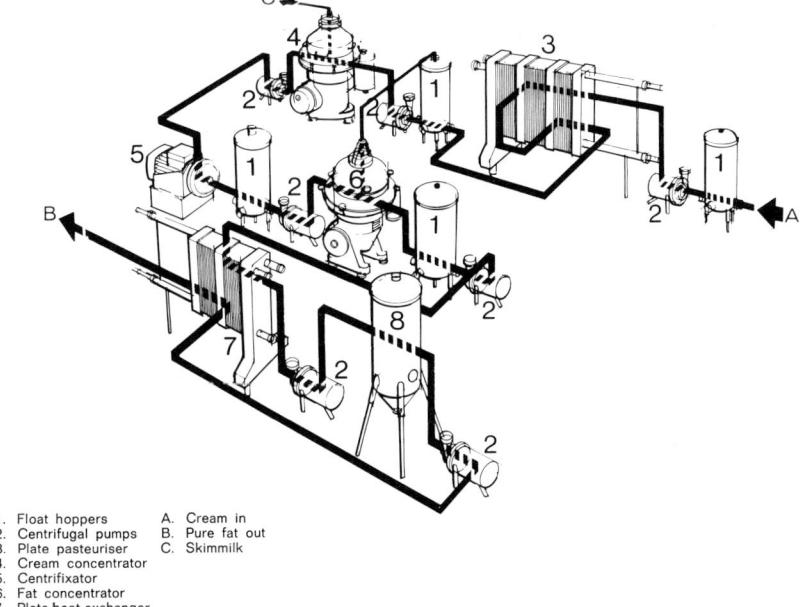

1. Float hoppers   A. Cream in
2. Centrifugal pumps   B. Pure fat out
3. Plate pasteuriser   C. Skimmilk
4. Cream concentrator
5. Centrifixator
6. Fat concentrator
7. Plate heat exchanger
8. Vacuum dryer

Fig. 2. Direct-from-cream plant for AMF manufacture (courtesy of Alfa Laval).

Further separation of the water-in-oil emulsion is carried out using an oil separator. Globular fat that passes through the phase inversion device unchanged leaves this separator with the protein and water through the heavy phase outlet and is returned to the cream concentrator feed tank. The light phase from the oil separator contains approximately 99·5% milkfat.

Phase inversion is possibly the most important of all the processing steps, as a loss of efficiency at this point has an impact on many other parts of the process.

The Principles of Phase Inversion

The expression 'phase inversion' is used to describe the process by which an oil-in-water emulsion (cream) is converted to a water-in-oil emulsion, such as butter at low temperatures or crude butteroil when the fat is molten. The first step in such a process is to rupture the fat globule membranes which in turn allows the globules to coalesce and form a continuous phase. A schematic diagram of the fat globule in the oil-in-water emulsion system of cream is shown as Fig. 3. The membrane layer which protects the fat from

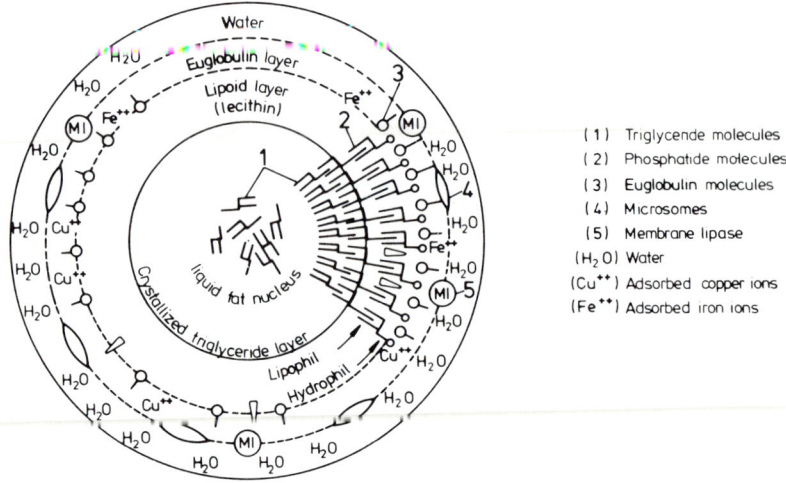

Fig. 3. The milkfat globule.

chemical and lipolytic action can be broken down by either chemical or mechanical means.

In continuous buttermaking processes, phase inversion is achieved at low temperatures (10–14°C) using a cream with a fat content of about 40%. The cream is foamed mechanically with the incorporation of air, and the fat globule membranes are ruptured by surface tension effects.

When AMF is produced directly from cream, phase inversion is made to occur in cream with a fat content of between 70 and 80% at a temperature of 55–60°C. The homogenizing device effects the rupture of the membranes, but in the absence of air. Figure 4 shows a diagrammatic representation of what occurs during phase inversion. The mechanism of the rupture process is not well understood, but it is believed that globules are split by the high shear fields in homogenizing devices.

Factors that influence emulsion stability also affect the efficiency of phase inversion. However, as phase inversion is a destabilizing of an emulsion, these factors must be optimized in the opposite direction to that which is considered desirable for normal cream handling.

The efficiency of phase inversion can be measured relatively easily. Mixtures of high fat cream, AMF and skim milk may be homogenized using a single-stage homogenizer at various temperatures and pressures. If the resulting emulsions are then centrifuged in a graduated tube, division of

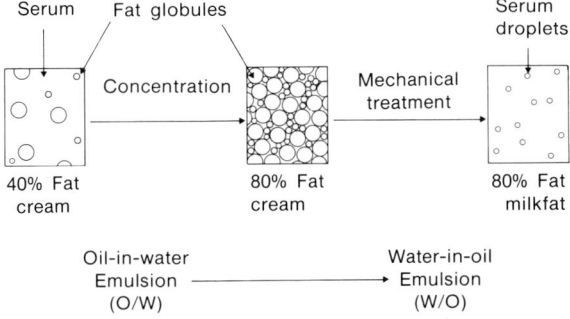

Fig. 4. Phase inversion in high fat cream.

the quantity of liquid fat in the top layer by the total fat content gives a measure of the efficiency of inversion.

Although multi-stage homogenization is possible, only single-stage is necessary. Two-stage homogenization can cause some of the free fat that is liberated in the first stage to become re-emulsified during the second stage.

The graphs in Figs 5, 6 and 7 (Watt, 1982) also illustrate that complete phase inversion is not possible by this means and that there must always be a certain percentage of intact fat globules present. Thus the process must be optimized to minimize the formation of new, stable fat globules.

The factors that influence the efficiency of phase inversion are listed below in order of importance:

(1) homogenizing pressure;
(2) fat content of the cream;
(3) free fat content of the cream;
(4) temperature.

Acidity in cream does not have any significant effect.

These factors are considered below, with their consequences on the direct-from-cream process.

The effect of varying the homogenizing pressure is shown in Fig. 5. At pressures below 1250 psi (8600 kPa), particularly if the fat content is low, the phase inversion efficiency drops rapidly as the pressure decreases.

In addition, if the pressure drops below 1250 psi (8600 kPa), small changes in the total fat content of a high fat cream result in large changes in the efficiency of phase inversion.

The smaller fat globules formed by homogenization are covered partly by original membrane material and partly by surface-active materials from

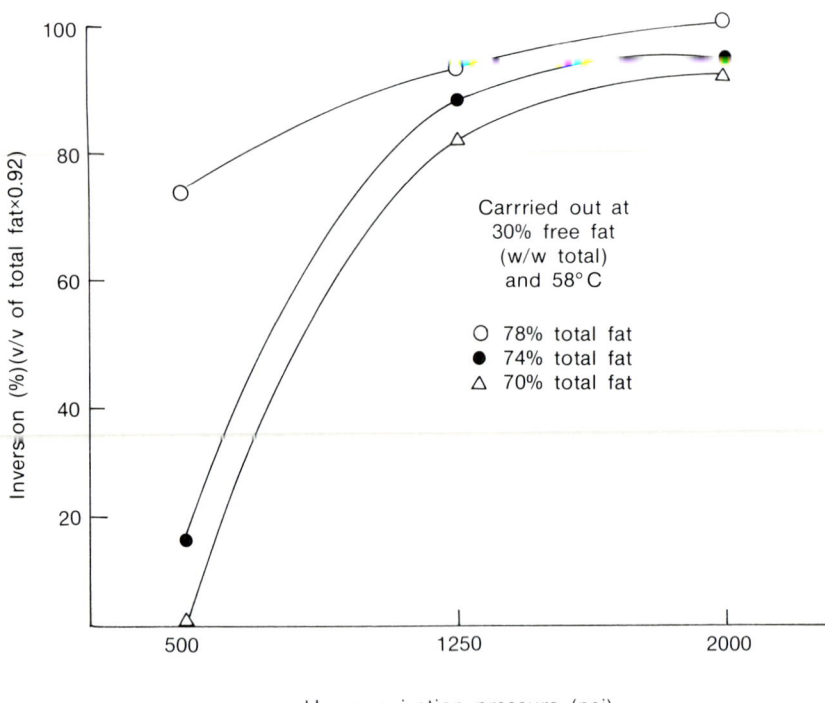

Fig. 5. Effect of homogenizing pressure on phase inversion.

the serum. The increased surface area requires more fat globule membrane material than was present in the original globules. Thus the formation of new globule membranes necessitates taking material, usually casein, from the serum. If as much as possible of the serum is removed during concentration of the cream to a high total fat content prior to homogenization, then fewer new globules can be stabilized and the phase inversion process is assisted.

Figure 5 also shows that the total fat content significantly affects the phase inversion efficiency at homogenizing pressures of less than 1250 psi (8600 kPa). Above this pressure, changes in the total fat content cause only minor variations in efficiency.

The phase inversion process has been found empirically to be improved greatly by the presence of 'free fat', obtained by recycling part of the phase-inverted stream back to the homogenizer feed tank. Free fat is then the oil present as large, unstable globules after inversion.

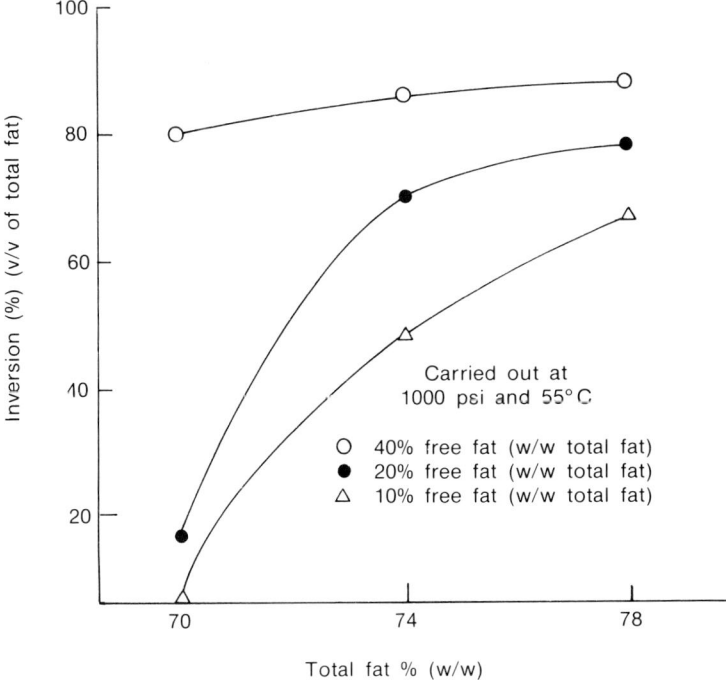

Fig. 6. Relationship between the total fat and free fat on phase inversion.

Figure 6 shows the relationship between total fat and free fat. If both are high, changes to one in relation to the other have little effect. If the total fat level is low, the quantity of free fat has a large effect on the phase inversion efficiency, with the effect diminishing once the overall fat content reaches 78% or higher. This relationship has clear consequences for the operation of direct-from-cream plants, particularly those that rely on low homogenizing pressures.

Free fat appears to 'catalyse' phase inversion by a mechanism that is still open to speculation. One theory is that the reduction in viscosity caused by the addition of free fat to a high fat cream results in better disruption of the fat globule membranes. Another is that free fat globules, which are unstable and readily separate out if agitation is stopped, are not surrounded by any protective natural milkfat globule membrane, and more readily coalesce with the natural fat globules ruptured during homogenization.

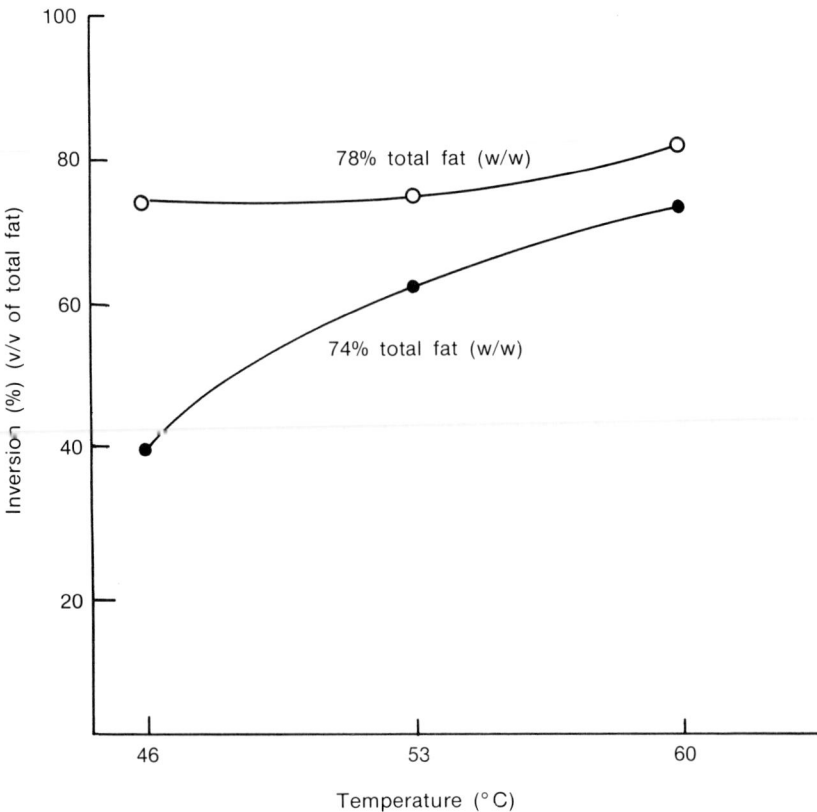

Fig. 7. Effect of temperature on phase inversion.

Whatever explanation is correct, the effect of free fat on the phase inversion efficiency is an essential element of the successful operation of Alfa Laval Centrifixator®[1] plants. Large quantities of free fat are cycled round the plant to ensure that high levels of both free fat and total fat can compensate for the low homogenizing pressure in a Centrifixator.

Although temperature has a lesser effect than other variables, the phase inversion efficiency increases as the temperature is raised from 40 to 60°C (Fig. 7). The effect is greater at low total fat levels.

Above 60°C, problems with direct-from-cream plants can occur owing to protein denaturation, resulting in the formation of stable, plastic creams

[1] Centrifixator® is a Registered Trademark of Alfa Laval.

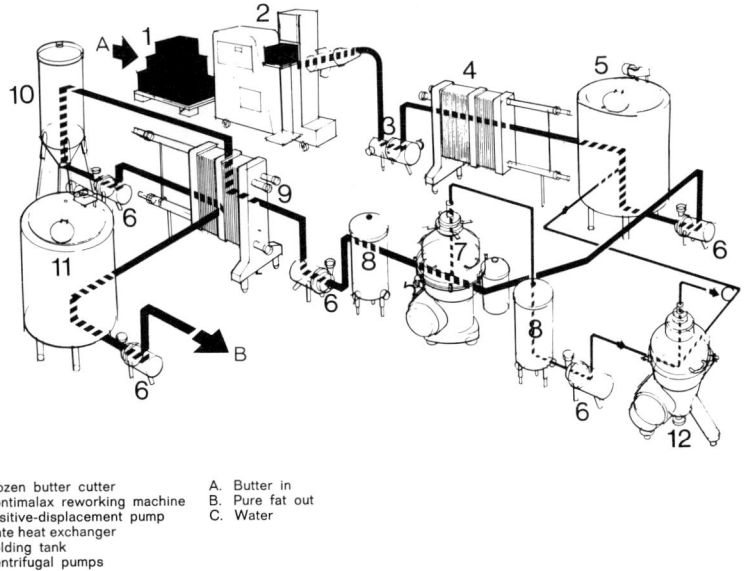

1. Frozen butter cutter
2. Contimalax reworking machine
3. Positive-displacement pump
4. Plate heat exchanger
5. Holding tank
6. Centrifugal pumps
7. Fat separator
8. Float hoppers
9. Plate heat exchanger
10. Vacuum dryer
11. Storage tank
12. Water phase separator

A. Butter in
B. Pure fat out
C. Water

Fig. 8. Plant for manufacture of AMF from butter (courtesy of Alfa Laval).

which cannot be inverted. As a result, direct-from-cream plants are operated at 55–60°C. The temperature must be maintained as high as possible to maximize separation and phase inversion efficiency while minimizing milk protein denaturation.

Manufacture from Butter

Both Alfa Laval and Westfalia supply plants that use butter that is either freshly made or stored for some time after manufacture as a feed material (Anon., 1970b). Figure 8 shows a typical plant layout. Cream is treated as for buttermaking, as described earlier. Fresh butter leaves the buttermaking machine through a jacketed pipe and is melted in a plate heat exchanger at 60–70°C. Alternatively, butter can drop into a melting funnel placed at the inlet of a positive pump. The latter method requires constant supervision to prevent air locks at the pump inlet.

The process from stored butter is similar to that from fresh butter except that some sort of butter melting device is required. Blocks of butter may be melted directly using steam- or hot-water-heated pipes or a rotating disc, but this tends to be a slow process (1 t/h), entailing the holding of large volumes of melted butter to enable continuous operation of the plant. The temperature is difficult to control and other disadvantages include burn-on of the non-fat milk solids and aeration of the melted butter.

Butter may also be shived using a rotating shiver drum connected to an auger-fed pump. Potable hot water or recycled melted butter sprayed on the auger in a way that minimizes aeration helps to speed up the process. Shivers have capacities of 3–5 t/h depending on the temperature of the butter, and can handle butter at low temperatures down to 0°C.

A third alternative is to slice blocks of butter, conditioned to around 10°C, into a number of smaller pieces (usually four) which then pass through a melting funnel into a butter reworker connected directly to a rotary positive pump. The funnel and pipes are jacketed with hot water at 70–80°C. Microwave heating can also be used for the melting of butter for AMF manufacture (Entremont et al., 1982).

AMF is separated from the melted butter using two, three or four self-desludging separators operated in series. The resultant fat stream is not as dry as that from the direct-from-cream process; thus the separators must be closely controlled to minimize fat losses in the serum without increasing the water content of the fat stream. An alternative to an oil-concentrating disc separator is a three-phase decanter, although this typically results in higher levels of moisture in the oil. As no sedimented solids are obtained from sweet cream butter, there is no discharge of a third phase. In processes using either three or four separators, serum from the first separator may be recovered as buttermilk.

When the feed material is salted butter, the buttermilk can contain up to 10% salt depending on the original salt content of the butter. Further processing of this buttermilk is limited. However, if citric acid is used to break the emulsion phase, the protein is denatured, as is the case when sour or cultured butter is the feed.

No emulsion phase is present when sour or cultured butter is processed. The pH of the serum phase is 4·75–5·2. The denatured protein can be concentrated to 20–50% solids in a decanter and converted to a stable, soluble state for uses such as fortification of milk (Lehmann et al., 1988).

The state of the protein in AMF manufacture has other implications. Low pH (4·5) and high pH (10–11) both weaken the hydrate envelope around the protein molecules that become denatured. In this state, protein molecules tend to clump into a flocculent precipitate by forming so-called hydrogen bridges. The phospholipids from the membrane are excluded and pass into the serum phase. Thus, if an AMF plant is intended to produce a buttermilk stream, the phospholipid content of this stream will depend very much on the pH of the serum.

Production of Ghee and Samna Balady

Ghee and samna balady are names also applied to versions of AMF, produced by traditional means in both the Indian sub-continent and the Middle East, for various purposes such as cooking and garnishing. A detailed description of their production is outside the scope of this chapter. Traditionally ghee production is a cottage industry (Parekh, 1970). The majority (80%) of the ghee produced in India is used in cooking and garnishing and the remainder is used in confectionery. Production is from milk that is allowed to stand and ripen overnight (Gupta et al., 1980), and is hand-skimmed; then the ghee is produced by boiling the cream to drive off the water. During the process, the milk solids-not-fat are caramelized by heat which imparts flavour and anti-oxidant properties to the ghee. Alternatively, the cream is churned to butter and the butter is heated to boil off the water. This method is also referred to as the pre-stratification method (Ray & Srinivasan, 1976). Factory-produced sweet cream or cultured butters may also be used as starting materials. The traditional processes have been adapted for factory manufacturing methods (Punjrath, 1974; Chakraborty, 1980). Methods to induce the traditional flavours of ghee using AMF as a starting material have also been proposed (Munro & Jebson, 1974; Wadwha et al., 1977).

PLANT AND EQUIPMENT USED FOR AMF MANUFACTURE

Phase Inversion Devices

Different types of phase inversion devices are available. All involve forcing the cream under pressure through devices with narrow orifices or clearances. The effect of this is to rupture the membrane surrounding the milkfat globule, producing free fat.

In the plant specified by Westfalia, high fat cream is forced through one or more homogenizer valve orifices at pressures of 100–150 bar (10–15 MPa). The high shear forces generated in the valve area of the homogenizer cause disruption of almost all the fat globules and result in a water-in-oil emulsion.

The Alfa Laval process offers a choice of equipment. The Clarifixator, shown in Fig. 9, performs a double duty (Fjaervoll, 1980). Firstly cream is concentrated to 70–80% fat in the separation disc stack of a centrifugal

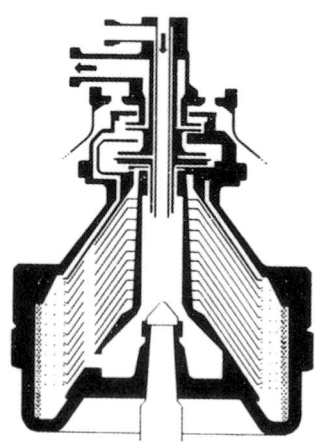

Fig. 9. Clarifixator (courtesy of Alfa Laval).

Fig. 10. Centrifixator (courtesy of Alfa Laval).

separator. Secondly, the milkfat globules are disintegrated, mainly by high shear caused when the cream passes over a toothed disc mounted on top of the light phase centrifugal pump. Clarifixators typically have capacities of 1000 kg/h.

The Centrifixator, shown in Fig. 10, is offered as an alternative (Joost et al., 1970). It consists of a toothed disc rotating at high speed with minimal clearance between the disc and its housing. Again, globules are disrupted by the high shear forces generated by forcing the cream through the device using a positive pump. The Centrifixator appears to have the same globule disruption efficiency as a conventional homogenizer operating at 50–60 bar (5000–6000 kPa), and has a capacity of 2000–7000 kg/h. Two such devices are often installed operating in parallel.

Separation Devices

Separation prior to phase inversion is required in a direct-from-cream process, to provide a cream of sufficiently high fat content for inversion by

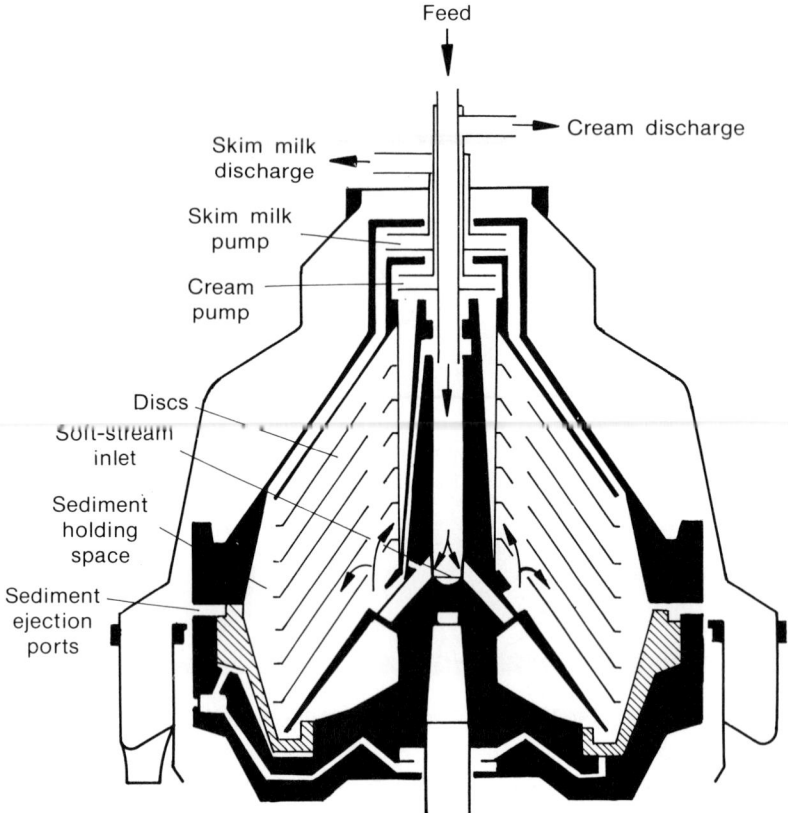

Fig. 11. Centrifugal separator shown in cross-section (courtesy of Westfalia).

high shear. Having achieved phase inversion, it is necessary to separate the phases. These phases differ in density and are termed 'light' and 'heavy'. The former is the oil or milkfat and the latter is the serum phase consisting of water and protein. These will separate fairly readily under gravity but the separation efficiency will be poor, with large fat losses, and will take a long time. In line with modern wet refining practice in the edible oil industry, centrifugal separators are used to effect rapid, efficient separation.

Separation is the basic operation in cream concentration, oil concentration or polishing and buttermilk/skim milk separation. The separators used for AMF manufacture, although basically similar in design principle as shown in Fig. 11, can be differentiated in terms of their function in the

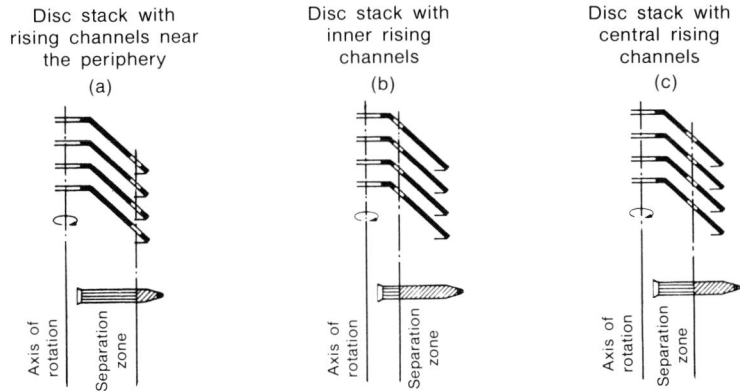

Fig. 12. Types of disc stacks available for separators (courtesy of Westfalia).

process. This is reflected in the position of the holes in the disc stack, as shown in Fig. 12. Both open-bowl and hermetic (sealed bowl) separation can be used.

The ideal cream concentrator should be capable of attaining high fat content in cream, with minimum fat losses to the heavy phase. Cream with 40% fat is fed into the inlet chamber via the central feed pipe at a velocity determined by the size of the plant. The kinetic energy of the cream flow is converted to pressure in the chamber. In an open-bowl type of separator, the feed to the disc stack is throttled in several stages to ensure firstly that the central distributor pipe is always in a flooded state and secondly that the flow to the disc stack is maintained at a constant rate. Unless this is the case, there is a danger that cavitation will occur which, if severe, can cause fat globule damage. Alfa Laval have overcome this problem by using a hermetically sealed separator which does not allow the ingress of air. If cavitation persists into the centrifugal pump at the top of the separator, this can become air-locked. Cream concentrators are equipped with central rising channels, as shown in Fig. 12(c). This arrangement guarantees a high cream concentration and gives a good separation effect with given phase ratios.

In the disc stack, the separation into concentrated cream and skim milk takes place with the light and heavy phases being pumped from the separator by centripetal pumps, also known as paring discs, sized to suit the throughput conditions of the concentrator. The separated solids collect on the surface of the bowl and an automatic time sequence ensures regular removal through the discharge ports. It should be noted that significant differences between high fat cream concentrators and normal milk separators also include the disc spacing and the relative sizes of the centripetal pumps on the light and heavy phases.

In separators used for oil polishing and separation of the oil-in-water emulsion, the holes in the disc stack are placed closer to the periphery than normal (Fig. 12(a)), because of the high oil concentration. This allows a wider separation zone for the light (oil) phase, and thus the zone can be adjusted to obtain an oil containing less than 0·5% serum (or moisture). The incoming product stream is guided into rising channels located towards the periphery of the disc stack. The separating zone is maintained at a constant position by means of a constant pressure valve fitted in the oil discharge line. Westfalia has developed a valve switching device to prevent oil losses during partial desludging. The separated streams are removed from the separator in a similar fashion to the streams from a cream concentrator.

Buttermilk and skim milk separators are very similar to those used for milk separation. In principle their operation is similar to those described previously. However, they have the holes in the disc stack closer to the centre to create the longest possible separating distance so the fat content of the serum phase is minimized (Fig. 12(b)).

Desludging, or removal of separated solids, plays an important role in the efficient operation of separators, particularly when the feed material is cultured or soured butter. Desludging should be accomplished in such a manner as to minimize fat losses to the serum phase. As an alternative to a conventional separator, Westfalia now offers a three-phase decanter in place of the polishing separator. This is a horizontally arranged separator that has a cylindrico-conical solid bowl with a built-in screw conveyor (Fig. 13).

The product stream is introduced into the feed chamber by means of a central feed pipe. Channels then guide the stream into the separation zone of the bowl where it is accelerated up to the bowl speed. Solid settles on the outer wall of the bowl, and is continually removed to the narrow diameter end of the bowl by the screw conveyor, which rotates at a slightly greater speed than the bowl. Here the solids are lifted from the liquid because of the conical shape of the bowl, and centrifugal force completes the removal of liquid. Solids are discharged through openings at the end of the bowl into a catching chamber, housed in the frame. A scraper ring effects the final ejection from the housing.

The liquid streams flow between the spirals of the screw conveyor

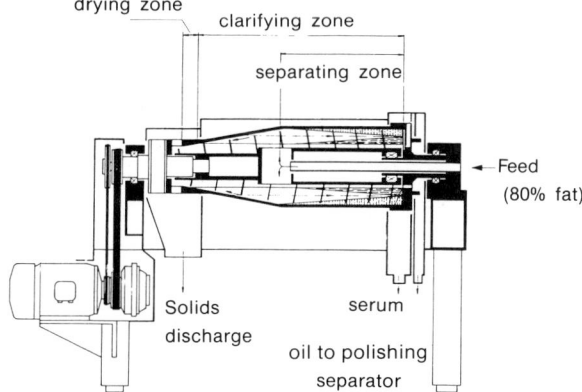

Fig. 13. Three-phase decanter (courtesy of Westfalia).

towards the cylindrical end of the bowl. They are separated into light and heavy phases in the separating zone, and are discharged via regulating rings and tubes.

The throughput capacity depends on the characteristics of the feed product, particularly the overall solids content and the permissible solids content of the clarified liquid. Between 5000 and 10 000 litres/h is possible depending on these factors.

The decanter seems better suited to processing cultured butter, which deposits more milk solids as a precipitate and causes frequent desludging and thus higher fat losses in a disc-type separator. However, the decanter has a lower separation efficiency than a disc separator and cannot reduce the moisture in the oil stream to the desirable 0·5%. It must therefore be followed by a disc-type polishing separator.

DEHYDRATION OF MILKFAT

The final step in AMF manufacture prior to packaging is the removal of water or dehydration. Removal of moisture and, at the same time, removal of as much dissolved oxygen as possible eliminate some of the likely causes of spoilage during post-dehydration storage. The low moisture content ($<0·1\%$) means that there is no possibility of microbial or enzymatic deterioration. The milkfat is heated to high temperature and introduced into a vessel under reduced pressure. Moisture and dissolved air are

Fig. 14. Dehydrator *in situ* in AMF plant (courtesy of New Zealand Co-operative Dairy Co. Ltd).

released from the fat and removed by a vacuum pump. It is important that the temperature used is not excessive and that the milkfat is not aerated afterwards as both promote oxidation.

Continuous vacuum dehydrators, similar to those illustrated in Figs 14 and 15, are steam-jacketed vessels and can be fitted with either a liquid-ring vacuum pump (Alfa Laval) or a water-jet condenser (Westfalia). These should be capable of achieving less than 7 kPa absolute. The jacket is usually heated with low-pressure steam during start-up and then milkfat at 90–100°C is introduced into the top of the vessel, through either a spray-ball (Alfa Laval) or a circular orifice ring (Westfalia). The dried, de-aerated milkfat is discharged continuously at the bottom of the vessel. The pipe connecting the dehydrator to the extraction pump is usually large, preferably with no constriction at the pump inlet, and is often equipped

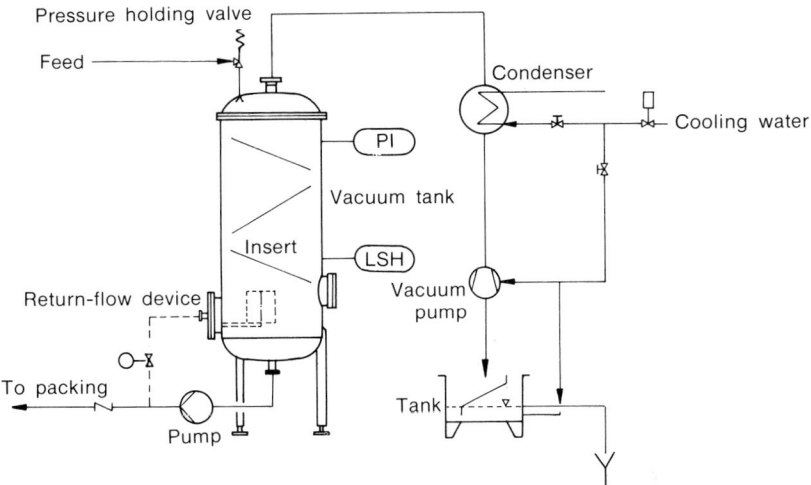

Fig. 15. Dehydrator shown in cross-section (courtesy of Westfalia).

TABLE 3
Solubility of water in milkfat

Temperature (°C)	Solubility (%w/w)
20	0·15
40	0·20
60	0·27
80	0·36
100	0·46

with a cooling jacket. In some plants, a by-pass pipe across the top avoids vapour locking of the pump inlet.

The dispersed water in milkfat can be almost completely flash evaporated when hot fat is introduced into a dehydrator under reduced pressure. The lower the pressure, the lower is the boiling temperature and hence the higher is the tendency of the water to evaporate. The boiling temperature of water at 20 kPa is 60°C. The theoretical evaporation is not always obtained, however, because moisture also appears as dissolved water in milkfat, which is not easily removed. Table 3 shows how the solubility of water varies with temperature.

Under constant operating conditions, the evaporation of dissolved water is affected by the diffusion of the water through the milkfat to the surroundings. This is primarily controlled by the distance the water must travel through the fat—the diffusion distance—and also by the residence time in the vacuum chamber, i.e. there are both thermodynamic and kinetic limitations.

In practice, in order to increase the removal of water below the solubility level, high temperature or, alternatively, low pressure is used. However, it is possible to dehydrate milkfat close to the water boiling temperature if the dehydrator is properly designed to maximize the diffusion rate. This necessitates improved methods of introducing milkfat into the vessel or increasing the dehydrator evaporation area.

As an alternative to the conventional dehydrator, it is also possible to use a stripping column similar in construction to a semi-continuous deodorizer as used in the edible oil industry for the final stage of refining. Such columns have a series of trays with perforations and are designed to expose a large surface area of oil to the vacuum. If dry steam is passed through the oil, some removal of volatile flavour compounds is possible.

Whatever design is used the dried oil is pumped from the bottom of the dehydration vessel, the level being controlled to prevent both it becoming too high in the vessel and increasing the chances of re-dissolution of water, and the discharge pump running dry.

MONITORING THE PERFORMANCE OF AN AMF PLANT

The monitoring of an AMF plant is confined to measurement of the fat content of the various heavy phases discharged from the plant and measurement of moisture in the final product. The final product is also monitored for flavour, PV and FFA. It may also be necessary in some applications to add specific levels of anti-oxidant, but these are usually confined to natural compounds such as mixed tocopherols, added at up to 200 mg/kg.

For monitoring the losses in a plant, both the production period and losses during any interruptions to production may be considered. The latter may include 'rinse losses' at the ends of production or losses occurring during CIP (clean-in-place). Production losses tend to have the most effect on a plant's efficiency rating because production times are disproportionately longer than other times. Thus loss monitoring is usually confined to production. Plant efficiency can be considered as the ratio of lost fat to total fat fed in per unit time and may be calculated (Lehmann et al., 1988). Depending on the initial raw material that feeds the plant and the design of the plant, losses should be between 0·3 and 1%.

The measurement of efficiency requires the fat contents of the various heavy phases to be determined accurately. These are influenced by the varying content of phospholipid in the various streams. Phospholipids cannot be separated from the serum by centrifugation because they tend to be colloidal. Thus the measuring method must allow for the varying levels, otherwise plant efficiency results will be erroneous.

A method that gives the 'real fat' content is recommended. Westfalia recommends the volumetric Gerber method (British Standards Institution, 1969), but gravimetric methods such as the Rose-Gottlieb (International Dairy Federation, 1987) are the accepted international reference methods. The former measures only triglycerides whereas the latter includes phospholipids. Whichever method is chosen, it should be carried out in a standard manner. It is then possible to relate one to the other.

QUALITY ASSESSMENT OF ANHYDROUS MILKFAT

Recommended methods of sampling and testing the constituents of AMF are as follows.

The stream from the production plant should be sampled using a standard method (International Dairy Federation, 1980a).

The moisture content of the AMF at any point in the production plant should be determined by the Karl Fischer method, in which the water (either bound or free) is converted into sulphuric acid and hydrogen iodide by titration of a weighed quantity of fat with a solution of sulphur dioxide and iodine in methanol and pyridine (International Dairy Federation, 1964a). The fat content of the AMF is determined after the water content measurement by measuring the non-fat residue left when the fat is dissolved in light petroleum and filtered (International Dairy Federation, 1964b).

FFA content as oleic acid is determined by titration against a standardized solution of potassium hydroxide in alcohol to a phenolphthalein end-point (Food and Agriculture Organization, 1969; International Dairy Federation, 1989). The number of milligrams of potassium hydroxide that neutralizes the FFA in 1 g of fat is the acid value. The FFA level is obtained by multiplying the acid value by 0·282.

If the peroxides in milkfat are reacted with a mixture of Fe(III) chloride and ammonium thiocyanate, they form a complex with the iron which can be determined photometrically to yield the PV (International Dairy Federation, 1980b). Similarly heavy metals such as iron (International Dairy Federation, 1986) and copper (International Dairy Federation, 1980c) can be measured photometrically. Iron is reduced to Fe(II) by reaction with bathophenanthroline and extracted into amyl alcohol as a red complex. Copper as Cu(II) is complexed with diethyl dithiocarbamic acid which gives a yellow colour in amyl alcohol.

Oxidation is the main cause of flavour deterioration in milkfat. It proceeds by a free radical mechanism catalysed by light, heat, heavy metals and other radicals. The primary intermediates are lipid hydroperoxides, measured by the PV test, which then decay into a range of unstable, off-flavour compounds including saturated and unsaturated aldehydes, lactones, fatty acids and alkanes. Tests for reaction intermediates such as the PV (for hydroperoxides), the thiobarbituric acid test (TBA) (for malonaldehyde) (American Oil Chemists' Society, 1990) and the anisidine value (AV) (for unsaturated aldehydes) (International Union of Pure and Applied Chemistry, 1987a), can be misleading and alone cannot give an adequate description of the quality of the product as received (Keogh & Higgins, 1986). However, when taken with a flavour assessment, particularly when the chemical tests are combined to give a 'Totox' value (Totox = 2PV + AV), the end-user can obtain a satisfactory assurance of product quality.

The effect of too high a moisture content during storage is to cause spoilage due to enzymic or microbial action. Deeth and Fitz-Gerald (1983) have shown that lipolytic rancidity is caused by the liberation of FFA from the triglyceride molecules. The characteristic flavours associated with this type of rancidity are, in the main, caused by the presence of free short- and medium-chain fatty acids which are abundant in milkfat. These have lower flavour thresholds than the long-chain fatty acids and thus their release has a more pronounced effect on the flavour. FFA levels must be controlled to preserve the milkfat against flavour deterioration. Similarly, PV is a measure of the oxidative deterioration of fats, although in the case of milkfat it does not always correlate with the flavour assessed by trained panels.

Iron and copper are both excellent catalysts for oxidation, which tends to be the primary cause of deterioration of oils and fats, and hence is a major factor in determining the shelf life of foods containing fats (Richardson & Korycka-Dahl, 1983).

It may also be necessary to check the purity of the AMF in relation to contamination by other animal fats. This is of particular importance when considering the Middle East as a potential market for the product. Food regulations, formulated according to religious preferences, demand that the producer gives adequate assurances that no contamination with pork fat is present. It is also most unethical to present a product such as milkfat as being pure when it may have other fats present. Some of the methods used for detection of other fats (e.g. Reichert-Meissl, Polenske and Kirschner values (International Union of Pure and Applied Chemistry, 1987b)) may seem tedious when compared with modern, rapid, automated, instrumental methods but it is worth remembering that newer methods may not be available to the prospective purchaser of the product. Vegetable fats may be detected either by the phytosteryl acetate test (International Dairy Federation, 1966) or by gas–liquid chromatography of sterols (International Dairy Federation, 1970). Gas–liquid chromatography of the triglycerides has also been used as a means of detection of adulteration (Timms, 1980).

PACKAGING AND HANDLING OF MILKFAT

Post-dehydration Handling

On leaving the dehydrator, the milkfat should be cooled to 40°C, using a plate heat exchanger, before being discharged into either a balance tank or storage tank prior to packing into drums or tins.

AMF leaving the dehydrator should contain less than 2 mg dissolved oxygen per kg. The rate at which oxygen is re-absorbed by milkfat from aeration during subsequent handling must be minimized in order to reduce milkfat oxidation during storage. The handling and packaging of milkfat should be done in a carefully controlled manner, using a procedure similar to that described next.

Balance tanks should be large enough to allow packing to be completed in one operation. The tanks should be filled and emptied from the bottom and should be fitted with a level control device.

Measurement of Dissolved Oxygen in Milkfat

As oxidation is the major cause of flavour deterioration in milkfat, it is important that the quantity of oxygen dissolved in AMF during processing and packaging is controlled and monitored. The rate of oxidation depends on the temperature and the dissolved oxygen concentration, as well as on the presence of pro-oxidants such as copper and iron. The degree of unsaturation of fats and the presence of anti-oxidants, either natural or synthetic, are other factors that need to be taken into account, because they also affect the rate of oxidation. Compared with many fats milkfat is relatively stable towards oxidation. It is quite high in saturated fatty acids and has a low polyunsaturated fatty acid content. Moreover, it contains low levels of natural anti-oxidants, mainly α-tocopherol. Thus it is necessary to monitor the level of dissolved oxygen, both immediately after manufacture and after packaging. To this end, for a period, the IDF set up a Committee of Experts (E57), since disbanded, to examine the various methods available for the measurement of oxygen content (Cant, 1985). The solubility of oxygen in milkfat is influenced by temperature and the physical state of the milkfat, and varies by more than 10% over the normal range of temperature at which milkfat is packed (30–40°C) (Jebson et al., 1973). When milkfat is allowed to crystallize, dissolved oxygen is excluded from the crystals, increasing the level in the remaining liquid (Timms et al., 1982). Several methods have been suggested for measurement of dissolved oxygen, including polarography (Ke & Ackman, 1973), gas–liquid chromatography and classic chemical methods (Keogh & Higgins, 1986). Polarographic detectors are the most convenient, offering the necessary sensitivity, reproducibility and convenience for use in monitoring the dissolved oxygen in milkfat during production and subsequent storage and handling. Polarographic probes respond to the partial pressure of oxygen. It is important therefore that dissolved oxygen is measured when all the milkfat is liquid, or high readings representative of only the liquid fraction will be obtained. In a closed drum, oxygen in the headspace and the liquid milkfat will equilibrate over a period of some hours, and will then give the same reading with the polarographic probe. A final standard method for measurement is still in preparation.

Packaging of AMF for Industrial Use

Traditionally AMF has been packaged in conventional oil drums holding 190–200 kg, as described above. The drums are lacquered on the inside, or have an inner plastic liner, to prevent the milkfat from coming in contact with the steel. Smaller metal containers—20 litres is a popular size—are usually tin plated. The design of these smaller containers varies from country to country, ranging from fairly flimsy kerosine-type cans (often with a cardboard outer) to more sophisticated drums with an integral pouring spout.

Drums require some headspace to control the pressure drop resulting from the volume reduction as milkfat crystallizes. Air needs to be excluded from the headspace, and this can be achieved by sparging the drums with nitrogen before or during filling. Alternatively, the headspace should be flushed with nitrogen gas to reduce the concentration of oxygen from 21% to less than 2%. Drums should then be pressurized to at least 20 kPa to prevent the partial collapse of the walls which can occur due to contraction of the milkfat as it cools and solidifies. The bungs on the drums should be tightened to specific torques. Similar precautions to exclude air should be taken when filling smaller containers.

Packaging of AMF in plastic-lined cartons, similar to those used for bulk butter, is now commonplace in the major manufacturing countries, whilst larger bulk packaging first appeared when Danish manufacturers introduced a bag-in-box pack with a capacity of 1 t. The AMF is held in a multi-wall plastic bag contained within a cardboard or fibreboard container which sits directly on a pallet. A valve arrangement allows the user to pump out the milkfat after melting. Similar packaging (950 kg capacity) has also been introduced by New Zealand (Fig. 16). A further alternative is the ISO-tank which holds nearly 20 t of oil (Fig. 17). This bulk tank sits in a frame of the same dimensions as a standard container. The tanks, which are fitted with coils for heating the contents for removal, can be pressurized and have been used successfully for the transport of other edible oils and fats. The precautions that are taken for filling the standard drums must also be taken for filling bulk containers.

The care taken in the packaging of milkfat is reflected in the stability of the fat during subsequent storage. It can be shown that the FFA level does not increase during storage under a variety of temperature and packing conditions. Oxidation occurs at different rates depending on the temperature of storage (Kehagias & Radena, 1973).

A word of caution is necessary here regarding the measurement of dissolved oxygen after prolonged storage of AMF. Whatever the dissolved oxygen content of the fat at the time of packaging, it must always be remembered that this oxygen is still a reactive species and can chemically attack unsaturated centres in the fat, forming peroxides, and be consumed in the process. It should come as no surprise, therefore, if a dissolved oxygen content of zero is found after prolonged storage. The condition of the fat at the time of opening will be indicative of the dissolved oxygen content at the time of packing.

Consumer Packaging of AMF

Consumer packs for ghee and clarified butter are usually 0·5, 1 and 2 kg conventional seamed cans, filled and sealed under vacuum. However, alternative can types have been adapted from other segments of the food

Fig. 16. 950-kg bag-in-box packaging for AMF (courtesy of New Zealand Dairy Board).

industry, such as ring-pull cans, cans with a push-on plastic lid for resealing or cans with a prise-off lid and with a vacuum seal of aluminium foil. The packaging of milkfat can involve controlled pre-crystallization to give large, grainy crystals, as are found in the traditional ghees of the Indian sub-continent. Such a texture is stable and tends to remain unchanged during storage at high ambient temperatures, because the crystals in the product can have individual melting points that are well above the overall melting point of the homogeneous fat. The alternative is to cool the product by scraped-surface heat exchangers to give a smoother textured product. This texture is preferred in some markets, e.g. Taiwan, where milkfat is used as an industrial baking fat.

Fig. 17. ISO-tank container used for bulk transport of AMF (courtesy of New Zealand Dairy Board).

In some markets, notably Europe and recently Australasia, AMF has appeared on supermarket shelves packed in foil-wrapped pats similar to the traditional consumer presentation for butter. This means that the milkfat must be processed through a conventional margarine plant with scraped-surface heat exchangers and pinworkers, as used for shortenings (Anderson & Williams, 1965), to crystallize, plasticize and render the fat suitable for packing in a patting machine.

USE OF AMF FOR RECOMBINING

AMF may be re-emulsified into buttermilk, skim milk or whey for the purpose of making butter and cheese (Schaap, 1973). However, in spite of the rapid development of recombining processes, particularly those for recombined butter, there is still a dearth of technical information in the literature. One of the principal reasons is that many developments have remained secret to maintain a manufacturer's lead in a very competitive area (Spieler, 1982).

Two seminars were organized by the IDF to discuss recombining techniques and some of the problems involved (Singapore, 1980; Alexandria, 1988). The use of AMF for recombining was discussed at great length at both, but, surprisingly, little new technology emerged between the two seminars.

The use of AMF in the recombining industry has caused doubts to be raised on the effectiveness of the FIL/IDF Standard 68A of 1977 (International Dairy Federation, 1977). The main areas of the world where recombined dairy products are manufactured are those developing areas that do not have a thriving local milk production industry that can satisfy the growing needs of the population, such as the Middle East and parts of South-East Asia (Kieseker, 1984; Al-Tahiri, 1987). AMF for the recombining industries must be imported from distant countries, mainly the EEC, Australia and New Zealand. Kieseker (1984) has pointed out that the often adverse storage conditions, which persist in these areas, make the use of anti-oxidants mandatory. The use of anti-oxidants has also been suggested as an alternative to temperature control and nitrogen flushing for limiting autoxidation of milkfat (Wade *et al.*, 1986). However, the IDF has not yet revised the standard, in spite of recommendations that anti-oxidants should be allowed in lower grade milkfats. As was emphasized at the Alexandria seminar by Norris (1990), the standard needs revision. The standard was set for AMF at the point of manufacture, not at the point of eventual use. Thus, there is no recommended colour specification in the standard—this is standard practice in the rest of the edible oil industry, being one of the quality criteria on which oils and fats are traded; headspace and dissolved oxygen are not yet included in the standard and no recommendations for tests for oxidative stability have been made, although the revision of the standard is expected to include this.

Recombined Butter and Other Spreads

It has been suggested that only top quality AMF should be used for recombined buttermaking. Such milkfat should comply with the FIL/IDF Standard 68A of 1977 (International Dairy Federation, 1977). However, this statement is rather sweeping and some further expansion on the influence of the quality of the AMF used for recombined butter on the quality of the final product is perhaps appropriate here. Generally, a recombined butter will be given a lower score by expert graders, than the original AMF from which it was manufactured (Rogers & Kieseker, 1985). A comparison of conventionally churned butter with recombined butter made from AMF manufactured from the same cream source—either direct or via melted butter granules—will show that the conventional butter has a higher initial grade score which will be maintained during storage at 2°C for up to 8 weeks. Both recombined butters will show rapid deterioration, with the PV rising.

This suggests that the milk solids-not-fat and membrane material removed during AMF manufacture have some anti-oxidant quality. Dehydration only, which retains this material, has been suggested as a means of producing an AMF source that will have a similar storage life to that of conventionally churned butter (Damerow & Kohl, 1971).

However, the other components making up the aqueous phase of recombined butter must also be considered. Thus the skim-milk powder used must be of the best quality (Sanderson, 1979); the salt should be of high quality, with no additives to assist free-flow characteristics (Jebson, 1979a), and, most importantly, attention should be given to the quality of the water. The water should be pure, both chemically and microbiologically (Leighton & Lawrence, 1979). If necessary the water should be boiled as this not only will destroy microorganisms but also may precipitate undesirable salts. Any other ingredients used such as emulsifying agents—usually mono-diglycerides and/or lecithin—should also be of the highest quality.

In considering the equipment required for the manufacture of recombined butter, a method for melting the AMF is of the first importance. It is necessary for the milkfat to be heated to above 40°C. This can be achieved for 20-litre containers by immersion in water at 80°C (2-3 h) or for 200-litre drums storage in a hot room (24-28 h) (Sanderson, 1982). The 950-kg bag-in-box and ISO-tank containers as described previously require no special facilities at the recombining factory. For this reason, their use is perhaps more economical in many circumstances.

The equipment used for the manufacture of recombined butter is virtually the same as that found in any margarine manufacturing facility. The product is passed through a scraped-surface heat exchanger and then a series of holding tubes and a gentle working device. The product is also recycled to the inlet side of the high-pressure pump without any heating and the recycle rate is often as high as three times the overall throughput (Jebson, 1979b). Butter made by this method does not have a good flavour and is not very plastic. Figure 18 illustrates the plant layout for manufacture in this mode. A method to overcome these disadvantages has been suggested, with a pinworker replacing the holding tube and gentle working device (Munro, 1982). Product recycle then occurs after the pinworker.

The alternative layout for this processing plant can be seen in Fig. 19.

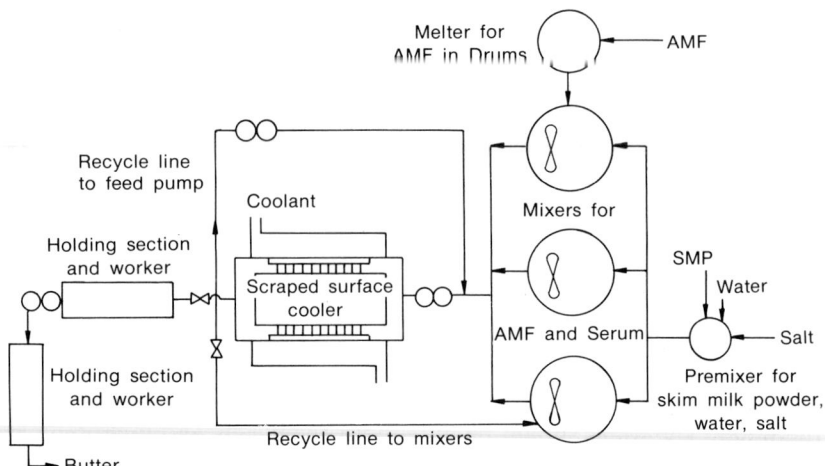

Fig. 18. Recombined butter plant layout as proposed by Jebson (1979b) (courtesy of New Zealand Dairy Research Institute).

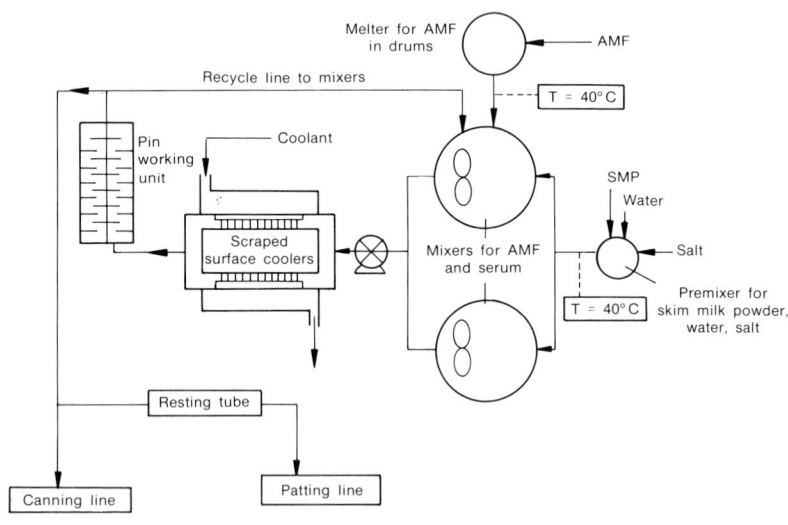

Fig. 19. Recombined butter plant layout as proposed by Munro (1982) (courtesy of New Zealand Dairy Research Institute).

The ingredients for the aqueous phase are combined separately and then added to the fat phase, which may contain emulsifiers, in the emulsion tanks. The emulsion, heated to at least 40°C, is pumped using a high-pressure pump through a suitable scraped-surface heat exchanger into a pinworker. The fat is rapidly super-cooled in the scraped-surface heat exchanger. Heat removed from the fat is sensible heat and only a small quantity of fat starts to crystallize. The crystallization is completed in the pinworker under conditions of mechanical working. Very small crystals of fat are formed; they remain discrete without forming the network that would build up under quiescent conditions. As a result of crystallizing the fat in this way, a large surface area of solid fat is produced, which allows the water droplets to be dispersed as a very fine emulsion. The latent heat given out when the crystals form in the pinworker causes the temperature to rise by about 5°C, although the rise depends on the residence time in the pinworker. Butter produced in this manner has a fine, smooth texture and shows no signs of leaking.

Use of AMF fractions in spreads

At the beginning of this chapter, it was noted that the physical properties of milkfat make it unsuitable for many applications, including baking. Butter also has some inadequacies as a spread, being hard from the refrigerator in comparison with margarine. Fractionation from the melt has in part provided a solution to some of these problems. The incorporation of fractions of milkfat in dairy and margarine spreads has been attempted on a number of occasions, mainly to produce a product that spreads from the refrigerator and that has the authentic butter flavour. Several products have appeared on the market, such as Danelite, marketed by ButterDane of Denmark, and briefly introduced to the UK market in 1987. That product was analysed by the authors and was found to have a fat phase based entirely on a low-melting fraction of milkfat. A product based on a low-melting fraction blended with milkfat appeared briefly on the New Zealand domestic market in the early 1970s prior to polyunsaturated margarines being approved for general sale. Other products based on low-melting fractions have also been developed, including a low calorie spread (Timmen, 1975; Verhagen & Warnaar, 1984).

High-melting fractions are now being used in margarine blends, where they fulfil the function of all or part of the usual hardstock (Lansbergen & Kemps, 1984), as well as in a spreadable water-in-oil emulsion in which the high-melting fraction is blended with liquid vegetable oil to the same end (Verhagen & Bodor, 1984). Balansia, a 60% fat spread incorporating milkfat hard fraction, was launched in Finland during 1992. Thus, it is a logical assumption that replacing the hydrogenated vegetable oil hardstock in dairy spreads with a hard fraction with a suitable melting point increases the flexibility that the recombiner has with regard to ingredient use and avoids problems with the increasing nutritional concerns about hydrogenated fats.

Low-melting fractions have also been used as the fat phase in low calorie spreads where the tendency for the product to 'slump' (the tendency for a cylinder of product to collapse when held at or just above ambient temperature) or 'oil off' (the tendency for liquid oil to leak from the product under the same conditions as for the 'slump' test) is much reduced because of the presence of powerful emulsifiers which 'stiffen' the product.

It should be emphasized here that, although several manufacturers now offer low-melting fractions for the recombining trade as an alternative to AMF, their usage is limited to those applications in which the physical state of the fat is non-critical. Thus they would be unsuitable for the manufacture of recombined butter or any other product that relies on the crystallization of the fat to maintain product structure and integrity at ambient temperatures.

Addition of hard fraction to milkfat has been suggested as a means of increasing the melting point and solid fat profile to the extent that the fat can be used in recombined butters for tropical markets. Such products, which are usually sold in cans, would not have the melting and 'stand-up' problems usually found with standard butter. However, it has also been suggested that this improvement should be able to be achieved by alteration of the processing conditions to change the crystallization behaviour of the fat (Hayes, 1982). To date, no such reliable process seems to have been made to work, presumably because of the slow crystallization behaviour of milkfat.

Other Blends of Milkfat with Vegetable Oils

The cost of domestic milkfat in Japan, coupled with enormous increases in demand for dairy products, has led to the establishment of a substantial industry for the manufacture of fat mixes or prepared edible fats. These are 70:30 blends of milkfat with hydrogenated vegetable oils—either soya or coconut—and are manufactured as either aqueous or anhydrous products. The aqueous products have a similar gross composition to that of butter. The composition is not dictated by any particular functional requirement; these products have become established as a means of importing milkfat into Japan. Volumes are strictly controlled by way of an import licensing system. The fat mixes are used in the manufacture of added-value products that may incorporate only small quantities of milkfat to give the product a prestige image. Currently, the majority of products are supplied from New Zealand, with the rest coming from the EEC, mainly from Belgium.

A new generation of fat mixes with the reverse composition is also beginning to emerge. These can be imported outside the licensing system under a special tariff which allows the free import of products that have less than 30% dairy solids on a dry basis. This has led to further development of products that have some functionality, although they are still mainly used as a means of obtaining milkfat. The more functional products mean that the vegetable oil component may be partially or wholly replaced by ingredients such as sugar, egg, coconut fines and flour.

Recombined Ice Cream

The same comments regarding ingredients, as were made for recombined butter, apply to ice cream, and the recommended plant is virtually the same as for the recombined butter with the exception that the pinworker is not required and some means of aerating the frozen product must be incorporated.

Again the use of fractions must not be ignored. As a result of the premium uses for hard fractions of milkfat in bakery applications, there is plenty of scope for the inclusion of soft fractions in ice cream in place of cream, often the preferred raw material. Soft fraction can be packaged exactly as for AMF with similar safeguards against deterioration. As such, manufacturers may find it an advantage to use soft fraction because it is easier to melt for incorporation into pre-mixes.

Recombined Milks and Creams

The setting up of recombining factories for milk can be used as a means of establishing indigenous dairy food production in developing countries. This may involve education of the process plant operators to use all the available milk as well as controlling the volume of milk and dried milk that is imported into the country (Thuraisingham, 1982). Such measures would benefit the dairy industry in a country such as Thailand, where the production of reconstituted milk products using imported ingredients presents a threat to the local milk producing industry (Maurer, 1979). Recombined milk industries are well established in Middle Eastern countries, with Iran (Anon., 1987) and Saudi Arabia (Salji *et al.*, 1984) being major producers.

Recombined milks are usually manufactured either from a mixture of

TABLE 4
Typical formulae for recombined milks (Zadow, 1982).

Ingredients	Recombined milk from SMP/AMF (kg/t)	Recombined milk from SMP/BMP/AMF (kg/t)	Recombined milk from WMP (kg/t)
SMP	89·5	80·5	—
BMP	—	9·7	—
AMF	34·1	33·3	—
WMP	—	—	123·7
Water	876·4	876·5	876·3
Composition of products (g/kg)			
Total solids	120	120	120
Fat	35	35	35
Non-fat solids	85	85	85

skim-milk powder (SMP), AMF and water or from whole-milk powder (WMP) reconstituted with water. It is quite acceptable to include a portion of buttermilk powder (BMP) in formulations based on SMP. The former ingredients were preferred until recently because of problems that occur on storage of WMP. The trend towards WMP does not mean that the problems have been solved, but there has been a significant swing away from a product mix of milkfat/butter/SMP exporting towards WMP/cheese exporting in countries such as Australia and New Zealand.

The manufacture of recombined milk involves the addition of melted AMF to a solution of the milk solids-not-fat in water, followed by homogenization. Typical formulae are shown in Table 4 (Zadow, 1982).

Recombining is also used for other milk products, in particular UHT milk and sweetened and unsweetened condensed milk (Kieseker, 1982; Newstead, 1982). The same criteria for the quality of the AMF apply as for other products. Many of the problems associated with these products, such as 'creaming' (the tendency for the cream layer to rise on storage), are due not to the use of AMF but to the fact that other vital constituents of milk, which contribute to the stability of the emulsion, are lost during AMF manufacture. Replacement of these by solids-not-fat as either skim milk or buttermilk does not restore the emulsion stability completely, and other emulsifiers are used in conjunction with them.

Other problems also depend on the fat. Sedimentation in UHT milk has been shown to increase with the age of the fat used for the recombining. This must be put into perspective with the storage conditions used for the

AMF prior to recombining. One factor that is of particular interest here is that the very process of recombining into an oil-in-water emulsion restores some oxidative stability to the fat. In the natural state, in milk, the milkfat is protected from degradation by the phospholipid-rich milkfat globule membrane (Mulder & Walstra, 1974). The same authors have suggested that the artificial milkfat globule membrane that forms during homogenization gives additional protection against oxidation of the fat in the recombined state. Further protection is provided during subsequent UHT treatment, provided that aseptic conditions are maintained. Heat activates sulphydryl groups (Taylor & Richardson, 1980) and the removal and exclusion of oxygen by the use of packaging with good oxygen barrier properties all add to the protection.

Milkfat fractions for recombined milk have also appeared on the market, particularly low-melting fractions. Although these fractions are marketed as being equivalent to AMF, in a standard formulation when emulsifier was used, recombined milks made with low-melting fraction were found to have similar cream rise to milks made from New Zealand AMF (McKenna, A. B., 1988, personal communication). When the emulsifier was left out, the AMF performed better than the low-melting fraction. High-melting fractions have been suggested for use in recombined creams to give better stand-up.

Milkfat in Other Recombined Products

AMF is also used in other recombined products such as cheese and yoghurts (Gilles & Lawrence, 1982). Again the same quality criteria as before apply, although oxidative deterioration is somewhat offset by the reducing environment in cheese manufacture. For white cheeses such as feta, a high colour in the AMF is a disadvantage. Manufacturers can offset this either by decolorizing the AMF using conventional edible oil refining techniques or by treating the recombined milk with a decolorizing agent such as benzoyl peroxide (Kuramoto & Jeseki, 1954; Kosikowski, 1975). This oxidizes and thus decolorizes the carotene and is in turn reduced to benzoic acid, a natural constituent of milk. The same process may also be used on milkfat or melted butter (Rachev *et al.*, 1979).

REFERENCES

AL-TAHIRI, R. (1987) Recombined and reconstituted milk products. *New Zealand Journal of Dairy Science and Technology*, **22**, 1–23.

AMERICAN OIL CHEMISTS' SOCIETY (1990) Method Cd 19-90. In *Official Methods and*

Recommended Practices of the American Oil Chemists' Society, 4th edn. American Oil Chemists' Society, Champaign, IL, USA.

ANDERSON, A. J. C. & WILLIAMS, P. N. (1965) Margarine, 2nd edn. Pergamon Press, London, UK.

ANON. (1970a) Description of the Alfa Laval plant, Lannarp, Sweden. Dairy Industries, 35, 775–779.

ANON. (1970b) Description of AMF manufacture ex butter. Dairy Industries, 35, 424–428.

ANON. (1987) Iranian dairy aseptically packs 100 tons of assembled milk/day. Food Engineering International, 2, 61.

ANON. (1992) Production figures for anhydrous milkfat. GATT Secretariat, Geneva, Switzerland.

BRITISH STANDARDS INSTITUTION (1969) British Standard BS 696. Specification for Gerber method for the determination of fat in milk and milk products. Part 2. British Standards Institution, London, UK.

CANT, P. A. E. (1985) Determination of oxygen in milkfat products. International Dairy Federation Document E-Doc 232, Annual Sessions, Auckland, New Zealand.

CHAKRABORTY, B. K. (1980) Industrial ghee production, trends and innovation. Indian Dairyman, 32, 737–742.

DAMEROW, G. & KOHL, H. (1971) Manufacture of anhydrous butter and its processing into other products. Chemie Ingenieur Technik, 3, 1121–1126.

DEETH, H. C. & FITZ-GERALD, C. H. (1983) Lipolytic enzymes and hydrolytic rancidity in milk and milk products. In Developments in Dairy Chemistry, Vol. 2, Lipids, ed. P. F. Fox. Applied Science Publishers Ltd., London, UK, pp. 195–239.

ENTREMONT, J., LEVARDON, R. & ENTREMONT S. A. (1982) Process and devices for separating fat from butter and products thus obtained. French Patent application FR 2 498 081.

FJAERVOLL, A. (1980) Anhydrous milkfat manufacturing techniques and future applications. Dairy Industries, July, 424–428.

FOOD AND AGRICULTURE ORGANIZATION/WORLD HEALTH ORGANIZATION (1969) FAO/WHO Standard B4: Determination of the acid value of fat from butter. Food and Agriculture Organization, Rome, Italy.

GILLES, J. & LAWRENCE, R. C. (1982) Manufacture of cheese and other fermented products from recombined milk. Bulletin, International Dairy Federation, 142, 111–117.

GUPTA, V. K., ARORA, K. L. & CHAKRABORTY, B. K. (1980) Ghee from curdled milk. Indian Dairyman, 32, 753–757.

HAYES, M. (1982) In discussion—Recombination of butter and ice cream. Bulletin, International Dairy Federation, 142, 140.

ILLINGWORTH, D., LLOYD, J. C. & NORRIS, R. (1989) Tailor-made industrial fats from milkfat, the New Zealand experience. Paper presented at Fats for the Future II, Auckland, New Zealand, 13–18 February.

INTERNATIONAL DAIRY FEDERATION (1964a) FIL/IDF Standard 23: Determination of water by the Karl Fischer method. International Dairy Federation, Brussels, Belgium.

INTERNATIONAL DAIRY FEDERATION (1964b) FIL/IDF Standard 24: Determination

of the fat content of butteroil. International Dairy Federation, Brussels, Belgium.
INTERNATIONAL DAIRY FEDERATION (1966) FIL/IDF Standard 38: Detection of vegetable fat in milkfat by thin layer chromatography of steryl acetates. International Dairy Federation, Brussels, Belgium.
INTERNATIONAL DAIRY FEDERATION (1970) FIL/IDF Standard 54: Detection of vegetable fat in milkfat by gas–liquid chromatography of sterols. International Dairy Federation, Brussels, Belgium.
INTERNATIONAL DAIRY FEDERATION (1977) FIL/IDF Standard 68A: Anhydrous milkfat, anhydrous butteroil or anhydrous butterfat, butteroil or butterfat, ghee: standards of identity. International Dairy Federation, Brussels, Belgium.
INTERNATIONAL DAIRY FEDERATION (1980*a*) FIL/IDF Standard 50A: Milk and milk products—guide to sampling techniques. International Dairy Federation, Brussels, Belgium.
INTERNATIONAL DAIRY FEDERATION (1980*b*) FIL/IDF Standard 74: Anhydrous milkfat: determination of the peroxide value. International Dairy Federation, Brussels, Belgium.
INTERNATIONAL DAIRY FEDERATION (1980*c*) FIL/IDF Standard 76A: Milk and milk products—determination of copper content—photometric reference method. International Dairy Federation, Brussels, Belgium.
INTERNATIONAL DAIRY FEDERATION (1986) FIL/IDF Standard 103A: Milk and milk products—determination of iron content—photometric reference method. International Dairy Federation, Brussels, Belgium.
INTERNATIONAL DAIRY FEDERATION (1987) FIL/IDF Standard 22B: Determination of the fat content of skim milk, whey and buttermilk by the Rose-Gottlieb gravimetric method (reference method). International Dairy Federation, Brussels, Belgium.
INTERNATIONAL DAIRY FEDERATION (1989) FIL/IDF Standard 6B: Determination of the acid value of anhydrous milkfat. International Dairy Federation, Brussels, Belgium.
INTERNATIONAL UNION OF PURE AND APPLIED CHEMISTRY (1987*a*) Method 2.504. In *Standard Methods for the Analysis of Oils, Fats and Derivatives*, 7th edn, eds C. Paquot & A. Hautfenne. Blackwell Scientific Publications, Oxford, UK, p. 210.
INTERNATIONAL UNION OF PURE AND APPLIED CHEMISTRY (1987*b*) Method 2.204. In *Standard Methods for the Analysis of Oils, Fats and Derivatives*, 7th edn, eds C. Paquot & A. Hautfenne. Blackwell Scientific Publications, Oxford, UK, p. 83.
JEBSON, R. S. (1979*a*) Salt. In *Monograph on Recombination of Milk and Milk Products (Technology and Engineering Aspects)*. International Dairy Federation, Brussels, Belgium, p. 50.
JEBSON, R. S. (1979*b*) Recombined butter. In *Monograph on Recombination of Milk and Milk Products (Technology and Engineering Aspects)*. International Dairy Federation, Brussels, Belgium, pp. 30–32.
JEBSON, R. S., EVANS, A. A. & COOKE, D. (1973) Continuous measurements of dissolved oxygen in anhydrous milkfat. *New Zealand Journal of Dairy Science and Technology*, **8**, 60–65.
JOOST, K., JOHANSSON, S., FORNAS, R. & HANSSON, B. (1970) Butter oil yield. *Svenska Mejeritidningen*, **26**, 3–10.
KE, P. J. & ACKMAN, R. G. (1973) Bunsen coefficient for oxygen in marine oils at

various temperature determined by an exponential dilution method with a polarographic oxygen electrode. *Journal of the American Oil Chemists' Society,* **50**, 429–435.

KEHAGIAS, C. & RADENA, L. (1973) Storage of butter oil under various conditions. *Netherlands Milk and Dairy Journal*, **27**, 379–398.

KEOGH, M. K. & HIGGINS, A. C. (1986) Anhydrous milk fat 1. Oxidative stability. *Irish Journal of Food Science and Technology*, **10**, 11–22.

KIESEKER, F. G. (1982) Recombined evaporated milk. *Bulletin, International Dairy Federation*, **142**, 79–88.

KIESEKER, F. G. (1984) Recombined dairy products. *Food Technology in Australia*, **36**, 34–36, 38.

KOSIKOWSKI, F. V. (1975) The whitening of Mozzarella. *Dairy and Ice Cream Field*, **58**, 52.

KURAMOTO, S. & JESESKI, J. J. (1954) Some factors affecting the action of benzoyl peroxide in the bleaching of milk and cream for blue cheese ripening. *Journal of Dairy Science*, **37**, 1241.

LANSBERGEN, G. T. J. & KEMPS, J. M. A. (1984) Hardened butterfat in margarine fat blends. Lever Brothers, US Patent 4 479 977.

LEHMANN, H. R., DOLLE, E. & UPHUS, A. (1988) Processing lines for the production of butteroil. Technical-scientific documentation No. 9, Westfalia Separator Company, Oelde, Germany.

LEIGHTON, F. R. & LAWRENCE, A. J. (1979) Water. In *Monograph on Recombination of Milk and Milk Products (Technology and Engineering Aspects)*. International Dairy Federation, Brussels, Belgium, pp. 46–48.

MAURER L. (1979) Increasing importance of the dairy industry in Thailand. *Ostereichische Milchwirtschaft*, **34**, 122–124.

MCDANIEL, M. R., SATHER, L. A. & LINDSAY, R. C. (1969) Influence of free fatty acids on sweet cream butter flavour. *Journal of Food Science*, **34**, 251–254.

MULDER, H. & WALSTRA, P. (1974) *The Milkfat Globule: Emulsion Science as Applied to Milk Products and Comparable Foods*. Commonwealth Agricultural Bureaux, Farnham Royal.

MUNRO, D. S. (1982) Alternative equipment for the manufacture of recombined butter. *Bulletin, International Dairy Federation*, **142**, 137.

MUNRO, D. S. & ILLINGWORTH, D. (1986) Milkfat based food ingredients: present and potential products. *Food Technology in Australia*, **38**, 335–337.

MUNRO, D. S. & JEBSON, R. S. (1974) Continuous manufacture of ghee from anhydrous milkfat. In *XIX International Dairy Congress*. International Dairy Federation, Brussels, Belgium, Volume IE, p. 679.

NEWSTEAD, D. F. (1982) Recombined sweetened condensed milk. *Bulletin, International Dairy Federation*, **142**, 59–62.

NIZO (1977) Method for producing an aromatic soured butter from sweet cream without acidifying it. Nederlands Instituut voor Zuivelonderzoek (Netherlands Dairy Research Institute), Netherlands Patent 7 513 464.

NORRIS, R. (1990) Milkfat for the production of recombined milk and milk products. In *Recombination of Milk and Milk Products*, IDF Special Issue No. 9001. Proceedings of Seminar, Alexandria, Egypt, 12–16 November 1988. International Dairy Federation, Brussels, Belgium, pp. 146–156.

PAREKH, J. (1970) Ghee and its technology. *Dairy Products*, **7**, 10–13.
PUNJRATH, J. S. (1974) New developments in ghee making. *Indian Dairyman*, **26**, 275–278.
RACHEV, R., SAKHANEKOV, KH. & KOSHEV, A. (1979) Study of decolorization of milk fat. *Nauchni Trudove, Institut po Mlechna Promishlenost*, **10**, 119–125.
RAY, S. C. & SRINIVASAN, M. R. (1976) Pre-stratification method of ghee making. *I.C.A.R. Research Series*, No. **8**.
RICHARDSON, T. & KORYCKA-DAHL, M. (1983) Lipid oxidation. In *Developments in Dairy Chemistry, Vol. 2, Lipids*, ed. P. F. Fox. Applied Science Publishers Ltd., London, UK, pp. 241–363.
RODENBURG, H. (1973) Use of subsidised EEC concentrated butter (as butter oil or dried butter) in bakery and confectionery. *Kakao und Zucker*, **25**, 496–499.
ROGERS, W. P. & KIESEKER, F. G. (1985) Observations on the manufacture of recombined butter. *Australian Journal of Dairy Technology*, **40**, 157–163.
SALJI, J. P., SAWAYA, W. N. & AYAZ, M. (1984) Fluid milk industry in the central province of Saudi Arabia. *Journal of Dairy Science*, **67**, 1054–1060.
SANDERSON, W. B. (1979) Dairy products. In *Monograph on Recombination of Milk and Milk Products (Technology and Engineering Aspects)*. International Dairy Federation, Brussels, Belgium, pp. 40–41.
SANDERSON, W. B. (1982) Recombination of milk and milk products. *Bulletin, International Dairy Federation*, **142**, 18–21.
SCHAAP, J. E. (1973) Re-emulsification of AMF into buttermilk, skim milk or whey for butter and cheese making. *Officieel Orgaan Koniklijke Nederlands Zuivelbond* (Official Bulletin of the Royal Dutch Dairy Union), **65**, 616.
SPIELER, R. (1982) Recombination of butter and ice cream. *Bulletin, International Dairy Federation*, **142**, 132–136.
TAYLOR, M. J. & RICHARDSON, T. (1980) Anti-oxidant activity of skim milk: effect of heat and resultant sulphydryl groups. *Journal of Dairy Science*, **63**, 1783–1795.
THURAISINGHAM, S. (1982) Recombination as a means to establish indigenous dairy food production. *Bulletin, International Dairy Federation*, **142**, 184–189.
TIMMEN, H. (1975) Milkfat fractionation to improve butter spreading using soft fractions and incorporation of vegetable oils to improve spreadability. *Die Molkerei-Zeitung Welt der Milch*, **29**, 1259–1261, 1264.
TIMMS, R. E. (1980) Detection and quantification of non-milkfat in mixtures of milk and non-milkfats. *Journal of Dairy Science*, **47**, 295–303.
TIMMS, R. E., ROUPAS, P. & ROGERS, W. P. (1982) The content of dissolved oxygen in air-saturated liquid and crystallized anhydrous milk fat. *Australian Journal of Dairy Technology*, **37**, 39–40.
TIRTIAUX, F. (1976) The industrial fractionation of fats by controlled crystallization—Tirtiaux method. *Oléagineux*, **31**, 279–285.
VERHAGEN, L. A. M. & BODOR, J. (1984) Spreadable water-in-oil emulsion based on a high melting butterfat fraction and a liquid oil. Lever Brothers, US Patent 4 438 149.
VERHAGEN, L. A. M. & WARNAAR, L. G. (1984) Low-calorie spread based on a low-melting butterfat fraction. Lever Brothers, US Patent 4 436 760.
WADE, V. N., AL-TAHIRI, R. & CRAWFORD, R. J. M. (1986) The autoxidative

stability of anhydrous milk fat with and without antioxidants. *Milchwissenschaft*, **41**, 479–482.

WADWHA, B., BINDAL, M. P. & JAIN, M. K. (1977) Simulation of ghee flavour in butteroil. *Indian Journal of Dairy Science*, **30**, 314–318.

WATT, G. N. (1982) The phase inversion of high fat creams. Dissertation, Diploma in Dairy Technology, Massey University, Palmerston North, New Zealand.

ZADOW, J. G. (1982) Recombined milks and creams. *Bulletin, International Dairy Federation*, **142**, 33–39.

5

Fats in Spreadable Products

D. P. J. Moran

Covington, Cambridgeshire, UK

SUMMARY

The fat spreads market shows considerable regional variations on a global basis. In Northern Europe, America and Australasia in particular, the market is strongly brands oriented; there has, however, been a changing emphasis from butter and packet margarine to products containing 'health' blends and reduced fat driven by increasing consumer awareness of the relationship between diet and health.

Within Europe the first low fat spread was launched in 1968 and many different types of yellow spreads are now commercially available including products with fat levels ranging from 80% to less than 5%, blends of milkfat with vegetable fat, and water continuous emulsions.

Until recently the microstructure of volume market spreads has been that of emulsions mainly of a fat continuous nature with dispersed aqueous drops. The structure of the product is dominated by the crystallisation characteristics of the fat (crystal size, shape and intercrystalline bonding). The stability of the product emulsion is significant during manufacture and spreading and also influences the perceived performance of the products especially those of the reduced fat type, which should predominantly invert to an oil in water emulsion on the palate. Bicontinuous phase spreads have also been reported.

Various methods for processing spreads are now available including traditional churning techniques as used for butter, controlled inversion of water continuous emulsions, crystallisation of fat continuous emulsions, and cold mixing of plastic products. Unlike the situation for traditional high fat products, control of the aqueous phase rheology of reduced fat spreads is a significant factor in successful processing and optimised palate response.

Methods for identifying key characteristics in spreads include the measurement of fat solids index in both fat blend and product and this last approach offers improved process control. Measurements of the hardness, spreadability, electrical conductivity, emulsion stability, size of dispersion, appearance, and a range of consumer detectable sensory properties are also employed.

Water continuous spreads are presently a relatively small section of the overall market and comprise soft and processed cheeses, spreading mayonnaise, peanut butter and the like.

Nevertheless water continuous yellow spreads can offer certain advantages over fat continuous emulsions in terms of reduced fat content, quicker flavour release on the palate, less demanding fat characteristics and simpler processing. One drawback may be a shorter microbiological shelf-life. Several methods are available for manipulating the structure of water continuous products. The implications of such structures into the yellow spreads area will have a major impact on product descriptions and legislative definitions.

Clearly with products containing up to 80% or more of aqueous phase the influence of this component on taste and emulsion stability is of increasing importance. Viscosity raising and gelling agents are added to provide structures which contribute to reduced droplet coalescence during processing.

Dispersions of microparticles of hydrocolloids which taste like fat are becoming of increasing importance in spreads technology as they can confer sophisticated rheological properties with consumer appeal to both fat and water continuous emulsions.

Looking to the future, it is clear that, with increased understanding of the functionality of spreads used both on bread and in the kitchen, the need for high fat traditional butter and margarine will diminish. The next generation of products will contain fat levels mainly in the 5–65% range.

Nutritionally adapted spreads will emphasise the trend towards 'positive' nutrition, with greater attention given to the biochemistry of fats consumed, and extensive processing of oils such as hydrogenation and interesterification may become undesirable. The consumer may come to accept the natural 'green' flavours resident in oils.

With improved technology the flavour quality of reduced fat spreads will closely match that of full fat variants and methods are already available for enabling ingredients lists to be shortened (in terms of added emulsifiers, etc.).

Multifunctional spreads should become available, particularly low fat spreads with reliable kitchen (baking) performance. Alternative methods of cooking, such as microwave heating, will also have an influence on the type of fat used.

More information on structured aqueous phases, particularly the behaviour when sheared, will lead to better control of product emulsion stability and, possibly by the 'shaping' of particles, offer interesting new textures.

Process studies will include more critical analysis of the factory unit operations currently in use, particularly in terms of fat crystallisation, aqueous phase structuring and emulsion phase stabilisation, leading to better control of product stability. Lower energy processing and line flexibility are also subjects likely to receive attention.

The blurring of the traditional technical lines of demarcation between the dairy and edible fats industries will continue and provide new opportunities for spreads. Increasing manufacturing efficiency combined with well-timed launches of new products with consumer appeal will be necessary to successfully compete in a market which is currently undergoing radical change.

INTRODUCTION

Until recently only two fat spreads were generally available in the developed world. Butter was an expensive luxury product whereas margarine was a rather down-market cheap substitute, but with more extensive usage in the kitchen. Several factors have now dramatically changed that situation, among which are pricing differentials and dietary recommendations concerning fat intake.

Within Europe commercial development of blended milkfat and vegetable fat spreads began about 1963 with the product Bregott in Sweden (Anon., 1969). The first low fat spread to be sold was Outline, marketed by the Van den Berghs company in the UK in about 1968. Related products had been patented some time earlier (Unilever Ltd[1]). Many different types of yellow spreads are now commercially available consisting of blends of milkfat and vegetable fat, products with fat contents varying from over 80% to less than 5%, spreads of a water continuous nature, and products containing significant levels of milk protein and hydrocolloid emulsion stabilising agents. Products with significant nutritional claims are becoming more popular. The major categories of spreads, together with typical ingredients, are shown in Table 1.

Consumption patterns vary widely between different groups of consumers within the European Economic Community. Butter consumption fell by some 13% during the 1980s, whereas margarine increased by about 1%. Some countries, however, such as the UK saw butter sales fall by about 50%, and newer spreads other than margarine or butter now account for a very significant and increasing amount of sales, a trend which is expected to continue.

RHEOLOGY OF FATS AND SPREADS

Fat Crystallisation

Edible fats consist of suspensions of crystals in a liquid oil. The crystals themselves are mainly triglycerides (although other minor components

TABLE 1
Typical composition of spreads

Spread type	Fat (%)	Protein (%)	Added emulsifiers or emulsifying salts	Stabiliser	Preservatives	Colour, flavour, vitamins
Butter	>80	0·3	–	–	–	+
Margarine	>80	0·2	+	–	–	+
Reduced fat	60–75	0·3	+	–	–	+
Low fat	38–40	0·2–6·5	+	=	+	+
Very low fat	20–25	0–8·3	+	=	+	+
Water continuous	5–12	12–20	+	=	=	+

= denotes an option

may be adsorbed or occluded). The properties of fats, such as hardness, work softening, plastic behaviour are highly dependent on how these crystals interact by reason of their number, size and shape, and the type of bonding force between them.

The crystallisation of fats in general is assumed to consist of a nucleation and a growth phase and both of these processes can exhibit a maximum value with respect to temperature.

Nucleation of crystallisation of the glycerides from the fat can be of the homogeneous or heterogeneous type, almost certainly the latter predominating in commercial fats due to the presence of trace impurities, high melting partial glycerides, etc. Thus in bulk fat systems degrees of supercooling will be relatively low, but in the globular fat state of dairy cream it is likely that degrees of supercooling will be greater due to the relative isolation within individual globules of impurities with the capability of seeding crystallisation. The nucleation rate in fat globules increases with decrease in temperature within the normal range of physical ripening temperatures for cream (Walstra & van Beresteyn, 1975).

The crystallisation of milkfat in bulk is said to be of the first order type with an activation energy of 11 kcal mol^{-1} (DeMan, 1963).

Fats can solidify in more than one crystalline polymorphic form, and these differ from each other in terms of melting temperature, density, heat of fusion and crystallisation, refractive index, etc. The alpha crystals are the least stable form, have the lowest melt point, and occur when melts are crystallised rapidly. Although each form has minor variants the three main forms are alpha, beta-prime and beta in ascending order of stability. The forward transitions between alpha and beta take place with the liberation of heat and are not reversible.

The various polymorphic forms are distinguished by X-ray powder spectroscopy which reflects the arrangement of the molecules within each lattice structure, although other techniques, such as differential scanning calorimetry, dilatometry, etc., can also be used. Fat blends for spreads production usually crystallise in the alpha form within the units of a continuous process line but under the influence of temperature rise and agitation rapidly transform to a beta-prime form. Of course, due to the variety of triglycerides present in each fat, several triglycerides are involved in such solid phases and the relationship between them is governed by phase rules. Further transition to the beta forms is much less likely as these require very close packing of the triglycerides and the complex nature of many spread blends (in terms of difference in length and shape of the fatty acid chains) make such close packing difficult even after a considerable time. Simpler blends, however, can crystallise in beta forms. The presence of molecules other than triglycerides, such as diglycerides, can inhibit the creation of more close-packed molecular arrangements, and thus act as beta-prime phase stabilisers. Such an effect has been noted in low erucic rapeseed oil blends (Hernquist & Anjou, 1983). The rejection of such molecules from the lattice could be an essential step in the process of further compaction to denser forms. The formation of high melting forms with melting or solution points too close to palate temperature can impair the melting performance of products. The process of the formation of solid phases in fats is described by solubility curves, and each polymorphic form has a different curve depicting the relationship between temperature and fraction of solute in the soluble/insoluble state.

X-ray diffraction studies have confirmed the existence of alpha, beta-prime and beta polymorphic forms in milkfat (DeMan, 1961).

In solidified globular milkfat it has been shown that beta crystals occur in the outer shells of the fat while beta-prime forms are correlated with lower melting triglycerides towards the centre, and also with the more liquid milkfat situated between the globules (Precht, 1980). It has been proposed that crystallisation is initiated at the surface of globules (Buchheim, 1970) and this is plausible due to the ordering influence on fat crystallisation of surface active molecules already aligned at the interface, and also the ability of the adjacent water phase to remove the latent heat of crystallisation comparatively quickly. However other observations suggest crystallisation can commence apparently randomly within globules (Precht & Peters, 1981). A major difference between bulk and globular crystallisation of fat is that the size of the resultant crystals is limited by the globule size in the latter and thus the formation of rigid extensive structures is inhibited. This probably contributes to the plastic texture of butter, which over a certain temperature range can consist of a dispersion of essentially non-interacting semi-solid globules and aqueous droplets in a relatively liquid fat continuous matrix.

When conventional butter churning practice is followed polymorphism of the fat is probably not a variable factor. However, this is not the case with votator and other types of processing as used for vegetable or blended products, which utilise a wide variety of different fat compositions.

Phase Behaviour

Milkfat like other fats consists of triglycerides which differ in both chemical composition and resultant different physical properties. Such triglycerides can form solid solutions with other triglycerides in the solid state mainly of the continuous solid solution or eutectic type, to a great extent depending on the molecular similarity in terms of chain length, unsaturation and molecular shape of the fatty acids composing the triglycerides. The most stable crystalline form, the beta form, having the highest molecular packing density, is least prone to form solid solutions particularly in the presence of mismatching fatty acid chain shapes and bulky molecular impurities such as phospholipids. These last have to be expelled from the lattice before higher packing densities can be achieved.

The phase relationships in crystallised milkfat in bulk and globular form have been shown to be affected by the temperature treatment such materials are subjected to (DeMan & Wood, 1959a). Such behaviour will be important in industrial applications such as the selection of butterfat fractions for use in spreads. The rate of crystallisation of milkfat can also affect both the compositon of the solid phase formed and the size of the resulting crystals. Very rapid cooling will produce many uniform crystals of small size which can retain a large amount of liquid oil due to their large interfacial area. Such blends will be comparatively firm in texture as a consequence of the forces of interaction between such crystals. The phase behaviour and mixed crystal formation of several fat systems have been studied for glycerides (Moran, 1963), non-milkfat systems (Timms, 1984), and milkfat systems (Cantabrama & DeMan, 1964).

Studies on the interaction of milkfat with other fats have shown strong evidence of eutectic and compound formation. The stable polymorphic form of both milkfat and tallow was found to be mainly beta-prime. Compound formation between these two fats in admixture resulted in a beta polymorphic form as evidenced from differential scanning calorimetry and X-ray diffraction measurements. Interesterification of the fats eliminated the beta form (Timms, 1979).

Fat Crystal Size

The dimensions of fat crystals are related to the manner of their formation. Rapid cooling below the solubility temperature of the alpha solid form gives a large number of imperfect crystals, whereas slow cooling at higher temperature results in larger crystals. Alpha crystals transform to more stable forms above the transition point but tend to retain their original size and crystal habit.

Structural relationships between larger fat crystals are said to be dominated more by primary than secondary bonds compared with smaller (0·1 μm) crystals (Nederveen, 1963). The shape of the crystals can be needle-like but are often thin platelets (as single crystals) of some tenths of a micrometre in dimension. Triglyceride molecules tend to add on during the growth phase in the form of successive layers.

In the bulk state milkfat crystals can vary in size from about 1 μm up to 40–50 μm (DeMan & Wood, 1959b) while clusters of spherulites can exist at up to several hundreds of micrometres in size.

The modification of fats by, for instance, interesterification usually results in a decrease in crystal size. Smaller crystals are difficult to detect with a light microscope and more specialised techniques such as electron microscopy or indirect methods such as permeametry have to be employed. Tempering of fats can also increase mean crystal size, especially if the fats contain significant amounts of those triglycerides capable of more extensive development such as symmetrical triglycerides. The size of the fat crystals can alter the appearance of the products in that usually the smaller the crystals the paler and more matt the appearance. Crystals which are small and do not melt quickly on the palate can give a rather slow melting sensation, in, for instance, margarines. Smaller crystals having a larger surface area which may not be smooth can retain a large amount of liquid oil in the network and the close approach of such crystals to each other tends to result in a tough strong lattice structure.

The Crystal Network

The three dimensional crystal network in plastic fats gives rise to both viscoelastic behaviour and a yield value. The inter-crystal forces which exist in the fat crystal matrix can be described as being of two types. Primary bonds are very strong and are formed by the crystallisation of glycerides between two adjacent crystals. Once broken they are not easily reformed and in this sense are irreversible. However, weaker secondary bonds of association are formed within seconds (Haighton, 1963) by van der Waals type bonding between crystals which are thus in a state of flocculation. Such bonds are relatively easily broken and are thus weak, can be reformed by reflocculation and are reversible. In actual practice a range of such primary and secondary forces probably exists in a typical network.

Individual crystals can be organised in fat continuous spreads into various structures. Crystals dispersed in an interstitial fashion between aqueous phase droplets contribute to product rheology. When clustered and aggregated around water droplets they can stabilise the emulsion by a Pickering type mechanism or by forming protective 'shells'. In butter they can be found in the globules dispersed in the fat phase, globule size being in the order of 4 μm (Fig. 1) (Juriaanse & Heertje, 1988). In higher fat content margarine and cooking fats spherulites can occur, which have diameters of

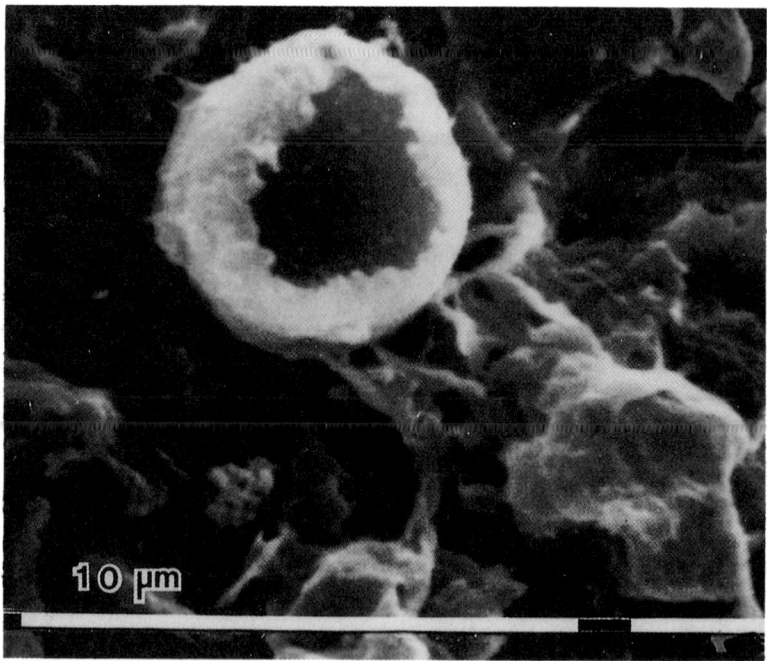

Fig. 1. Fat globules in churned butter (Juriaanse & Heertje, 1988). (Reproduced with permission from *Food Microstructure*.)

tens of micrometres, organised in three dimensions with long crystals radiating from a central nucleus. These are strong structures likely to increase the strength of the network. Sometimes grains can be seen, which appear to consist of loose agglomerations of crystals of similar size to spherulites. When they exceed about 25 μm in size they may be recognised as a distinct texture on the tongue, variously described as powdery, gritty or sandy.

The Aqueous Phase

With the more traditional high fat spreads the drops of aqueous phase normally consist of milk components, optionally cultured, which provide flavour to the product. In more recent lower fat variants the flavouring ingredients have been extended to include whey and ultrafiltered whey, caseinates, butter serum and concentrates of fat globule membrane proteins. Post-manufacture culturing of the aqueous phase within the packed product is also feasible for the development of more natural flavours.

As the fat level is reduced in products the influence of the aqueous phase on taste and emulsion stability becomes much more important. Stabilisers are added to the water phase to provide structure and assist stabilisation of the water in oil emulsion. Four different types may be identified:

—viscous (by use of high levels of milk protein or polysaccharides such as alginates or carrageenans),
—gelling (based on gelatin or carrageenan modified by appropriate cations),
—phase separating (by selection of suitable mixtures of biopolymers such as protein (gelatin) and a polysaccharide (maltodextrins),
—synergistic (the viscosity being increased by interaction of casein with starch, modified starch, or monoglycerides).

All these systems contribute to reduced coalescence of the aqueous droplets during product processing of lower fat spreads. Some can also give sophisticated rheological characteristics, which may confer advantageous attributes to products in terms of mouthfeel, texture on spreading, etc. The gelling systems are particularly favoured for very low fat spreads since they can mask emulsion breakdown when the product is spread.

Particles of hydrocolloids (less than 8 μm in size) can be added either within the aqueous phase drops themselves, or in the continuous phase of water continuous spreads. Examples are heat denatured proteins from milk or egg white, or mixed biopolymer coacerate particulates. These are claimed to enhance the fat-like taste.

The Structure of Spreads

Historically most spreads have been of the fat continuous type with aqueous phase droplets in the range 2–4 μm (margarine) to 4–80 μm (for low fat spreads). However, with the move towards even lower fat spreads, water continuous types are being marketed. High internal phase emulsion technology indicates that fat continuous emulsions with as little continuous phase as 5% are possible, so that emulsions of this type with 10–20% are certainly feasible. Commercially, however, they may offer different characteristics to those of higher fat spreads and their stabilisation systems are possibly unique. With highly structured aqueous phases a true emulsion system may not be necessary, and simple mixtures of fat and shaped pieces of structured water are feasible. The 'shapes' may be significant in terms of optimised consumer properties. Bicontinuous phase systems are also possible. Clearly such systems will exhibit both water and fat continuous characteristics as indeed some butter has been shown to do. Butter may be seen to consist of compressed wet fat globules, so clearly a continuous network of aqueous phase is possible, in addition to recognisable drops of emulsified water. Limited inversion spreads (see later) also show both water and fat continuous characteristics.

The implications of such subtle structures in novel reduced fat spreads on product descriptions and legislation definitions are clearly of some significance.

In margarine and many reduced fat spreads of the fat continuous variety the water droplets are situated in a continuous oil phase, with fat crystals in a protective 'shell' acting as a barrier to coalescence. In general the higher the water content and the higher the fat solids level, the greater will be the emulsion instability. The bonds between the crystals will range from strong (primary) to weak (secondary) according to the fat type, process and storage regime adopted. By contrast churned butter (made by part crystallising fat globules in cream of about 35% fat content) contains globules which are protected by a strong membrane at the surface some 0·01 μm thick consisting of a tough elastic complex of milk proteins (on the aqueous side), phospholipids, cholesterol and fat crystals. During the cooling (physical ripening) process a layer of higher melting triglycerides crystallises and gives further support to the membrane, which can thicken to dimensions up to 1 μm thick. On churning the introduction of air bubbles into the cream causes a partial destabilisation of the membrane, and the mechanical action results in some globules losing the entrained liquid part of the fractionated milkfat. This then becomes the continuous phase of the product. Hence the structure of butter, and also some churned reduced fat spreads, is mostly a continuous oil phase containing small free fat crystals with suspended aqueous droplets and more or less undamaged globular fat, together with some empty globular residues. There is thus only limited opportunity for the higher melting milkfat crystals to form a rigid three dimensional network of primary bonds, as can occur in votated products, which can be more brittle in structure as a result. However, some continuous water phase structure may also be possible in certain types of butter for the reason explained above.

The partial disruption of the milk globule membrane can begin at the stage of physical ripening of the cream when clustering of droplets takes place. A greater degree of destruction of globules during churning releases more free fat, and results in harder products at lower temperatures, and also more oiling off when warm (Knoop, 1965). The strength of the globules is enhanced by a shell of crystalline fat around the periphery together with further crystals within the interior of the globule (Precht & Buchheim, 1979).

From a rheological standpoint plastic deformation can be correlated with systems which contain a certain minimum phase volume of suspended particles, these particles being able to move relatively freely with respect to each other. The presence of fat globules and possibly water droplets with associated fat crystals on the surface suggests butter has such a structure which can be both tough and viscoelastic on deformation. Non-churned spreads by contrast have structures which tend to be dominated by the nature of the intercrystalline bonds between fat crystals, and the presence of associations of fat crystals. Reduced fat versions of such spreads can have higher residual viscosities on the palate (after the fat crystals have almost all melted) by reason of the high phase volume of suspended water droplets. When the fat crystals are very small the preponderance of secondary bonds results in relatively more plastic structures. High cooling rates and much mechanical working during processing favour such structures, but these are not always possible with lower fat spreads because such processing conditions will favour emulsion instability.

The Emulsion

During processing the water droplets in fat continuous spreads (when the fat is in the melted state) are initially stabilised against coalescence by a surface film of adsorbed emulsifier, usually monoglycerides, fatty acids or phospholipids, which may have in part originated from the oil phase itself. At a later stage fat crystals contribute to the stabilisation mechanism by a combination of being situated at the oil/water interface and a rigid network which helps keep the droplets apart. A good margarine will have a mean water drop size of less than 2 μm, but individual drops can range up to 20 μm in diameter. In reduced fat spreads water drop sizes can be much larger and range up to 80 μm or so. Whereas with higher fat products water droplets do not have a marked effect on product consistency (Haighton, 1965) with lower fat spreads this is not the case. In fact with 40% fat products the final emulsion can have a pronounced viscosity on the palate in the almost melted state, unless emulsion inversion has already taken place. Butter has been described as having an aqueous phase droplet distribution similar to margarine but with a more continuous aqueous phase, probably linked to the manufacturing process in which wet grains of butter are compressed together following the churning process. Also relevant may be the fact that some of the milk fat membrane (and associated water molecules) surrounding the globules can exist as collapsed fragments in the fat phase of butter.

The amount of water phase and the nature of the distribution of drop size can affect the appearance, flavour release and microbiological keepability of spreads. Clearly this last factor is a greater hazard in those spreads with a higher water content. Smaller drops tend to reflect more light due to the increased interfacial area so spreads with such a structure tend to be light in colour and matt in appearance, and the size of the fat crystals can have a similar effect. The Ostwald ripening effect can apply to both water drops and fat crystals, so that both tend to increase on storage.

The size of the water drops is also increased by the presence of milk proteins, which tend to destabilise the water in oil emulsion, and such larger drops increase the flavour impact of the product when melt down occurs on the palate.

In normal processing high rates of shear such as in votator chilling units favour a reduction in drop size, whereas low shear conditions tend to promote droplet coalescence.

Other factors which favour a smaller drop size are higher levels of certain emulsifiers. Normally mono/diglycerides of low HLB number are used to stabilise the water in oil emulsions in margarine, and distilled monoglycerides are used for reduced fat spreads. The more saturated high melting types are effective in that they can both nucleate crystallisation of triglycerides and themselves crystallise at the water/oil interface. In addition monoglycerides can, by surface adsorption, alter the contact angle of the fat crystals which collect at the interface. This makes such crystals partially wettable by water and increases their ability to stabilise the emulsion by a Pickering type mechanism. By contrast milk proteins tend to produce oil in water emulsions and increase droplet size, and lower the stability of water in oil products under shear. Such shear is applied to the products during the later stages of processing and tub filling, when spread by the consumer, and during mastication on the palate. Under these last conditions the fat crystals melt, and the emulsion is rapidly inverted in saliva (at least in part) to a water continuous state. This emulsion breakdown promotes the rapid release of flavours and salt and hence accentuates the taste sensation. In terms of surface activity lecithins are considered to behave similarly to proteins when fat crystals are absent, but can adsorb on the surface of fat crystals in a similar fashion to distilled monoglycerides and thus modify the contact angle and emulsion stabilising properties of these crystals.

The stability of higher water content fat continuous emulsions is usually increased by introducing structuring agents into the water. These can include high levels of milk proteins (in excess of about 12%), or lower levels of polysaccharides such as sodium alginate. In addition gelling agents, including gelatin, carrageenan, pectin, etc., have also been used. The final drop size in the product emulsion is the result of a balance between those factors promoting droplet disruption, and the reverse influences favouring coalescence.

Although structuring agents probably retard the back coalescence of water drops during processing, an additional major function is to restrict droplet coalescence on spreading, and in this regard the gelling agents are particularly effective. They may be used in the process in a fluid state, which gels later when the product is packed, or pregelled before the water in oil emulsion is formed. Very often the structure in the aqueous phase is matched to that of the accompanying fat phase (i.e. related to the solid fat content) for the best effect in terms of product characteristics.

Clearly with the increasing popularity of lower fat spreads the influence of the aqueous phase is of increasing importance. Particularly significant are the rheological characteristics of water structuring systems of biopolymers under the influence of shear, both during the manufacturing process and on spreading/mastication.

Functions of Surface Active Lipids

Surface active components may make several contributions in spreads technology. Some of these may be summarised as follows:

1. At the initial stages of spreads processing molten emulsions of both oil and water continuous type are stabilised by partial glycerides, lecithins and/or fatty acids absorbed into the interface, thus lowering the interfacial energy.

2. Monoglycerides, particularly those from saturated acids, can be hydrated at certain concentrations and temperatures to liquid crystalline phases with highly structured properties. Such viscous phases are known to stabilise air cells in liquids and thus assist in foam creation in cake batters. It is also feasible that structured aqueous phases using such ingredients can be used as alternatives to biopolymers in low fat spreads production.

3. When molten emulsions are crystallised, the highly oriented molecular films of surface active lipids at the interfaces can act as nucleation sites for triglyceride crystallisation. Such a process would be assisted by the adjacent water phase acting as a heat sink for the latent heat emitted.

4. Monoglycerides have higher crystallisation temperatures than their

parent triglycerides and thus assist nucleation of the latter within supercooled oil droplets or bulk fats.

5. Partial glycerides such as diglycerides can inhibit phase transformations in triglyceride lattices and thus stabilise fats in the beta prime polymorphic form.

6. By using selected blends of surface active agents of the low and high HLB variety, combinations can be utilised which are effective in both stabilising fat continuous products at the manufacturing stage, and yet assisting in inverting such emulsions on the palate. This results in an enhanced flavour release.

7. Very specific partial esters, such as erucic acid esters, sucro-esters and polyglycerol esters have been claimed to be particularly effective in stabilising lower fat spreads.

8. Surface active lipids may alter the contact angle of fat crystals at the water/oil interface and thus confer enhanced emulsion stabilising properties of both water and fat continuous character, on such crystals.

It is clear more investigations are needed in the above areas to broaden the extent of technology available, if optimised reduced fat spreads are to be realised in terms of flavour release, required emulsion stability under key operating conditions, and multifunctionality in use (including for instance baking properties).

Defining Spreads Structure

Spreads are visco-elastic materials which can be distinguished by a variety of measurable physical characteristics, typically yield value, elasticity, viscosity and work softening, all of which contribute to the perceived properties on spreading and chewing. General flow behaviour may be described in terms of Newtonian, Bingham and pseudo-plastic materials. Liquid oil is usually of the Newtonian type but crystallised emulsions behave as pseudo-plastic materials. The viscosity of such materials can only be measured after extensive mechanical working. An overview of the general aspects of texture may be found in DeMan *et al.* (1976).

The consistency of butter, margarine and reduced fat spreads can often be distinguished by means of alterations in yield value after mechanical working, viz. work softening. The yield values for products after a standard amount of work softening often show that those products with a high degree of secondary bonds retain proportionally more structural hardness than others with essentially primary bonded structure. These latter will slowly recover structural hardness but not regain their original values unless the fat crystals have been allowed to melt and recrystallise. Naturally

TABLE 2

Yield value (g/cm^2)	Panel assessment
Less than 50	Very soft, almost pourable
50–100	Very soft, not really spreadable
100–200	Soft, spreadable
200–800	Satisfactory, spreadable
800–1000	Hard, but spreadable
1000–1500	Too hard, almost unspreadable
Greater than 1500	Too hard

the basic emulsion structure of the product should not be altered by the test conditions.

Lower degrees of work softening can be correlated with plastic spreads and high degrees with brittle spreads (Haighton, 1965). For instance butters show values between 50 and 55% whereas margarines have values of 70–75%, where the figure represents the proportion of initial hardness lost. Puff pastry margarines should show low degrees of work softening for preference.

Although butter and some margarines can have similar values for hardness (yield value) and work softening, they can still be different in terms of elastic recovery after deformation. Butter usually shows higher values than margarines but values vary with the hardness of the samples. Some animal fats can produce spreads with similar elastic recovery to butter. When considering the impression of spreadability of a product with a knife a combination of the above factors is involved, namely hardness, work softening, toughness and elastic recovery. Clearly a range of values for these factors can be set for acceptable spreadability as recognised by a panel of consumers. This is normally confined in practice to limits for yield values when measured at certain temperatures.

For accurate measurement good temperature control of the samples is very important. Some typical values are given in Table 2 (Haighton, 1959).

SPREADS PROCESSING

General Aspects

Fat based spreads can be divided into two groups, fat and water continuous. By far the largest of these in volume terms is the fat continuous group.

TABLE 3
Fat continuous spreads process options

	1	2	3
Processing route	Inversion of precrystallised oil in water cream	Inversion of a cream crystallised continuously in the line	Processing of wholly fat continuous emulsions
Option A	Churning with air (emulsion partly destabilised by air bubbles)	Cream made by continuous injection of oil in the unit	Emulsion made by continuous injection of water phase in the unit
Option B	Shear churning (no air present)	Premixed cream	Premixed emulsion

Fat continuous spreads

Altogether six general methods of processing fat continuous spreads can be identified, as shown in Table 3. Other methods are possible such as cold blending of crystallised emulsions. Method 1(A) is widely used for the production of butter involving typically the Fritz type churning procedure. Method 1(B) can be identified with such processes as the Alfa and New Way. Margarine manufacture in recent times has tended to concentrate on variations of method 3(B). The other techniques, including the use of mixed emulsions of milkfat and vegetable fat, have all been described for the preparation of new types of spreads with fat contents varying from about 70% down to 20% or so. In particular methods 1(B), 2(A) and 2(B) all relate to the necessary collision and subsequent coalescence of oil drops. The kinetics of such processes has been studied by earlier workers (Smoluchowski, 1917).

When drops collide the question as to whether they flocculate, coalesce or simply redisperse is dependent on a number of factors which all have to be considered in the process of inversion of emulsions to fat continuous spreads. Fat droplet coalescence is generally favoured by high fat content in the initial emulsion, higher shearing rates (related to the mixing unit geometry), and a critical range of solid content (related to temperature) within the fat droplets, larger drop sizes, low continuous phase viscosity under shear, the absence of soluble proteins, the presence of emulsifiers such as lecithins or monoglycerides, a low temperature rise during the inversion process, etc. Very often several of these factors must be set off against others to achieve efficient emulsion inversion and stable final fat continuous products.

Lower fat spreads which contain relatively high viscosity aqueous phases (i.e. 12% milk protein or more) can be processed by option 3(B). Those with lower viscosity aqueous phases will show some reluctance to maintain a fat continuous state during processing and are best processed using option 2(A) or 2(B). Rework streams, which will tend to re-invert to an oil in water emulsion on heating/pasteurisation, may then be conveniently fed back to an early point in the process line.

Processing routes 2 and 3 use very similar units (scraped wall crystallisers and pin mixers), one operating difference being that pin mixers employed for the emulsion inversion step in route 2 must be capable of generally higher rotor speeds than customary in standard margarine processing practice.

Water continuous spreads
The processing of water continuous fat spreads is usually more straightforward and often involves controlled homogenisation as the texture modifying step before packaging. By careful control of formulation a variety of products can be obtained, the final texture being controlled by means of a combination of aqueous phase texture (as contributed by hydrocolloids) and fat droplet association. Methods for fat droplet association include protein bridging, fat crystal bridging (Unilever NV[1]) and limited inversion (Lever Bros[1]). The influence of polysaccharides (Unilever NV[2]) and milk proteins (Alfa Laval) has also been described. Scraped surface units can be used for preparing such products. Apart from fresh, processed and analogue cheese variants and spreading 'mayonnaises', the commercial exploitation of water continuous spreads, particularly of the yellow type, has so far been limited.

The following discussion will be limited to the two currently most widely used process options, viz. the churning of precrystallised cream and the use of scraped wall heat exchangers.

Churning of Butter Type Products
Products with fat contents in excess of about 70%, including butter, can be prepared by emulsion inversion of concentrated milk. Higher test creams (80% fat or more) can be converted by shear alone to fat continuous emulsions without any loss of buttermilk and this is practised in such processes as the Cherry-Burrell, Creamery Package, Alfa Laval and Meleshin. In Western Europe processes such as the Fritz are employed

capable of operating at up to 16 t h^{-1}, in which creams of 30–45% fat content after pasteurisation are part crystallised and tempered before churning. The churning process involves the use of the air–serum interface to assist in destabilising the milkfat membrane around the fat droplets. The butter grains are drained and washed and then optionally salted. Buttermilk is a by-product and generally contains 0·1–0·6% milkfat partly in the form of lipoprotein originating from fragments of disrupted membrane.

Normally the centrifugation of the cream is carried out at 40–50°C, and pasteurisation at 81–98°C for between 1 and 25 s. The physical ripening is designed to crystallise the fat into major fractions, and a much favoured regime is that known as the Alnarp process, involving various holding times at for instance 8, 21 and 12°C. This will effectively raise the melt point of the fat fraction which structures the globular fat. Such globules are carried through into the fat phase of the final product, and the tempering process restricts possible mixed crystal formation.

During churning air is whipped into the cream and the resulting air–serum interface collects the milkfat globules the surfaces of which are destabilised as a result. Released fat promotes the agglomeration of other fat globules and crystals from broken globules and also helps to collapse the foam. The procedure continues until granules of fat continuous butter several millimetres across are grown, containing both entrained buttermilk and intact and broken globules.

If soured cream butter is required then during the physical ripening a few per cent of starter culture may be added. The resultant production of lactic acid reduces the pH to about 4·7. This results in a lower level of retained fat in the buttermilk (about 0·2%). After draining of the butter granules they are washed with cool water and then sufficient brine/aqueous phase is added to produce an aqueous phase content of 16%.

The product texture is achieved by kneading in which worm conveyors firstly remove excess water and then reduce the drop size of the aqueous phase and plasticise the fat.

For the preparation of high fat spreads (more than 70% fat) much effort has gone recently into ways of improving the efficiency of the churning process. Developments have included ways to improve the temperature control of the cream (VEB Komb. Fortschritt), mechanical handling of cream (Butter Cheese Industry Research) and evaporation of cream (Creamery Cheese Industry; Butter and Cheese Industry) as part of the churning process.

A flowing loop system has been used to eliminate traditional cream ripening. Variation in the time of resting of cream between cooling and churning has been studied, improved ways of texturising the butter grains from churning described (Krasheninin *et al.*) and a process developed for destabilising cream not involving actual churning (Provision Machinery Research). The churning of cream to butter is said to be assisted by using direct current at 6 V to destabilise the cream (Leningrad Soviet Trade).

Microwaves have been used to speed up product temperature stabilisation. Computer directed monitoring and control of beater speed, water supply and the cream pump output have also been described.

Further innovations include the direct collection of butter fat globules from evaporating milk (Provision Machinery Construction), use of compressed air (at 10 bar) to destabilise cream and reduce loss of fat through mechanical action, arrangement of two separate units in line operating at 6° and 17°C to make a spreadable product continuously (VEB Komb. Fortschritt), dual kneading equipment for better product spreadability and reduced air content in the butter, kneading under reduced pressure for better texture (Ahlborn), and the use of equipment designed to reduce fat losses in the waste buttermilk (Westfalia Separator AG[1]).

Fat can be obtained directly from milk by adding liquid fat to a rotating churn to initiate agglomeration of the globules followed by an extended holding time of the fat rich fraction in the apparatus.

Various developments have taken place in the design of continuous buttermaking machines involving for instance specially designed orifice plates which are said to stabilise butter composition and increase the yield of the final product.

Use of Scraped Wall Heat Exchangers and Stirred Crystallisers

The general principles of margarine processing are given in many reference books, typically by Anderson & Williams (1965). The aqueous phase of margarines consists of skim milk or whey proteins, and the keepability may be extended by a selection of approaches involving heat treatment, small water drops, and the use of additives such as acidifying agents, anti-mould agents and salt.

In a similar way to butter processing a low pH can be achieved by fermentation over about 10 h at about 22°C using a few per cent of starter cultures in the protein containing aqueous phase.

Other minor ingredients often added include emulsifiers (0·1–0·5%

monoglycerides), lecithin (0·3–0·4%) (to avoid spattering of higher fat products and promote browning during frying), flavours (such as diacetyl, short chain fatty acids and ketones of the nature identical type), colours (such as annatto or beta-carotene), vitamins (principally A and D), and antioxidants (such as BHA, BHT) which can enhance the effect of the naturally occurring tocopherols.

Margarine processing unlike churning is carried out in the absence of air, generally by means of votator type units, in which by using a sequence of scraped wall chilling units, and mixer/crystalliser pin mixers, the emulsion ingredients are mixed, part crystallised and textured. The older open process involving a cooling drum followed by vacuum plodding is microbiologically less sound and more labour intensive and is rarely used nowadays.

In a typical process line dosing pumps proportion the fat and water phases to the first A-unit which is cooled by evaporating ammonia. The scraper blades (on a shaft rotating at up to 800 rev min^{-1}) remove crystallising fat from the cold walls and so maintain the heat transfer characteristics of the unit while promoting nucleation of crystallisation. The emulsion normally leaves the unit in a supercooled state so that after cooling the products are allowed to approach an equilibrium state with respect to crystallisation in a following crystalliser unit. This results in mainly higher melting fractions crystallising. The process is continued in following pairs of A-units and crystallisers, and products can be further worked by in-line mesh screens before packing.

Votator A-units themselves offer residence times to the product of only seconds due to the narrow gap between the rotor and the walls of the cooling surfaces of some 5–10 mm. Several such units can be assembled in sequence to increase the cooling capacity further.

Crystalliser units provide much longer residence times (of several minutes) and subject the product to intensive working by means of a series of intermeshing pins fixed to both the static walls and rotor shaft. The volumes of such units can be as large as 150 litres, with rotor speeds up to 200–500 rev min^{-1} or even higher for special purposes. Higher speeds can be used for inverting oil in water emulsions. Other uses for this type of unit include adjusting the hardness of the product flow immediately before the packaging machines.

A rework system is usually provided to allow material from just before the packaging machine or final shaping device to be remelted and stored in a buffer tank. This normally is led back into the process line at a steady low rate. Such systems allow the votator line to continue operating without over-crystallisation of any product halted within the units in the event of a breakdown in the packaging machine.

Nucleation of fat crystallisation can be promoted, and the size of crystals influenced, by either precrystallisation or recirculation. Precrystallisation is effected by introducing units before the first A-unit. In recirculation a stream of the part crystallised product is recirculated back into the uncrystallised fresh product flow through a recirculation loop.

Flowing fats can be rapidly matured (or crystallisation stabilised) by microwave heating (Mitsubishi Heavy Ind. KK) and also by the use of high/low temperature zones in a heat exchanger which is said to convert fat solids to a stable form (SCM Corp.[1]). The development of crystallisation can be monitored by nuclear magnetic resonance (New Correspondence Poly.).

Relationship of Processing and Product Structure

During the process, factors such as product residence times in the units, the shear conditions as defined by rotor speeds and blade/pin design, and the viscosity of the emulsion phases result in an equilibrium water drop size. The overall emulsion viscosity is itself influenced by the drop size and the amount of solid phase in the fat. This last is influenced by the temperature, shear and residence times. The earlier units in the process line operate with high rates of shear on a relatively low viscosity emulsion and tend to reduce drop size. In the later units the viscosity is higher, and drop sizes tend to increase. The final drop size in the product is the result of the competitive processes of drop size reduction and re-coalescence.

The development of structural hardness in spreads can be associated with the formation in particular of primary bonds between fat crystals. These are generally favoured by undisturbed crystallisation. Hence crystallisation after the processing units should be restricted, and residence times and temperatures, especially relative to storage temperatures, adjusted accordingly. Working the crystallised mass can also be assisted by the use of mesh screens in the lines. To avoid structure build-up, temperature control in votators should allow for a gradual cooling regime, and there should be limited temperature cycling during storage.

The anticipated performance of blends during processing may in part be predicted from a knowledge of their crystallisation rates as derived, for instance, by pulse NMR spectrometer/time readings for solidifying fat, or from cooling curves. In this way the correct amount of solids can be approximately forecast for a given blend in different parts of the process line consistent with acceptable texture, emulsion stability, packability, and oil exudation in the final product.

The functions of the chiller units are to mix the two phases, oil and water, and to abstract heat and so crystallise the fat. Such factors as rotor speed, heat transfer coefficients and residence time are all of consequence. Usually scraped wall jacketed heat exchanger units are used, the efficiency with which nucleated higher melt point glycerides are scraped from the wall and tumbled back into the mix being determined by the geometry and placing of the knives mounted on the rotor. The design also determines the balance between back mixing and plug flow in the annular space. The chilling units should give sufficient cooling capacity to nucleate even difficult blends. The volume of each chilling unit is related to the product residence time and higher residence times increase the amount of crystallisation which takes place within the unit. Occasionally unusual designs are met with such as eccentric shaft units, and these can increase the efficiency of fat nucleation. Figure 2 shows the effect of shear on aqueous phase drop size (Juriaanse & Heertje, 1988).

Crystalliser units are normally larger than chiller units with volumes of many tens of litres. They consist of intermeshing pins fixed to both wall and rotor. Crystallisers are usually situated after chilling units, so that crystals of fat nucleated in the latter continue to grow and transfer to the more stable beta-prime polymorph in the crystalliser unit. At the same time crystallisers can mechanically work solidifying blends and break up primary bonded structures, so that softer, more plastic products can result. Final emulsion structures are also developed (see Fig. 3) (Heertje et al., 1988). Occasionally crystallisers are used to precrystallise blends in higher fat content products before the chilling stage, which usually results in fewer but larger crystal nuclei. The overall effect is a general softening in the product, with higher values for oil exudation and quicker melting behaviour.

Following the crystallising stage, with higher fat products resting tubes are sometimes employed for allowing crystallisation to proceed further, and mesh screens are used to work the product. This allows product consistency to be aligned to the requirements of the subsequent packing machines.

With lower fat spreads modified crystalliser units are often used to ensure the inversion of the primary water continuous emulsion to the final fat continuous product. This allows for pasteurisation (and therefore

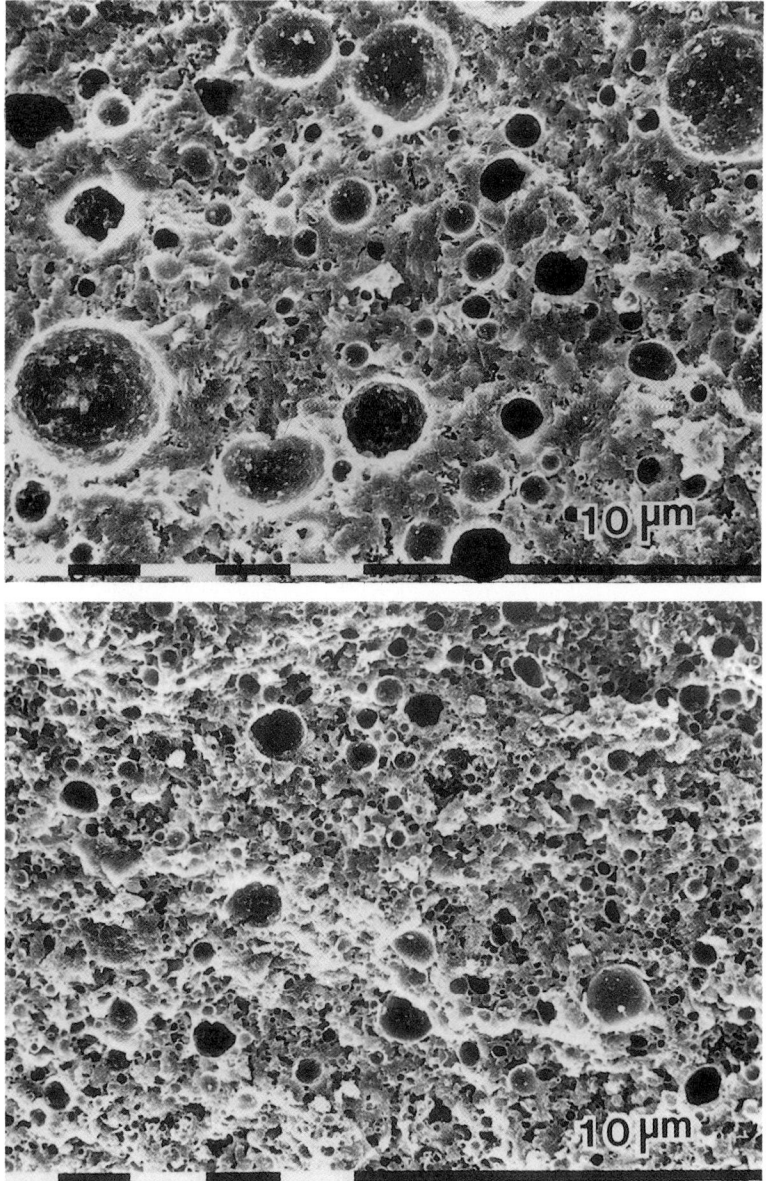

Fig. 2. Reduction in water drop size in margarine as a result of shear during processing: (top) low shear; (bottom) high shear (Juriaanse & Heertje, 1988). (Reproduced with permission from *Food Microstructure*.)

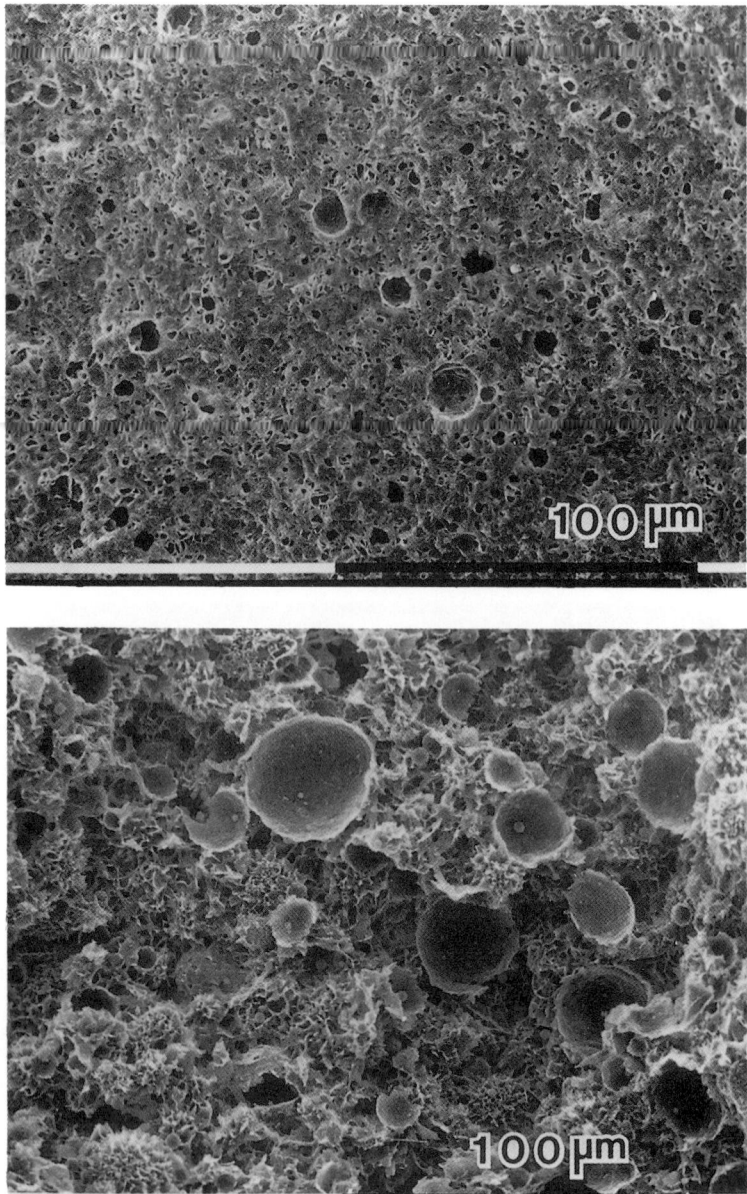

Fig. 3. Influence of crystalliser units: (top) no shear; (bottom) with shear (Heertje *et al.*, 1988). (Reproduced with permission from *Food Microstructure*.)

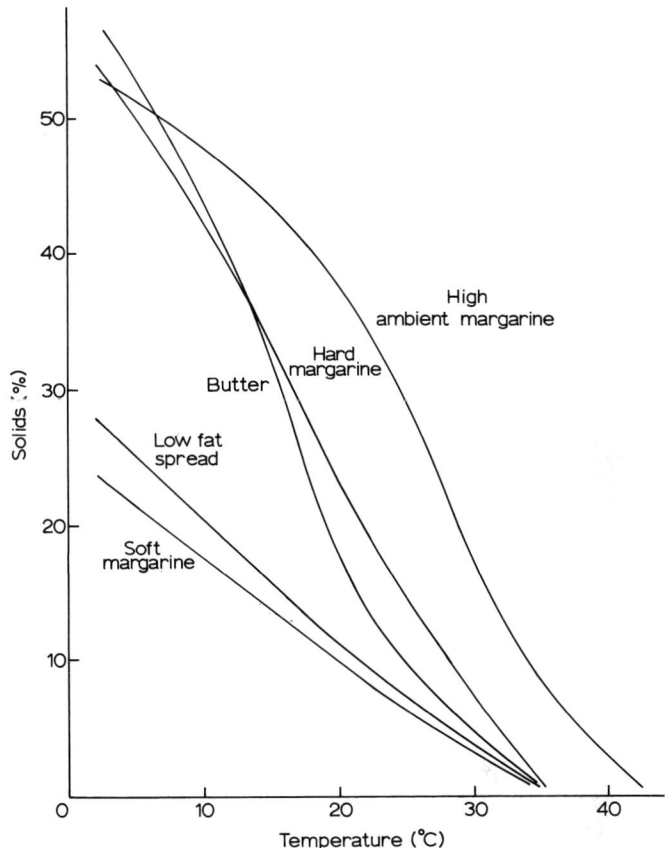

Fig. 4. Typical fat solids indices for spreads products.

consequent inversion) of rework product to be employed which may then be fed back to an early stage of the processing.

Alterations in the Properties of Spreads

Products of a fat continuous nature can be altered in structural properties by a variety of methods involving the fat, including blending, interesterification (including enzymatic interestification), fractionation and, to a degree, modification of votator conditions. Thus the blends can be selected to vary in solid fat level (Fig. 4), crystallisation rate and rate of melting, and processing conditions can be optimised to take account of these dependent on the product required. By such means considerable changes in product hardness, spreadability and coolness on the palate can be achieved.

Butter itself may be modified in physical properties either by pretreatment of the cream or by working post-manufacture. Variations of the Alnarp type cold–warm–cold treatment of the cream before churning is claimed to result in a softening effect (Dolby, 1953). Reduced fat butter spreads using milkfat only can additionally be modified in texture by alterations in fat level, granulation of the fat phase, and processing conditions to control drop size.

The post-manufacture plasticising of higher fat spreads can be carried out most efficiently at low shearing stresses, resulting in maximum bond breakage. Working at higher temperatures (15°C) apparently gives similar results to those at 4°C, when measurements are taken at 15°C (DeMan, 1969).

Simple blending of milkfat with other fats, mainly vegetable in origin, showed there were few constraints, the resulting physical properties being intermediate between those of the components of the mixture (Timms & Black, 1978).

Interesterification by an enzyme route with liquid vegetable oils offers the possibility of raising the polyunsaturated fatty acid content of milkfat (Kalo et al., 1987).

Fractionation of milkfat has been used to produce spreads with improved spreadability compared with butter. Churned products have been reported involving adding a liquid fraction of milkfat to cream (Kaukare & Antila, 1974). Further developments using milkfat fractions may be inhibited by the relative imprecision of the dry fractionation process for milkfat, the cost of the alternative solvent fractionation, the loss of flavour during fractionation, and the resultant change in milkfat composition which may cause problems with product description in some countries.

High fat products are available which contain a blend of 80% milkfat with 20% vegetable oil, although at which stage the oil is added does not appear very critical (Dixon et al., 1980). Milkfat can also be blended with other animal fats such as tallow (Timms, 1979).

Apart from blending milkfat with other fats in simple mixures, both fats may be incorporated into a structured globular form by adapting the standard churning process. At least three ways of doing this have been described. In one, liquid oil is added to dairy cream before ripening and churning, with up to 30% of the fat being non-dairy in character (Olsson). In a second method a filled cream produced from liquid and partly hydrogenated bean oil with dairy cream is churned after pasteurising and ripening to a final product containing at least 35% non-dairy fat (Milk Marketing Board). Yet a third method involves the production of a cream from crystallisable vegetable fat, blending two physically ripened creams, and churning the mixture with liquid vegetable oil to a product with a non-dairy fat content of about 50% or more of the fat phase (Svenska Mejeriernas). The different structures resulting from these approaches is probably due in part to the presence of both dairy and non-dairy fat globules, and also the further dilution of globules with liquid oil. It can be anticipated that both plasticity and palate melt down will be affected. The fat content of products involving these approaches is usually in excess of 70% of product weight, and such products are claimed spreadable from the refrigerator. The introduction of granular fat structure into low fat blended spreads via a precrystallised cream has also been described.

Process Control

Computer control of the buttermaking process is possible and can monitor signals from the beater speed, water supply pump, and also the cream pump (VEG Nahrungsguterm).

Improved monitoring of the churning process includes measurements of the conductivity of the emulsion and the use of high frequency signals to provide information on butter globules.

Infra-red determination of the fat content is now known and measuring the dielectric constant of both the water and fat phases extends process control information. Automatic control of fat, non-fat milk solids, and salt is now possible and water and fat content can be checked continuously (Westfalia Separator AG[2]; Leningrad Refrig. Ind. TE; VEB Nahrungsguterm). Liquid flow monitoring has been used to control both the cream feed and also the addition of minor ingredients such as culture and salt.

PERFORMANCE OF SPREADS

Measurement of Solids

Nuclear magnetic resonance
Protons subjected to a strong magnetic field can absorb electromagnetic energy at particular frequencies. Wide line nuclear magnetic resonance or NMR allows an estimate to be made of the proportion of liquid in a fat sample by comparison of the absorption of the unknown sample with that

of a completely liquid sample at the same temperature (Samuelsson & Vikelsoe, 1971). The alternative pulsed NMR technique exposes the fat sample to a pulse of electromagnetic radiation and measures the decay of the response to the pulse.

The decay is related to the mobility of the protons and thus the proportion of solid fat phase can be measured (Waddington, 1986). As far as products are concerned protons from other ingredients apart from fat can also give a response to NMR, such as water, protein, carbohydrates, etc. Although paramagnetic ions (such as Mn^{2+}, Ca^{2+}) can be used to relax the water phase signal, this is hardly of use in edible product systems. Nevertheless NMR can be used as both a static and in-line technique for estimating solid/liquid ratios in fat product systems. The performance of the wide line and pulse techniques has been compared and both can be used at fat solids content close to 100% (Mills & van der Voort, 1981), and also directly on spread type emulsions containing an aqueous phase.

Ultrasonics

This developing technique is based on measurements of the velocity of high frequency sound through solid and liquid fat, and experiment shows a reasonable correlation between NMR and ultrasonic velocity techniques for solids content determination (Hussin & Povey, 1984; Miles *et al.*, 1985). Ultrasonics could offer a significantly cheaper method than NMR, and with modification holds out prospects for in-line monitoring of possibly emulsion drop size and flow rates in addition to solid fat estimation. When combined with in-line conductivity measurements all data necessary for monitoring and controlling the manufacture of low and very low fat spreads would be available.

Differential scanning calorimetry

A popular thermal method for determining solid levels in fats or products containing fats is differential scanning calorimetry or DSC. The method is time consuming and the equipment is expensive, but when carried out carefully can give an accurate trace of the melting curve, the area of which is related to the melting of the solids phase. The result of course is subject to polymorphic transitions and the heat of fusion varies with the nature of the triglycerides and therefore the melting temperature (Timms, 1978).

Dilatometry

This much older method relies on the fact that, when the solid phase of fat melts, the melting can be followed by the change in volume that takes place

(Hannewijk & Haighton, 1957). Measurements are usually made on fats themselves, and a standard method has been described (American Oil Chemists Society, 1973). As with other methods valid results can be obtained only if samples are reproducibly prepared in terms of melting out and recrystallisation before measurement.

Dilatometry is being superseded by NMR (nuclear magnetic resonance) due to the easier procedure and more rapid evaluation time for the NMR technique. Dilatometry, DSC and NMR have been assessed on a comparative basis (Walker & Bosin, 1971).

While measurement of the solids content of fats is of value in assisting in the formulation of products with certain required characteristics, and also provides guidance towards setting process parameters, the relationship between such values and product structure is not always a direct one. Nevertheless the incorporation of such measurements for in-line process control purposes can provide useful feedback criteria for establishing self-monitoring process routines.

Tests for Spreads Structure

Penetrometry

A common method for measuring the strength of the structure of spreads is by determination of the yield value. One convenient method is cone penetration. In this method a cone suspended on a shaft is released from a starting position with its tip on the surface of the sample. After release the cone penetrates the surface and its descent can be arrested, conveniently after 5 s. A practical equation relating yield value (c) to penetration (p) is:

$$c = k \frac{w}{p^{1.6}}$$

where w is the weight of the moving parts. The constant k depends on the angle of the cone. Convenient apparatus can be purchased for carrying out this test with preset facilities for stopping the cone and time of descent. Under a given set of conditions yield values can be correlated with observed penetration. It has been proposed that the apparent yield stress can be calculated from measurements made using penetrometry. Penetrometry is probably the most routinely used of the various methods for measuring physical texture.

Recent studies have indicated that for butter the yield stress can be used as an indicator of spreadability (Mortensen & Danmark, 1982).

The yield value in spreads has been found to be inversely proportional to the second power of the fat solid content (Haighton, 1976) and, although solids level is not a complete description of the hardness of a spread, clearly with many fat blends it is significant in terms of the ease of spreading.

Extrusion method
The spreadability of butter is often expressed by the pressure necessary to extrude the product through a nozzle. Measurements are made at constant extrusion speed and temperature using for instance the FIRA-NIRD extruder (Prentice, 1943).

Cutting or sectility method
In this, typically, the consistency of butter is measured by drawing a wire 25 mm long by 0·3 mm diameter through the butter sample at 15°C with a speed of 0·1 mm s^{-1}.

Oil exudation
The exuding of oil from spreads is thought to be brought about by a reduction in the volume of the fat crystal network due to the formation of primary and secondary bonds, and also by the influence of external pressure. The fat crystal network can show variability in ability to contain liquid oil, particularly in terms of the size and shape of crystals and the capillary forces at play. Oil exudation, which is undesirable, can be limited by attention to such factors as correct selection of fats (particularly hardstocks), controlling processing to produce smaller crystals, higher solids content values, lower ambient temperatures during transport, and, where relevant, lower stacking heights in stores. Empirical tests can be devised to measure the defect whereby the oil lost from specimen samples under controlled pressure is measured by absorption into preweighed paper, etc.

Microscopy
Both optical and electron microscopy have been used for identifying spreads structure. With optical microscopy magnification factors of 400 × and 1000 × are normally sufficient for quantifying emulsion drop size and the type of crystal network present. Care has to be taken when preparing sample slides to ensure minimum damage is done to the sample to achieve reliable results as especially with lower fat spreads the emulsion structure can be unstable under such conditions. It is not usually necessary to resort to staining techniques.

Electron microscopy (using for instance staining or freeze fracture techniques) is employed for recognising structural features in more detail such as the location of individual fat crystals, protein micelles and, under certain conditions, surface active materials. Confocal microscopy has the advantage of being relatively non-disruptive while allowing specimens to be examined to some tens of micrometres in depth. It is especially useful for studying less stable reduced fat spreads.

Sizing of emulsions
An accurate knowledge of the droplet size of creams prior to processing can be useful for predicting subsequent process conditions to ensure emulsion inversion to spreads. In the final product the distribution of water drop size will be an indicator of product stability in terms of spreading, emulsion breakdown on the palate, and microbiological shelf-life.

For creams, ordinary light microscopy using counting techniques when required is normally sufficient. However, several types of commercial analysers are available for sizing stable flowing emulsions using such techniques as light scattering and electrical capacitance.

In fat continuous products light microscopy is useful for obtaining a general impression of the range of drop sizes but suffers from the problem that the eye appears less sensitive to smaller drops even within the range of vision. A more recent technique has involved pulsed NMR where a measure is obtained of the mean free path of protons in the water phase (Packer & Rees, 1972; Callaghan & Jolley, 1983). Sizes are derived by comparison with the performance of model emulsions.

Electrical conductance
For lower fat products the tendency for the emulsion to revert to the water continuous state during processing has led to the use of conductivity as a measure of product stability. Using suitably designed cells measurements can be carried out either on the final product as an indicator of product quality, or during in-line processing, to guide the operator as to the state of the emulsion at significant points in the process. Values can vary from 10^{-3} to several thousand μS cm^{-1} (Unilever Ltd[2]).

Appearance of Spreads

Fat crystals, water droplets and occluded air bubbles all influence reflected light from the surface of, and indeed some depth into, a spread. In addition the larger the reflecting internal surface area the more matt, opaque and pale the appearance. The refractive index of water and air are sufficiently different from oil to alter the optical appearance. The CIE system can be used to describe colour in terms of standard components, and also the overall intensity or saturation (Wright, 1969). The onset tint of a sample can be described in terms of a dominant wavelength and the saturation.

In general winter butters are paler than summer butters at refrigerator temperatures. Differences between butter and vegetable fat products are in part due to the different minor components in the oils such as carotenes and chlorophylls. Lower fat spreads can be much paler and more matt in appearance than full fat products if the aqueous phase drop size is sufficiently small, due to the greater internal reflecting surfaces present. Larger drop sizes deepen the emulsion colour significantly.

Other factors in addition to colour which are used to describe the optical appearance are gloss and translucence. Gloss can be measured instrumentally as the amount of light reflected from the surface, which normally varies between 10 and 50%. In general butter shows low values and reduced fat spreads containing mainly water in the aqueous phase are only slightly higher. The more liquid oil in the blend and the larger the water drops the higher the measured gloss factor. The translucency of a spread is the proportion of light which can penetrate a thin (< 1 mm) sample. In general butter shows higher translucency values than other spreads. Dehydration of the surface of spreads is particularly relevant to low fat spreads and water continuous spreads (for instance processed cheese). The optical effect is to deepen the colour where it occurs.

Thus the optical appearance of spreads can be controlled using a combination of factors such as added colour, size and amount of fat crystals and water drops, and by the use of close contact between packaging material and product surface, sealed containers, etc., to restrict dehydration effects. Commercial apparatus is available to measure several of the optical properties of spreads.

Effect of Fats on Product Texture

Due to recrystallisation phenomena occasionally problems can arise in product texture. If crystallisation rates during processing are too high, lower melting mixed crystals can form which can melt at higher storage temperatures and from which higher melting components can recrystallise. Accompanying this, polymorphic transformation can take place whereby the more usual beta-prime form is changed to beta forms. Finally, by the Ostwald ripening effect, very small crystals can grow into larger ones. The overall result is that either large individual crystals can form some tens of micrometres in size, or spherulites of associated crystals may appear some hundreds of micrometres in diameter. The average human palate can distinguish particulates above about 25 μm in diameter, so products with such defects can have a rough unpleasant feel on the palate. Such effects during storage can result in both the product firmness and emulsion structure varying. The latter may give rise to problems in terms of microbiological safety.

Clearly some blend components should be avoided to minimise the risks of the above happening and processing conditions can be used to minimise the effect. Interesterification of at least part of the blend components can also be beneficial, and in general temperatures should be kept low to inhibit recrystallisation.

On some occasions, due to delayed crystallisation during processing, irregular shapes can develop in the packed product. Such problems can be reduced by using mesh screens towards the end of the process line.

Insufficient crystallisation during processing can lead to completion of the process, unstirred, in the packed product. This can give an unexpectedly hard and brittle result.

With lower fat products a too high fat solids development while processing may result in wet products due to 'overworking' and emulsion destabilisation.

Palate Behaviour of Spreads

The function of a fat spread is multiple and includes:

lubrication of bread when eaten,
energy source,
flavour carrier,
vitamin transport,
a source of essential fatty acids,
a coolness/taste contribution during eating,
provides product structure.

If the product is also intended for kitchen use other functions such as batter aeration, shortening power and heat transport in frying will be relevant. When warmed on the palate fat crystals melt, water droplets coalesce and the resultant emulsion is diluted by saliva. The process usually takes 5–12 s. The result is normally an emulsion inversion at least in part to an oil in water emulsion of sufficiently low viscosity to trigger the swallowing process. At the same time volatile flavours are released by migration from the oil phase to the aqueous phase and then into the oral cavity, whilst contact tastes, such as salt, impinge directly on the tongue (Fig. 5).

The various relevant properties that can be measured in the laboratory

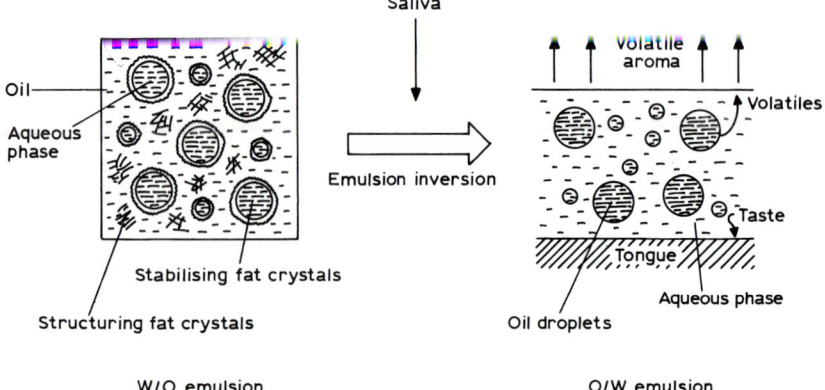

Fig. 5. Emulsion inversion on the palate.

include the quickness of the melt down/emulsion inversion process, the coolness of the product (as heat is withdrawn from the palate by the melting fat), and the viscosity of the final emulsion (problems can result from possibly incomplete inversion in the case of lower fat spreads, and/or residual fat crystals if their melting point is too high).

Although quickness can be measured experimentally, trained panellists are the most reliable method. One feature controlling the quickness of melting of a spread is the solid fat index at appropriate temperatures, and also the fat crystal size. In reduced fat spreads the use of water structuring/gelling agents also has a measurable effect on melt down.

The viscosity of the final emulsion on the palate can also be judged by panellists but meaningful values can be measured using a suitable viscometer operating at about 100 s^{-1} and 33–35°C. Usually values higher than about 170 mPa s are considered thick. The situation is complicated by the variable dilution effect of saliva, the presence of bread on the palate, etc.

The rate of emulsion destabilisation on the palate governs the rate of flavour release, in terms of both volatile flavour and contact tastes (such as salt). Ingredients should be selected so that the emulsion inversion process takes place below palate temperature and the corresponding rate of destabilisation of the product water in oil emulsion can be measured experimentally. Factors which affect emulsion stability include the amount and type of emulsifying and stabilising agents, the emulsion drop size, and the amount and type of fat crystals especially those protecting the water drops.

The coolness impression of spreads on the palate is a function of water content, emulsion structure, and the amount and heat of fusion of solid fat melting on the palate. Normally reduced fat spreads taste cooler than higher fat products of similar components due to the specific heat of the increased water content. The rate of release of such water is also a factor. In terms of the fat, the difference in solid fat index between selected appropriate temperatures can be used as an approximate measure of coolness. Values for milkfat, lauric fats and palm oil mid-fractions tend to be higher than other vegetable blends.

Other Functions of Spreads

The possible defects in spreads which have to be taken into consideration include spattering during shallow frying, oil exudation and emulsion stability at ambient temperatures. The relative importance of these factors varies with geographical conditions and local cooking habits. In some localities spreads may be used for shallow frying and minimal spattering is an important factor. In the distribution chain undue pressure in stacking may cause products to exude oil, and exposure to warm kitchens can have a similar effect. In tropical markets emulsion stability at high ambient temperatures has to be within acceptable limits, particularly for reduced fat spreads. Problems concerned with poor oil exudation or emulsion stability characteristics are best approached by controlling the percentage of high melting fat, selecting appropriate fats for the quite separate functions of emulsion stabilisation and product structuring, and the amount and type of emulsifier, as far as fat continuous products are concerned. Spattering problems are usually overcome by the use of lecithin based emulsifying systems which produce fine water in oil emulsions and allow the release of the aqueous phase as steam in a more gradual fashion.

HIGHER FAT SPREADS (PATENT REVIEW)

Structure Control

Improved butter-like properties in spreads can be brought about using esterification products of mono-diglycerides of carbon number between 16 and 20 with butyric acid, at less than a 20% inclusion level in the fat (Henkel KG Auf Aktion). The glycerides involved probably contain both long and short fatty acid chains, symmetrical and unsymmetrical in type. A similar approach involves esterifying fatty acids of carbon number 2–8 with higher fatty acid monoesters where up to 30 parts of the ester are added per 100 parts of a main fat blend. This is suggested as giving better palate melting properties and spreadability.

Control of the palmitic/stearic acid content of specific blends in the presence of C16–C24 unsaturated *trans*-fatty acids are related to improved cool melting properties, spreadability, mouthfeel and decreased graininess (Unilever NV[3]).

Softer oils can be produced by ester exchanging (for instance) palm and soya oil using immobilised enzymes from *Mucor miehei* at about 70°C in the absence of a solvent.

Particular types of triglycerides are claimed to enhance product performance in terms of spreadability and heat stability. For example fractions from palm oil with solid fat contents between 67 and 80% at 50°F and with defined proportions of trisaturated and disaturated/mono-unsaturated (both symmetrical and unsymmetrical) glycerides are said to be useful in this respect (Procter and Gamble[1]). A similar approach proposes esterifying long chain monoglycerides with short chain fatty acids.

Ester interchange of fats brought about using enzymes in the presence of a lipase activating agent such as a trihydric lower alcohol with an insoluble carrier is claimed to have certain advantages in terms of a shorter time for the process and thus minimal hydrolysis of the fats (Kao Corp.).

Interesterification can help to avoid the formation of fat grains following processing as exemplified by blends of Colza 38°/42° and palm oil 50° (mixed in a ratio of 60/40), processed at 80°C under nitrogen gas using sodium methoxide catalyst (Gerscheld). Graininess can also be inhibited by using behenic acid at a level between 40 and 60% based on the fat composition.

The structure of butter can be made smoother without the formation of large crystals on slow cooling if a system for vibrating the product at 45–55 Hz is used (Sibe Dairy Ind. Res.).

The blending of milkfat with other fats, mainly of a vegetable character, is now widely practised in spreads formulation. Enzyme catalysed ester-exchange between milkfat and vegetable oil has also been described (Kalo *et al.*, 1986).

Fresh milkfat for spreads can be obtained as part of a continuous process by precrystallising cream, breaking the emulsion by pressure, re-melting the separated grains together with selected vegetable fats and processing to a spread using added protein (Arla Ekonomisk Forening). Yet another method involves concentrating dairy creams to 80% fat, homogenising at 161°F and 1400 psi to invert to a water in oil emulsion, then mixing with vegetable oil in a ratio of 40:60 (Land 'O' Lakes Inc.[1]). Several other similar inventions concern homogenising vegetable fats with milk and churning to products (Land 'O' Lakes Inc.[2]). The maintenance of a granulated fat structure in blended spreads typical of churned butter has been referred to earlier (Olsson; Milk Marketing Board; Svenska Mejeriernas). It is claimed that softened butter (surface sterilised with alcohol) and margarine can be cold blended to preserve much of the original product structure (Kannegafuchi Chem. KK).

Emulsifiers

Selected emulsifiers are said to have specific effects in spreads. Apart from the generally accepted mono/diglyceride type, sucrose fatty acid esters of HLB greater than 11 have been claimed to give improved texture. Highly substituted sucrose fatty acid esters of HLB less than 3 at a level of 0·1–1·0% can contribute to a stable viscosity in emulsions containing high solid blends over considerable periods of time (Miyoshi Yushi KK).

Mono/diglycerides from rapeseed oil high in erucic acid are claimed to be efficient in high water content emulsions intended for bakery applications (Grindsted Products A/S). Polyglycerol esters of ricinoleic acid with a degree of polymerisation of 2–3 are said to be useful for stabilising water in oil spreads (Asahi Denka Kogyo[1]).

Double Emulsion Products

Double emulsion spreads of both water in oil in water and oil in water in oil types have been described in many inventions. Typically lecithin fractions are claimed to be useful in the former (Meiji Milk Products KK) and a mixture of sorbitan fatty acid esters in the inner oil phase, sucrose esters combined with polyglycerol esters in the water phase, and lecithin/monodiglycerides in the outer oil phase have been shown to be useful for stabilising the latter type (Snow Brand Milk Products[1]). Advantages can lie in selecting different fats for inner and outer oil blends (Unilever Ltd[3]).

Miscellaneous Developments

Convenient forms of handling fats have been described including pourable particles, fluidised margarine, and shortenings in particulate form (Procter and Gamble[2]). Fat spreads can be preserved against oxidation by using liposomes prepared from lecithin and unsaturated monoglycerides containing tocopherol (Larsson VK).

Helium gas has been used at small levels (3–15 volume %) to improve the spreadability of spreads with up to 45% of dispersed aqueous phase and is also claimed to improve frying properties (Unilever NV[4]).

Health Aspects

The nutritional properties of oils are receiving increasing attention, in particular the claimed beneficial dietary effects of specific unsaturated fatty acids. New eicosapentaenoic glycerides can be synthesised from sardine oil to give oils valuable in cooking or as constituents in blends (Nisshin Flour Mills KK).

Claims are made for spreads containing marine oils high in eicosapentaenoic acid and docosahexaenoic acid which are free of fish oil smell (QP Corp.) and blends of olive and corn oil are said to have cholesterol lowering effects. The effects claimed for EPA and DHA may be enhanced when combined with plant sterols (Asahi Denka Kogyo[2]). Cholesterol can be continuously removed from oil by using absorption on carbon (New Zealand Dairy Research Inst.).

Oils containing large amounts of gamma linolenic acid can be protected against oxidation using cyclodextrin or degradation products of starch which contain cyclodextrin (Nitto Electric Ind. KK).

A replacement for human milkfat in infant feeding can be made by maintaining at least 50% of the fatty acids in the 2-position of triglycerides as saturated, while randomly distributing mainly unsaturated fatty acids in the 1,3-positions, using a ratio specific enzyme lipase (Unilever Plc[1]). Improved digestibility in infant feeds can be achieved with quite conventional blends of for instance palm oil, olive, coconut, and corn oils plus lecithin (Wyeth J. Ltd). Vegetable fats very high in arachidonic and docosahexaenoic acids in a ratio of 2·5 to 1 plus high cholesterol concentrations are claimed to improve cell membrane functions in infants (Miln).

Fat Substitutes

The physical states of fats in spreads can include the bulk, an oil in water emulsion, double emulsion and a bicontinuous emulsion. In terms of functions fats provide structure, energy and taste (including creaminess), a carrier for flavours and vitamins, lubrication of bread during eating, a physiological feeling of satiety, and a heat transfer medium when used for shallow frying. As yet no fat substitute can adequately fulfil all these functions. Developers can, however, manipulate ingredients to partly replace fats from the point of view of structure, creaminess, flavour retention, body, and, with the ester/ether type of substitute, heat transfer characteristics.

In general there are four recognised methods for substituting fats. The first of these utilises materials with ester/ether structures which have physical properties very similar to normal fats. The best known of these are sucrose polyesters and polyglycerol esters, but others which have received attention include sterically hindered esters, polysiloxanes, glycerol di-ether monoesters and jojoba oil. Medium chain triglycerides also offer some calorie reduction. Some of this group, notably sucrose polyesters, reputedly contribute no calories to the diet by reason of their non-absorption in the digestive system. However, with this class of materials possible problems may exist to the extent that very small traces may be absorbed (which could imply long term toxicity studies), vital fat soluble ingredients such as vitamins and drugs may be lost with indigestible substitute, and it may not be certain that such waste substitutes can be adequately broken down in the environment. However, these materials readily lend themselves to be used in spread products either alone or blended with normal fats.

The remaining three methods for fat substitution are all based on modification of the structure of the aqueous phases. The first of these involves simply raising the viscosity and thus changing the flow characteristics on the palate. Typical well-known examples in spreads technology include the use of galactomannans, alginates, glucose polymers and caseinates. The second method involves carbohydrate based gelled systems. Polysaccharides such as carrageenan have been employed but commercial systems are available where dextrins or maltodextrins form the main structuring agent. These do contribute calories during digestion but give heat sensitive gels at about 25% dry matter, so offering only 4 kJ g^{-1} to the spread's total energy content.

Finally there is a method based on particle engineering. The principle here is to utilise particles of protein or polysaccharides, less than about 8 μm in diameter. Better known examples include the use of particles of milk proteins and/or egg albumin, vegetable zein protein, and polysaccharide coacervates. Again these materials are used at relatively low levels and calorie contributions are of the order of 5–6 kJ g^{-1}.

Several of the above approaches have been employed in formulating low calorie spreads and appropriate references are given in the relevant sections. Some specific examples of the non-digestible type of substitute are given below.

Sugar esters from base materials such as erythritol, xylitol, glucose, sucrose, and common fatty acids can replace fat at the 10–100% level, and are poorly absorbed during digestion (Procter and Gamble[3]).

Sucrose fatty acid polyesters, suitable as low calorie fats, can be prepared by transesterification of edible oil methyl esters with sucrose at 100–140°C in the absence of toxic solvents and without the formation of difficult to handle contaminants (Procter and Gamble[4]). Alternatively similar ingredients can be reacted at 75°C, but in the presence of water and isopropanol, giving a yield of 91% of hexa-, hepta- and octa-esters (Ethyl Corp.).

Another type of substitute which is claimed to inhibit hypercholesterolaemia is composed of non-absorbable polyol fatty acid polyesters having at least four fatty acid ester groups (Procter and Gamble[5]).

Dialkyl glyceryl ethers with a specification based on unsaturated and branched alkyl groups are claimed as fat substitutes which are not absorbed, and closely simulate the functional properties of the fats they replace (Swift Co.). Hydrogenated jojoba oil at a concentration greater than 10% on fat is only slightly absorbed during digestion and thus can act as a fat substitute (Société des Produits Nestlé[1]). Non-hydrolysable polyorganosiloxanes are also claimed to replace fats and oils in foods and such products are claimed to differ only in terms of minor visual appearance factors from the parent foodstuffs (Dow Corning Corp.).

Polyol fatty acid esters of polycarboxylic acids and fatty alcohols such as 8–22 carbon number fatty acid sucrose esters can give specific viscosity, yield point and thixotropy characteristics which are claimed to be unique (Procter and Gamble[6,7]).

Other examples of fat substitutes are dioleyl dihexadecyl malonate (Frito Lay Inc.[1]), glycerol tri-alpha-oleyl oxylaurate (Procter and Gamble[8]), hexadecyl dioleyl malonate (Frito Lay Inc.[2]), and 2,3,4 tri-acylhexose as extracted from the exudate of the lycopersicon penellii plant (Atlantic Richfield Co.).

Hollow gelatin spheres (less than 0·2 mm in size) containing air cannot be tasted as such in spreads on bread, and their tendency to solubilise in water can be reduced by cross-linking. They have been used to reduce calorie content (Schaefer).

LOW FAT SPREADS (PATENT REVIEW)

Fat Continuous Products

Processing
In terms of lower fat products, many disclosures are concerned with the manufacture of 45–60% fat low calorie butter spreads (Bahr *et al.*; Institut für Milchforschung der DDR). Caseinate solution may be added to 80% fat cream during phase inversion (Dairy Ind. Res. Des. Inst.[1]). Fat products (60%) can be made by first cooling cream in two steps at 1·5° and 5°C per minute with up to 15 min between each stage (Meat Cheese Ind. Res.), while 45% fat spreads can be prepared by intensive shearing of high fat cream after cooling to 15°C (UKR Meat Dairy Inst.). Non-churned (i.e. non-

inverted) low fat blended spreads can be made by homogenising high fat cream (80% fat) followed by mixing with vegetable oils (Land 'O' Lakes Inc.[1]) Double fractionated butterfat when made up into creams can be churned to a butter-like product (UKR Meat Dairy Ind.).

Creams may be converted to low fat spreads by concentration with cooling and mechanical working (FA Dr Oetker), also by controlled two stage chilling (Lithuanian Butter Cheese), and low pressure evaporation of water (Rieder).

For producing spreads with reduced levels of additives the continuous incorporation of cream into a fat in water emulsion until inversion takes place is said to require no emulsifiers and is amenable to processing in established apparatus (Balla[1]). Alternatively, the ripening step normally used before inversion of dairy cream can be eliminated by careful temperature control of the emulsions in a flowing loop system or intermediate container (Balla[2]).

Spreads can be produced by temperature conditioning and working of cream between 6°C and 17°C in two separate units followed by further processing in the beater of a buttermaking machine (VEB Komb. Fortschritt).

Reduced fat butters of about 60% fat content can be made by churning cream containing 43–45% fat after culturing (Bahr *et al.*), and by processing cream of 80–85% fat with caseinate solution (Dairy Ind. Res. Des. Inst.[2]).

Interesting developments include the use of deep cooling conditions ($-10°C$) to destabilise cream over 24–36 h before votating to a product with a thickened aqueous phase (Unilever NV[5]). Partly inverted cream can form the basis of a spread with an improved flavour release compared with normally churned products made by a controlled churning process monitored by the change in viscosity, the final product containing both a continuous and discontinuous aqueous phase (Unilever NV[6]). Hydrophobic surfaces on process units have been demonstrated to assist the inversion of dairy cream to a fat continuous spread which requires no additives (Unilever NV[7]).

The gradual addition of aqueous phase to fat under agitation is a way of retaining a water in oil structure. One variation involves pumping a water in oil emulsion through a cooling unit in a cyclical fashion, while adding further ingredients before each cycle. In this way the temperature is reduced from 24°C (max.) to 4°C (min.) in a stepwise fashion while the line pressure rises to between 100 and 200 psi (Corn Products Co.). The supply rate of the aqueous phase can be adjusted downwards once the amount added has reached certain critical amounts, and the speed of the mixing blades is maintained so that the power is 150–250 kg s^{-1} m^{-2}, so maintaining the water drops at 0·08–0·1 mm (Snow Brand Milk Products[2]).

Multistream processing involves the use of votator units in parallel fashion, with the final product comprising a blend of two water in oil emulsions, or a water in oil emulsion with an aqueous phase, cream, etc. It has been mainly used for reduced fat spreads development. Typical inventions in this direction include the incorporation of both gelled and ungelled aqueous phases in low fat spreads (Unilever NV[8]) and a process based on the handling of individual component phases with a blending only at the final (small volume) mixing stage. This is said to reduce waste of the product through minimised rework during packaging machine breakdown, etc. (Unilever NV[9]). Mixing of butter with other components often does not give the expected result particularly in terms of butter flavour, if the overall product fat level is less than about 50%.

Double emulsion low fat spreads

In recent years so-called duplex spreads have been developed. These mainly consist of emulsions which are of oil in water in oil type, in which the identity of the two separate oil phases can be different.

Sucrose fatty acid esters of HLB 10–17 can be used for stabilising the cream phase (50 fat: 50 water) of a duplex type water in oil low fat spread (Asahi Electrochem. Ind.). Lecithin in combination with such sucrose esters and a polyhydric alcohol fatty ester is said to produce an oil in water in oil emulsion having oil granules of 0·5–2 μm size. Alternatively dairy-like ingredients can be used for stabilising the inner oil phase (Lever Bros.[2]).

One invention concerns the inversion of a bimodal oil in water emulsion in which the intended fat continuous phase is initially incorporated as large oil drops in a cream, amenable to inversion during the process (Unilever NV[10]), and whey proteins can be employed as the stabilising agent for the cream phase (Unilever NV[11]). Smaller oil drops invert less easily during such processing and so remain as a discontinuous phase within the product, and thus release a pleasant creamy taste sensation during inversion on the palate.

Proteins

The use of low levels of protein in the aqueous phase of reduced fat spreads poses significant problems in terms of emulsion stability during manufacture, although the overall quality of products in terms of taste is improved.

The viscosity of the aqueous phase is of increased significance in reduced fat spreads and is normally raised to fairly critical limits for stabilising the products. However, the influence of milk proteins on destabilising the emulsion is also a major factor.

Inventions pertinent to blended products include the use of the viscosity raising properties of membrane-filtered skim milk, buttermilk or sweet whey, the process being carried out preferably above pH 5·8. Subsequent elevation of the temperature to 90–95°C also coagulates the whey protein. The protein concentrate is then treated with complexing agents such as calcium phosphates/calcium citrates, and processed at a concentration of 10–11% (on final product) with a mixture of butter oil and bean oil (Mjolkcentralen Ekonomisk Forening).

Enhanced viscosity in the aqueous phase of proteinaceous low fat spreads can be obtained with casein and calcium caseinate giving a viscosity of 3800 cP at pH 7·3 the spreads being prepared by mixing of butter oil and such aqueous phase at a temperature of 25°C (Société des Produits Nestlé[2]).

By carefully controlling the calcium ion to casein ratio, emulsions with fat levels of 10–60% with a very stable texture may be prepared. The possibility of raising the viscosity of low concentrations of protein solutions exists as for instance heat treatment can gel milk (Andrews, 1975). Controlled reactions of this type could offer substantial savings in spreads manufacture by offering high viscosity aqueous phases without the necessity for stabiliser usage.

Stabilisers

Hydrocolloids other than milk proteins can both modify the viscosity for product structure purposes, and also the palate breakdown and taste. The influence of such additives on possible binding of flavour components, and also on the emulsion dynamics affecting flavour release on the palate is clearly of significance. Many thickening agents have been used to stabilise reduced fat spread emulsions including sodium alginate, propylene glycol alginate (Unilever Ltd[1]), gelatin (Unilever Ltd[4]) and blends of carrageenan and locust bean gum (Unilever NV[12]). In some instances very specific selections of stabilisers with proteins are preferred (Unilever Ltd[5]).

Other materials useful for structuring the aqueous phase of low fat spreads include gelatin with polysaccharides such as xanthan and locust bean gum (Konin Brinkers Marg.) and carrageenan modified by the addition of a cation to control the gel properties (Unilever NV[13]). Instead of defining the stabiliser system by type, the required characteristic behaviour of the aqueous phase such as softening/melting behaviour has been described (Lever Bros.[3]).

Emulsifying systems
Problems of legislation and the nature identity concept suggest that broadening the chemical emulsifier base beyond what is currently used may be difficult. The use of monoglycerides in reduced fat spreads is well established. An alternative to these are a solid-particle stabilised (Pickering) system utilising mixtures drawn from hydrated phospholipids, fatty acids, propylene glycol esters and sorbitan esters (Spitzer *et al.*). Triglycerol monostearate, hexaglycerol distearate and decaglycerol distearate have been found to be useful for preparing oil in cold water emulsions suitable for low calorie foods resembling butter or margarine (Drackett Co.).

Erucic acid esters of glycerol or sorbitan are said to improve the processibility of water in oil type low calorie spreads without causing an unwanted increase in stability on the palate. Acetylated polyglyceride esters are also useful for stabilising low fat spreads.

Water in oil type emulsions produced from a specific emulsifier system, consisting of polyglycerol condensed ricinoleic acid ester and glycerol fatty acid ester, are stable for a long time in melted or liquid condition. The use of monoglycerides which are predominantly from unsaturated fatty acids have been recommended, the saturated portion being mostly from stearic acid. The blend is said to inhibit coalescence of water drops.

Combinations of monoglycerides, propylene glycol or sorbitan fatty acid esters with polyoxyethylene sorbitan monostearate or oleate have been employed for preparing soft spreads with a fat content of 10–30% having a gum stabiliser in the aqueous phase at a level of 0·1–3·0% (Standard Brands Inc.). Monoglyceride type emulsifier added to dairy cream at 70°C assists subsequent inversion in a votator to a low calorie butter provided a thickening agent is used, selected from alginate, gelatin or caseinate (Gay Lea Foods Corp.). Fat bridges can be generated between partly coalesced oil drops of hydrogenated palm kernel oil and sorbitan polyoxyethylene ester emulsifiers to produce spreadable structures from oil in water creams (Unilever NV[1]).

By using lipoprotein type emulsifiers (from animal or fish sources), oil in water emulsions can be churned to water in oil type products in a butter-making machine. The resultant products may be of the duplex type and contain discrete fat particles in the fat phase (Asahi Denka). Similarly protein–lipid complexes from a variety of vegetable, fish or egg sources, plus added surface active agents have been described for use in mayonnaises and spreads (Akad der Wissenschaft DDR).

Fat blends

Although proteins in reduced fat spreads can cause stability problems certain fats are said to facilitate the use of proteins. In these inventions 2–8% beta phase tending hardstocks with an iodine value below 12 are a feature. Fat crystal sizes are below 10 μm, and increased consumer appeal is said to result from a uniform solids content over a wide temperature range (Procter and Gamble[9]). Similarly a product of improved spreadability and stability incorporating protein is claimed by employing a blend with 5–30% of a beta tending hardstock containing triglycerides from C_{10} to C_{16} fatty acids.

Cream prepared by emulsifying high or low melting point butter oil fractions with a low fat milk fraction can be churned to prepare a butter-like product spreadable at refrigerator temperature (Bratland). Products of improved crystallisation properties with a tendency towards limited post-crystallisation can be obtained by randomising palm oil alone or together with small amounts of fish and whale oils.

Palm mid-fraction fats with a specified steep solids/temperature line can improve the inversion of low fat spreads on the palate giving enhanced flavour release. Characteristics of the blend are a difference in solid content between 10° and 20°C of not more than 5%, and a solids level at 35°C of less than 0·1%. The low fat spread has a measured emulsion phase inversion temperature (PIT) of 28°C (Unilever Ltd[2]).

Butterfat fractions, particularly butterolein (typically $N_{10}=36$, $N_{20}=6$, $N_{30}=0$), can be votated with sodium caseinate containing aqueous phases to give softer products than low calorie spreads from unfractionated milkfat (Unilever NV[14]). A similar invention based on mixing butterfat stearin with soft oil has also been disclosed (Unilever NV[15]). The use of polyglycerol fatty acid esters (mainly di-, tri- and tetraglycerol) at a level of 0·5–3% on fat can reduce the tendency to graininess formation in spreads with blends of the SUS:SSU type (Unilever NV[16]).

Microbiological stability

Problems with low fat spreads relating to shelf-life are more pronounced than for higher fat spreads. Usually with the fat continuous type this is on account of larger aqueous phase droplets, or even because of water continuous regions being present. As a general rule droplets should be as far as possible below 10 μm diameter to retard the development of spoilage micro-organisms, but in practice drops or lakes of 40–80 μm can be detected in commercial products. Recognised aids to increasing shelf-life include lower pH values (<5), significant levels of salt, and the presence of certain organic acids. Sorbates and benzoates are often used to retard mould spoilage.

One patented approach has involved clustering effective levels of inhibiting agents together with vulnerable materials such as proteins in a relatively few of the total drops of aqueous phase in the product, so raising the effective level of the agents in a localised sense but without affecting the overall flavour balance of the product. In this way lower levels of usage of such agents can be achieved (Unilever Plc[2]).

Methods for extending shelf-life could be based on for instance naturally occurring lactoperoxidase systems.

Water Continuous Products

The water continuous spreads market is rather disparate, being divided up into fresh cheeses, processed cheeses, peanut butter, yellow spreads, spreading mayonnaise and sweet flavoured spreads based on chocolate, etc. Products tend to have advantages and disadvantages compared with fat continuous spreads. Advantages include a more rapid flavour release, relatively simple and cheap processing and a texture less dependent on fat type. Disadvantages could be shorter microbiological keeping times when opened, often more expensive packaging (using sealed containers for instance), and possible dehyration problems when opened.

Nevertheless many different types of water continuous spreads have been described whose spread-like properties are based on the use of agents which structure the water phase such as polysaccharides and enhanced levels of water binding milk proteins, or mechanisms which promote the bridging of oil drops by protein molecules or fat crystals, and also by the part coalescence of oil drops.

Processing

In general reduction in the oil drop size of creams can raise the product viscosity to a spreadable consistency if the fat content is high enough or water binding hydrocolloids are present. A process including the steps of fractionation of cream followed by dehydration of the low fat fraction, addition of the concentrate to the high fat fraction, and homogenisation of the mixture is described as giving a spreadable consistency (Molkerei Zentralen Westfalen[1]).

A water continuous spread containing 41% of milkfat together with other milk components has been prepared by homogenising the mixture at 350 bar and 25°C and cooling the subsequent product to 12°C before packaging (Ruhoff). Agents for further raising the viscosity include milk protein and gelatin to give a product containing 39–41% butterfat (Molkerei Zentralen Westfalen[2]).

Non-dairy emulsions which could form the basis of low fat spreads can be prepared by intensive mixing at 10 000 rev min^{-1} of sunflower oil in soured skim milk, followed by centrifugation to give a 50% fat and 6% protein containing product (Kroll). Other products can be made by stabilising cream from lard, beef fat or hydrogenated oil, using phospholipids from buttermilk and a double homogenisation process (Ulich Butter and Cheese).

Proteins
A very much higher amount of protein can be incorporated into water continuous spreads than fat continuous spreads, within the limits of palatability and spreadability. Fat contents can be comparatively low, and soluble protein concentration in water can be as high as 18% or more. This offers the possibility of better balanced products in a nutritional sense. Up to 65% of vegetable fat can be stabilised by a blend of whey protein and vegetable proteins after homogenisation (Ralston Purina).

Either butterfat or vegetable fats can be blended with milk protein (ultrafiltered) and heat-precipitated buttermilk protein and water to give a final product containing 40% fat and 12% non-fat solids into which gas can be injected before cooling (Alfa Laval Berg Eisen). A product structured mainly by ultrafiltered buttermilk with homogenised fat gives a rich smooth taste on the palate (Alfa Laval).

Stabilisers
Many inventions have been published describing the use of thickening agents in dairy creams or dairy analogue creams for producing spreads often with a mayonnaise-like texture. The agents used have included xanthan, seed gums, cellulose and starch derivatives, etc., and, in addition, cellulose fibres. Inventions involving the use of cellulose derivatives include producing vegetable oil based butter-like spreads by the use of cellulose crystallite aggregates (American Viscose Corp.), and the use of swellable aerated gels from cross-linked gelatin with cellulose added (Battista).

Fibrous cellulose of less than 40 μm length added at a level of 10–85% on product weight has been used in combination with low levels of gum (at least 0·1%) selected from guar, carrageenan or locust bean, etc., and 1–50% of a gelling agent, e.g. gelatin, alginate, agar, giving a base for low calorie spreads with a fat content of 50–60% (Contrelle Kitchens).

A vegetable oil based low fat spread can be stabilised by 0·25–1% hydroxypropyl methyl cellulose and 0·25–1% of a blend of microcrystalline cellulose plus sodium carboxymethyl cellulose at a ratio of 8:1. The final viscosity is stated to be above 100 000 cP at between 2 and 32°C (SCM Corp.[2]).

Xanthan gum has been used as the major thickening agent together with minor amounts of gum arabic, locust bean gum or methyl cellulose in a 10% coconut fat product, also at a level of 0·2–2% together with lesser amounts of guar, tragacanth or tamarind for preparing transparent stable mayonnaise type spreads.

Vegetable gums at an amount equivalent to 0·15–0·5% of product weight can be dissolved in sweet cream (optionally with vegetable oil added) to give a spread-like texture after homogenising (Better Spreads Inc.), whereas 0·3% gums consisting of alginate:locust bean:guar gum in a ratio of 2:1:1 are preferred for a solid product from dairy cream with a final fat content of about 30% (Graves Stambaugh Corp.). Gelatin and sodium carboxymethyl cellulose are preferred for a 25–40% milkfat product (Hall Sandford). Anhydrous butterfat can be reconstituted into a xanthan:guar gum:sodium alginate blend of thickening agents in a ratio of 3·5:71·5:25 in water at 50°C, resulting in a mayonnaise-like product with a smooth texture (Merck & Co.).

Emulsifying systems

There are many emulsifier options for stabilising oil in water type spreads such as high HLB non-ionic or lecithin types; however, the proteinaceous variety offers the added benefit of improved mouthfeel. In addition functions have been claimed such as cold water active emulsifiers for preparing spreads from dry ingredients in a domestic kitchen.

Fat blends

In terms of fats with particular characteristics beta phase solid triglycerides are said to produce oleaginous gels of uniform consistency on quickly cooling from 55° to 7°C. When dispersed in water at pH 2–6·5 as particles having a diameter of 10 μm or less at a level of 40–50%, spreadable products result (Procter and Gamble[10]). Fat replacer systems of the oil in water type can be based on structured particles of an appropriate size prepared from proteins (Labatt Ltd) or polysaccharides (Unilever NV[17]).

VERY LOW FAT SPREADS

Products with Less than 40% Fat

Very low fat spreads can be of the fat or water continuous or even bicontinuous emulsion type. Fat continuous products with less than 10%

fat are possible provided suitable methods of preparation appropriate to high internal phase emulsions are used. One process involves the production of emulsions suitable for low fat spreads with more than 75% aqueous phase (Petrolite Corp.). Levels of fat between 15% and 25% in fat continuous spreads are claimed with aqueous phases consisting of water only (Unilever Ltd[6]). Very low fat products of the fat continuous type have special processing problems. Due to the comparatively low level of liquid oil present adequate separation of the water droplets is increasingly difficult as the water/oil interface expands with higher water contents. Consequently fat blends are chosen with low solids levels, and short process lines adopted to avoid overworking the relatively unstable emulsion. The disruptive effect of scraped wall heat exchangers may be minimised by using very cold water in the formulation which has the advantage of assisting the crystallisation of the fat. This relates to the now outdated method of making margarine using ice crystals. Pregelation of the aqueous phase prior to emulsification is also possible and this assists in stabilising the water in oil emulsion by limiting back coalescence and the appearance of visible water in the product. Ultimately methods could be available for simply distributing small particles of structured water in the fat phase to make the product, and the size and shape of such particles could have a significant influence on the properties of such products. Products with distributed water in the region of 80% or more clearly offer the possibility of controlling overall texture and performance by means of the aqueous phase. In addition it is often preferable to structure the water to enhance emulsion stability with fat continuous products.

Viscosity enhancement in the aqueous phase can be grouped into four broad types:

1. A homogeneous gel system using, for instance, carrageenan (Unilever NV[18]).
2. Phase separating systems based on, for example, gelatin/maltodextrin (Unilever NV[19]).
3. Synergistic viscosity using components such as milk protein with modified starch (St Ivel Ltd[1]), monoglycerides (Laude Int. Sweden AB), or gelatin (St Ivel Ltd[2]).
4. One component viscous systems such as starch hydrolysates (Grain Processing Corp.).

In addition systems which involve suspensions of structured aqueous particles in an aqueous medium based on single polysaccharides or coacervates of gelatin/gum arabic (Unilever NV[17]), or heat denaturable milk proteins (Labatt Ltd), may also be used. Phase separating mixed biopolymer systems are potentially very useful for application in the area and could provide interesting rheological/melting characteristics (Clarke, 1987).

Very low fat water continuous products offer some advantages over the fat continuous type in terms of even lower fat content (well under 10%), and higher protein levels. Examples may be based on phase separating biopolymer systems (Unilever NV^2), microfragmented particles of proteins (Labatt Ltd) or polysaccharide/protein complexes (Kraft Inc.). Bicontinuous systems of fat and water may also provide interesting textures suitable for very low fat spreads and can offer a combination of the characteristics of both fat and water continuous types of product emulsion.

THE FUTURE

Nutritionally Adapted Spreads

Although legislation in terms of standards of identity may well limit the marketing of specific products in the future, nevertheless certain technical developments are in progress. It is clear that with increasing emphasis on healthy eating, and a drive towards 'positive nutrition' rather than diets which only respond to impending or actual health problems, greater attention will focus on the biochemical impact of the fats consumed. Spreads constitute a significant portion of visible fat intake and will be a vehicle for controlling the amount and type of fatty acids in the diet.

Reduced intake of saturated fatty acids has been widely recommended. Polyunsaturated fats are well established in spreads, especially those containing omega-6 (linoleic) fatty acids. Omega-3 fatty acids differ in structure, metabolism, physiological and biochemical effects (Kinsella, 1988) and apart from being claimed at least as effective in lowering serum lipid levels as omega-6 acids are said to have additional benefits in terms of reduced blood pressure, certain arthritic conditions, etc. An increase in omega-3 acids intake with respect to omega-6 has been suggested (Lands, 1986). Products with a balance of these two constituents, containing blends of fish and vegetable oils, could well find a market. Omega-9 monounsaturated fats may also be of benefit in terms of reduced risk of heart disease (Grundy, 1987).

Adverse comments have been expressed in terms of *trans*-fatty acids especially those from non-natural sources (i.e. *trans/trans*-acids from hydrogenation processes). Hydrogenation of edible oils will probably decline, in addition to interestification processes. Suitable hardstocks for use in spreads will probably be obtained by improved fractionation or controlled enzymatic synthesis.

Technical developments will enable even more of the functions and properties of current 80% fat products to be matched in lower fat variants. This will result in spreads with fat contents between 5 and 65% becoming the norm, to the eventual exclusion of higher fat products. Water continuous spreads will come to occupy a significant market niche as taste and keepability problems are resolved and the benefits of their very low fat contents appreciated. More products based on fat replacers can also be anticipated, probably originating in the USA.

Butter and Milk/Vegetable Fat Products

Many attempts have been made to produce refrigerator temperature spreadable butter by means of blending winter and summer butters, churning fractionated milkfat, improvements in the churning procedure, mechanical working, and gas incorporation into the product. Only moderately successful results have been achieved so far. Milkfat can be interesterified with oleic and linoleic acids using specific enzymes for churning to spreads with nutritional appeal, but flavour will be impaired. Milk from herds fed protected oil has also been used for butter production. Convenience forms of butter, for instance granulated with maltodextrins, have appeared.

Cholesterol levels in butter have been lowered using supercritical carbon dioxide extraction and low fat variants of butter are available but can lack the full flavour impact of the parent product. Efforts to improve this situation will also benefit spreads from blended milkfat and vegetable fat. Such products can be made by both churning and scraped wall heat exchanger (including emulsion inversion) routes. Other factors which can contribute to the quality of butter flavour in the final product include the level of fat, the concentration of milkfat in the blend, and the ratio of water in oil to oil in water emulsions which exist within the product. Further work will undoubtedly see improved products of this type emerging.

Multifunctional Spreads

Current low fat products cannot be reliably used for baking and shallow frying purposes. Poor aeration performance, however, could be improved by the judicious use of selected emulsifying systems. In addition, additives such as humectants and viscosity raising agents may assist in modifying the imbalance to recipes that low fat spreads contribute. Such approaches will accelerate the eventual substitution of margarine and butter by reduced fat spreads.

Natural Spreads
A preference for products with a reduced 'processed' content among some consumers is reflected in a trend towards shorter ingredients lists and the use of fats which have not undergone hydrogenation or interesterification processing. Better utilisation of the residual levels of emulsifiers in natural oils and the controlled development of partial glycerides during refining will assist in excluding such materials as added ingredients. High viscosity forms of milk protein may obviate the need for additional thickening agents. Reduced refining operations may result in products carrying a certain level of natural flavours such as olive or peanut oil.

Novel Textured Spreads
The high proportion of fat globules in churned butters contribute a texture and mouthfeel notably different from, for instance, soft margarines. Globular fat can be included in spreads from both milk and vegetable fats. Similar effects could be introduced into both fat and water continuous spreads using particularised gelled water. The structuring of the aqueous phase of spreads will be receiving increasing attention in terms both of giving plastic fat-like consistencies, and in other particulate forms, for simulating the taste/flow properties of fats. The shaping of gelled aqueous particles under flow within an aqueous phase of different composition can be achieved by careful control of rates of setting, interfacial tension, etc. This could offer interesting possibilities for alternative chewing sensations in spreads.

Processing
The process operations in spreads manufacture include mixing, cooling, crystallisation, emulsion phase stabilisation, texturisation and filling. Future developments will give better independent control of these operations particularly in respect of fat crystallisation, water phase viscosity manipulation (for instance gelling at the most efficient point in the process line), and emulsion phase stabilisation. Improved control over the process of emulsion inversion to stable products will result.

Lowered energy consumption can be achieved by substituting simpler heat exchangers for scraped wall units, especially for the processing of lower viscosity oil in water emulsions prior to inversion. Microfiltration will be adopted where suitable for low temperature pasteurisation of ingredients with delicate flavours. Flexibility of process lines, with one line having the capacity to handle products with different fat contents and pack sizes, will help reduce costs.

Research and Development

The trends outlined above will only come to fruition if supported by development programmes with well-defined objectives. Typical areas for research include the influence of non-fat ingredients on flavour impact, and deeper insight into the rheology of aqueous phase ingredients under the shear fields occurring during processing, spreading and on the palate.

More work is needed on understanding the role of fat in baking if spreads with truly multifunctional characteristics and desirable nutritional qualities at low fat contents are to become a commercial reality. Different methods of food preparation, such as microwave heating, will also have an influence on the type of fat product used. Overall a better understanding of the factors controlling emulsion stability in a product environment, more refined analysis of factory unit operations, and accurate correlations of identified consumer preferences with product properties and the impact of the process, all require further long term background study.

The rapid changes in the yellow fats market seen over the more recent past will continue as competitors introduce increased manufacturing efficiency combined with well-timed and appealing new products.

REFERENCES

AHLBORN, E., GMBH. German Fed. Rep. Patent 252 4452.
AKAD DER WISSENSCHAFT DDR. German Dem. Rep. Patent 150539.
ALFA LAVAL BERG EISEN. German Fed. Rep. Patent 2245814.
ALFA LAVAL. Euro. Patent 0082 105.
AMERICAN OIL CHEMISTS SOCIETY (1973) *Official and Tentative Methods*, CD10–57, Illinois.
AMERICAN VISCOSE CORP. US Patent 3023104.
ANDERSON, A. J. C. & WILLIAMS, P. N. (1965) *Margarine*, Pergamon Press, Oxford.
ANDREWS, A. T. (1975) *J. Dairy Res.*, **42**, 89.
ANON. (1969) *Svenska Mejeritildningen*, **61**, 438.
ARLA EKONOMISK FORENING. Euro. Patent 185 631 A.
ASAHI DENKA. GB Patent 2080325.
[1]ASAHI DENKA KOGYO. Jap. Patent 6 1015 646 A.
[2]ASAHI DENKA KOGYO. Jap. Patent 6 1015 647 A.
ASAHI ELECTROCHEM. IND. US Patent 391 7859.
ATLANTIC RICHFIELD CO. Euro. Patent 194 154 A.
BAHR, N. *et al.* German Dem. Rep. Patent 939 18.
[1]BALLA, A. German Dem. Rep. Patent 131 714.
[2]BALLA, A. German Dem. Rep. Patent 131 339.
BALLA, A. German Dem. Rep. Patent 131 714.
BATTISTA, O. A. US Patent 440 1682.
BETTER SPREADS INC. GB Patent 1102928.

BRATLAND, A. GB Patent 2087211.
BUCHHEIM, W. (1970) *XVIII Int. Dairy Congress*, **IE**, 73.
BUTTER AND CHEESE INDUSTRY. German Fed. Rep. Patent 2500 159.
BUTTER CHEESE INDUSTRY RESEARCH. Soviet Union Patent 401 338.
CALLAGHAN, P. & JOLLEY, K. (1983) *J. Coll. Interf. Sci.*, **93** (2), 521.
CANTABRAMA, F. & DEMAN, J. M. (1964) *J. Dairy Sci.*, **47**, 32.
CLARKE, A. H. (1987) *Food Structure and Behaviour*, Academic Press, New York, p. 13.
CONTRELLE KITCHENS. German Fed. Rep. Patent 2709297.
CORN PRODUCTS CO. GB Patent 118 72 32.
CREAMERY CHEESE INDUSTRY. Soviet Union Patent 495 064.
[1]DAIRY IND. RES. DES. INST. Soviet Union Patent 427 692.
[2]DAIRY IND. RES. DES. INST. Soviet Union Patent 888 449.
DE MAN, J. M. (1961) *J. Dairy Res.*, **28**, 117.
DEMAN, J. M. (1963) *Milchwissenschaft*, **18**, 67.
DE MAN, J. M. (1969) *J. Text. Stud.*, **1**, 109.
DE MAN, J. M. & WOOD, F. W. (1959a) *J. Dairy Res.*, **26**, 17.
DE MAN, J. M. & WOOD F. W. (1959b) *XV Int. Dairy Congress*, **2**, 1010.
DE MAN, J. et al. (1976) *Rheology in Food Quality*, AVI, Westport, CN.
DIXON, B. D., CRACKNELL, R. H. & TOMLINSON, W. (1980) *Austr. J. Dairy Technol.*, **35**, 43.
DOLBY, R. M. (1953) *J. Dairy Res.*, **20**, 201.
DOW CORNING CORP. Euro. Patent 205 273 A.
DRACKETT CO. US Patent 3936391.
ETHYL CORP. GB Patent 2061941.
FA DR OETKER, A. German Fed. Rep. Patent 3324 821A.
[1]FRITO LAY INC. US Patent 4582 927 A.
[2]FRITO LAY INC. US Patent 4673 581 A.
GAY LEA FOODS CORP. US Patent 4307125.
GERSCHELD, D. French Patent 2 570 388 A.
GRAIN PROCESSING CORP. US Patent 4 615 892.
GRAVES STAMBAUGH CORP. US Patent 3314798.
GRINDSTED PRODUCTS A/S. Res. Disclosure 271 007 A.
GRUNDY, S. M. (1987). *Am. J. Clin. Nutrit.* **45**, 1168.
HAIGHTON, A. J. (1959) *J. Am. Oil Chem. Soc.*, **36**, 345.
HAIGHTON, A. J. (1963) *Fette Seifen Anstr.*, **65**, 479.
HAIGHTON, A. J. (1965) *J. Am. Oil Chem. Soc.*, **42**, 27.
HAIGHTON, A. J. (1976) *J. Am. Oil Chem. Soc.*, **53**, 397.
HALL SANDFORD AND CO. French Patent 2005794.
HANNEWIJK, J. & HAIGHTON, A. J. (1957) *Neth. Milk Dairy J.*, **11**, 304.
HEERTJE, I., VAN EENDENBERG, J. & CORNELISSEN, J. M. (1988) *Food Microstruct.* **7**, 189.
HENKEL KG AUF AKTION. Euro. Patent 37 017.
HERNQUIST, L. & ANJOU, K. (1983) *Fette Seifen Anstr.*, **2**, 64.
HUSSIN, A. B. & POVEY, M. (1984) *J. Am. Oil Chem. Soc.*, **61**, 560.
INSTITUT FÜR MILCHFORSCHUNG DER DDR. German Fed. Rep. Patent 2241 236.
JURIAANSE, A. C. & HEERTJE, I. (1988) *Food Microstruct.*, **7**, 181.
KALO, P., VAARA, K. & ANTILA, M. (1986) *Fette Seifen Anstr.*, **9**, 362.

KALO, P., KEMPINNEN, A. & ANTILA, M. (1987) *J. Am. Oil Chem. Soc.*, **64** (9), 1263.
KANNEGAFUCHI CHEM. KK. Jap. Patent 5 9130 135 A.
KAO CORP. Jap. Patent 6 20 61 591 A.
KAUKARE, V. & ANTILA, V. (1974) *XIX Int. Dairy Congress*, **1E**, 671.
KINSELLA, J. A. (1988) *Food Technol.*, October, 124.
KNOOP, E. (1965) *Kieler Milchwirtschaftlicke Forschungsberichte*, **17**, 73.
KONIN BRINKERS MARG. GB Patent 2205 849 A.
KRAFT INC. Euro. Patent 0340 035 A2.
KRASHENININ, R. F. *et al.* German Dem. Rep. Patent 99 720.
KROLL, J. German Dem. Rep. Patent 106777.
LABATT LTD. Euro. Patent 0250 623 A1.
[1]LAND 'O' LAKES INC. US Patent 4447 463 A.
[2]LAND 'O' LAKES INC. US Patent 4425 370 A.
LANDS, W. E. N. (1986) *Nutrit. Rev.*, **44**, 189.
LARSSON VK. Dutch Patent 8402 867 A.
LAUDE INTERNATIONAL SWEDEN AB. Euro. Patent 0312 514 A2.
LENINGRAD REFRIG. IND. TE. Soviet Union Patent 1163 815.
LENINGRAD SOVIET TRADE. Soviet Union Patent 1292 701 A.
[1]LEVER BROS CO. US Patent 4540 593.
[2]LEVER BROS. CO. US Patent 4071634.
[3]LEVER BROS CO. US Patent 4414236.
LITHUANIAN BUTTER CHEESE. Soviet Union Patent 1178 388.
MEAT CHEESE IND. RES. Soviet Union Patent 618 089.
MEIJI MILK PRODUCTS KK. Jap. Patent 6 1174 938 A.
MERCK AND CO. INC. Euro. Patent 45158.
MILES, C. A., FURSEY, G. A. T. & JOWS, R. C. D. (1985) *J. Sci. Food Agric.*, **36**, 215.
MILK MARKETING BOARD. Euro. Patent 0 106 620.
MILLS, B. L. & VAN DER VOORT, F. R. (1981) *J. Am Oil Chem. Soc.*, **58**, 776.
MILN. Euro. Patent 231 904 A.
MITSHUBISHI HEAVY IND. KK. Jap. Patent 5 9041 396 A.
MIYOSHI YUSHI KK. Jap. Patent 5 9042 842 A.
MJOLKCENTRALEN EKONOMISK FORENING. Belgian Patent 832 934.
[1]MOLKEREI ZENTRALE WESTFALEN. Belgian Patent 809469.
[2]MOLKEREI ZENTRALE WESTFALEN. German Fed. Rep. Patent 2600028.
MORAN, D. P. J. (1963) *J. Appl. Chem.*, **13**, 91.
MORTENSEN, B. K. & DANMARK, H. (1982) *Milchwissenschaft*, **37**, 530.
NEDERVEEN, C. J. (1963) *J. Coll. Sci.*, **18**, 276.
NEW CORRESPONDENCE POLY. Soviet Union Patent 1206 662 A.
NEW ZEALAND DAIRY RESEARCH INST. Euro. Patent 0174848.
NISSHIN FLOUR MILLS KK. Jap. Patent 6 2153 249 A.
NITTO ELECTRIC IND. KK. Jap. Patent 6 2084 041 A.
OLSSON, I. T. H. GB Patent 12 17 395.
PACKER, K. & REES, C. (1972) *J. Coll. Interf. Sci.*, **402**, 206.
PETROLITE CORP. GB Patent 2194 166 A.
PRECHT, D. (1980) *Fette Seifen Anstr.*, **82**, 142.
PRECHT, D. & BUCHHEIM, W. (1979) *Milchwissenschaft*, **34**, 745.
PRECHT, D. & PETERS, K. (1981) *Milchwissenschaft*, **36**, 673.
PRENTICE, J. H. (1943) *Lab. Pract.* **3**, 186.

¹PROCTER AND GAMBLE CO. US Patent 4447 462 A.
²PROCTER AND GAMBLE CO. Euro. Patent 106 393 A.
³PROCTER AND GAMBLE CO. US Patent 360 0186.
⁴PROCTER AND GAMBLE CO. US Patent 396 3699.
⁵PROCTER AND GAMBLE CO. US Patent 4034083.
⁶PROCTER AND GAMBLE CO. Euro. Patent 233 856 A.
⁷PROCTER AND GAMBLE CO. Euro. Patent 236 288 A.
⁸PROCTER AND GAMBLE CO. US Patent 4582 715 A.
⁹PROCTER AND GAMBLE CO. GB Patent 1124084.
¹⁰PROCTER AND GAMBLE CO. US Patent 3425843.
PROVISION MACHINERY RESEARCH. Soviet Union Patent 599 788.
PROVISION MACHINERY CONSTRUCTION. Soviet Union Patent 441 898.
QP CORP. US Patent 4764 392.
RALSTON PURINA. US Patent 3843828.
RIEDER, J. German Fed. Rep. Patent 3304 108A.
RUHOFF, K. German Fed. Rep. Patent 2545607.
SAMUELSSON, E. & VIKELSOE, J. (1971) *Milchwissenschaft*, **26**, 621.
SCHAEFER, P. German Fed. Rep. Patent 2701361.
¹SCM CORP. US Patent 4391 838 A.
²SCM CORP. US Patent 4284655.
SIBE DAIRY IND. RES. Soviet Union Patent 1050 637 A.
SMOLUCHOWSKI, M. V. (1917) *Z. Phys. Chem.*, **92**, 129.
¹SNOW BRAND MILK PRODUCTS. Jap. Patent 6 1074 540 A.
²SNOW BRAND MILK PRODUCTS. Jap. Patent 71843 183.
¹SOCIÉTÉ DES PRODUITS NESTLÉ. Euro. Patent 673 58.
²SOCIÉTÉ DES PRODUITS NESTLÉ. Belgian Patent 862 264.
SPITZER, J. G. *et al.* US Patent 336 0378.
¹ST IVEL LTD. Euro. Patent 256 712 A.
²ST IVEL LTD. Euro. Patent 2 208 29 AA.
STANDARD BRANDS INC. Belgian Patent 886241.
SVENSKA MEJERIERNAS. Euro. Patent 0 155 246.
SWIFT CO. Canadian Patent 1106681.
TIMMS, R. E. (1978) *Austr. J. Dairy Technol.*, **33**, 130.
TIMMS, R. E. (1979) *Austr. J. Dairy Technol.*, **34**, 60.
TIMMS, R. E. (1984) *Prog. Lipid Res.*, **23**, 1.
TIMMS, R. E. & BLACK, R. G. (1978) Dairy Res. Report No. 22, CSIRO.
UKR MEAT DAIRY IND. Soviet Union Patent 468 319.
UKR MEAT DAIRY INST. Soviet Union Patent 961 634.
ULICH BUTTER AND CHEESE. Russian Patent 793554.
¹UNILEVER LTD. British Patent 1094268.
²UNILEVER LTD. British Patent 1564 801.
³UNILEVER LTD. GB Patent 1091593.
⁴UNILEVER LTD. British Patent 1564800.
⁵UNILEVER LTD. GB Patent 1450269.
⁶UNILEVER LTD. GB Patent 2035 360B.
¹UNILEVER NV. Euro. Patent 0 059 510.
²UNILEVER NV. Euro. Patent 0298 561 A2.
³UNILEVER NV. Euro. Patent 186 244 A.

[4]UNILEVER NV. Euro. Patent 285 198 A.
[5]UNILEVER NV. GB Patent 208 12 94.
[6]UNILEVER NV. Euro. Patent 986 63.
[7]UNILEVER NV. Euro. Patent 986 64.
[8]UNILEVER NV. Euro. Patent 11891.
[9]UNILEVER NV. Euro. Patent 101104.
[10]UNILEVER NV. Euro. Patent 76548.
[11]UNILEVER NV. Euro. Patent 76549.
[12]UNILEVER NV. Belgian Patent 848 067.
[13]UNILEVER NV. Euro. Patent 271 132 A.
[14]UNILEVER NV. Euro. Patent 62938.
[15]UNILEVER NV. Euro. Patent 63389.
[16]UNILEVER NV. Euro. Patent 70080.
[17]UNILEVER NV. Euro. Patent 0011345.
[18]UNILEVER NV. Euro. Patent 279 499 A.
[19]UNILEVER NV. Euro. Patent 237 120 A.
[1]UNILEVER Plc. Euro. Patent 209 327 A.
[2]UNILEVER Plc. Euro. Patent 0101105.
VEB KOMB. FORTSCHRITT. German Dem. Rep. Patent 143 203.
VEB NAHRUNGSGUTERM. German Dem. Rep. Patent 221 626A.
VEG NAHRUNGSGUTERM. German Dem. Rep. Patent 211 268 A.
WADDINGTON, D. (1986) *Analysis of Oils and Fats*, Elsevier Applied Science, London, p. 341.
WALKER, R. C. & BOSIN, W. A. (1971) *J. Am. Oil Chem. Soc.*, **48**, 50.
WALSTRA, P. & VAN BERESTEYN, E. C. H. (1975) *Neth. Milk Dairy J.*, **29**, 35.
[1]WESTFALIA SEPARATOR AG. German Fed. Rep. Patent 3039 807.
[2]WESTFALIA SEPARATOR AG. GB Patent 2091 912.
WRIGHT, W. D. (1969) *The Measurement of Colour*, Hilger, Bristol.
WYETH J. LTD. Euro. Patent 129 990 A.

6

Fats in Bakery and Kitchen Products

J. Podmore
Research & Development Centre, Pura Food Products Ltd,
Merseyside, UK

SUMMARY

Fats have been used for many years in food preparation to provide structure, flavour and nutritive value. Climate and agricultural practices have influenced the fat used in food preparation, e.g. northerly climates using plastic fats like butter and suet, whilst in more southerly climates liquid oils, like olive oil, are more popular.

Economic conditions and population growth led to the invention of margarine as a substitute for butter, and developments in refining technology and fat modification techniques allowed the use of a widening range of fats in margarines, and as alternatives for lard and suet.

Progress in the technology of butter production, oil refining and modification, and margarine and shortening manufacture has led to the food processor having available a very wide range of fats and oils with differing functional properties that will meet his product and process needs.

The structural and crystalline properties of fats can control their functionality in food and this is readily illustrated in the manufacture of baked products like short pastry, cake and puff or flaky pastry. Advances in the knowledge of the properties of emulsions, and emulsifier systems, have been applied to bakery products to give improvements in bread volume and shelf-life, as well as leading to recipe balance in other baked products, and altering the requirement for plastic fats so that fluid and liquid shortenings could be used. The use of powdered fat and fat powders can add convenience in a number of food sectors, e.g. prepared cake mixes, toppings, bread improvers, etc.

Ghee, the clarified and crystallised butter from buffalo or cow's milk, is an extremely important culinary fat for a very large sector of the world's population and its method of production has developed from a domestic method to continuous industrial manufacture. Economic pressures again have led to the development of the hydrogenated vegetable oil alternative vanaspati. Ghee and vanaspati can be used as a spread, but are used principally in cooking, for deep and shallow frying or basting.

Salad oils and cooking oils are now used very extensively. These oils must either be highly characteristically flavoured, e.g. olive oil, almond oil, or completely bland, but in both cases must show a high degree of oxidative and flavour stability.

Salad oils are used directly or as the base for a sauce or dressing like mayonnaise. Cooking oils are used in the preparation of many dishes to add texture and flavour or control the cooking process.

An area of cooking where the high temperature stability of fats and oils is widely exploited is frying, both in shallow and deep fat frying. In shallow frying salad oils and butter and margarine can be used; however, in deep fat frying different conditions apply and knowledge of the conditions means the producer can modify the oil to give the best performance qualitatively and economically.

Nutritional demands and eating habits continue to change and so the oils and fats manufacturer will continue to develop products to meet the changing needs.

INTRODUCTION

Fats in their natural state have been a culinary and bakery ingredient for many thousands of years improving the palatability and nutritive value of food. Climate and agricultural practices have influenced the fats used in food preparation. Traditionally in Europe butter, lard and suet have been extensively used whilst in the warmer Mediterranean area liquid vegetable oils, particularly olive oil, are the traditional culinary oils.

Increasing industrialisation and population expansion in the later part of the nineteenth century led to a shortage of the traditional fats. These conditions stimulated the invention of the first substitute food—margarine. Margarine was invented in 1869 by a French chemist, Mège Mouriès, and the invention was exploited by Dutch butter exporters. Starting from modest beginnings margarine has developed into an important and sophisticated food processing industry. Additionally it has had important repercussions on the agricultural industry for as margarine production has expanded it has stimulated an expansion in the production and export of tropical oils and oilseeds and these now represent a substantial proportion of world trade in agricultural products.

During this period butter production has seen improvements in agricultural techniques and processing methods combined with a more scientific appreciation of milk and milkfats which has led to a consistent and highly characteristic product which endows cooked products with distinctive flavours and characteristic textures.

Progress in the understanding of the function of the ingredients in food now means the fat processor and food manufacturer can work together to improve the food products available to the consumer. The crystalline form and product consistency have a profound influence on the performance of the fats in food, particularly in baked products. Thus an understanding of physical properties like crystallisation behaviour, polymorphism and

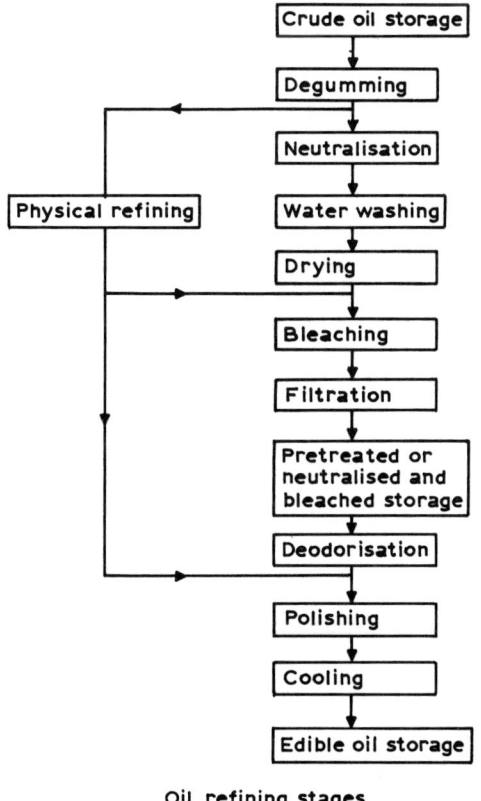

Fig. 1. Block diagram of the stages of refining oils and fats.

crystal structure in fats is necessary to control production processes such that they can be 'tailor made' to suit particular applications.

PRODUCTION OF MARGARINE AND SHORTENING

The modern processor has available bland oxidatively stable and low coloured edible oils of vegetable and animal origin, achieved by the processes shown in Fig. 1. These natural oils can be modified by the processes of hydrogenation, interesterification and fractionation used either singly or in combination to produce fats that bear no relation to the original material.

Blending oils and fats to achieve the required solid to liquid ratio is a major part of the processor's skill as it is critical to the firmness and texture of the finished product. Added to this is the influence of the crystal habit of the oils and fats selected and their polymorphism. Thus the processor requires an understanding of these characteristics when preparing blends for margarines and shortenings. Blend selection has now become highly sophisticated and is becoming defined on the basis of fatty acid and triglyceride types rather than on oil types in order to achieve the specified physical and performance properties. This knowledge must be combined with the technology of crystallisation and particularly in the application of scraped surface heat exchangers which are now extensively used to chill and crystallise fat based products.

CRYSTALLISATION BEHAVIOUR

In common with all other long chain molecules, fats and fatty acids exhibit polymorphism, that is the ability to exist in more than one crystalline form and so possess multiple melting points. Triglycerides occur in any one of three basic polymorphs designated α, β' or β (Bailey, 1950):

- α (alpha) the most loosely packed arrangement and hence the least stable and lowest melting.
- β' (beta prime) more stable than α but transforms irreversibly to
- β (beta) the most closely packed and highest melting polymorph.

Work by Timms (1984) describes the behaviour of a monoacid triglyceride where it can be shown that as it cooled rapidly the α form is obtained which on slow heating melts to resolidify and give the β' form. After further slow heating it melts and then resolidifies in the β form. Most fats possess an α form which is so unstable that it can be ignored, some also possess both β' and β forms and others just a stable β' or just a stable β form. Some examples are shown in Table 1.

X-ray studies on tristearin have shown that the triglycerides pack side by side in separate layers. The triglyceride molecules form the shape of a chair and the molecules are arranged in pairs, head to tail. Figure 2 shows the packing arrangements possible in pairs of two or three fatty acids. Figure 3 is a schematic diagram showing the main features of the molecular packing of the three polymorphs of tristearin. It can be seen in the α form the fatty acid chains are perpendicular to a basal plane (that plane containing the methyl end groups). In the β' form the fatty acid chains are

TABLE 1
Crystallisation performance of some natural edible oils

β' form	β form
Cottonseed oil	Soyabean oil
Palm oil	Sunflower oil
Tallow	Groundnut oil
Butter fat	Coconut oil
High erucic acid rapeseed oil	Palm kernel oil
	Lard
	Low erucic acid rapeseed oil

tilted at an angle to the basal plane. Each fatty acid has its hydrocarbon chain in regular zig-zag formation in a plane perpendicular to its neighbour, whilst the β form which also has the fatty acids tilted to the basal plane has the zig-zag planes of the hydrocarbon chains in parallel in the same plane. These descriptions show an increasing closeness in packing and hence increasing melting point and stability.

Where there is a wide variety of molecular size and type of triglyceride, for example coconut oil and beef tallow, the β' form rather than the β form predominates because it is more able to accommodate the distortion of the chain packing necessary for a solid solution.

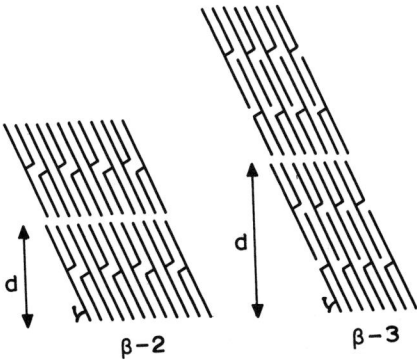

Fig. 2. Schematic arrangement of triglycerides in the β-2 and β-3 polymorphs. (From de Jong as given in Timms, 1984.)

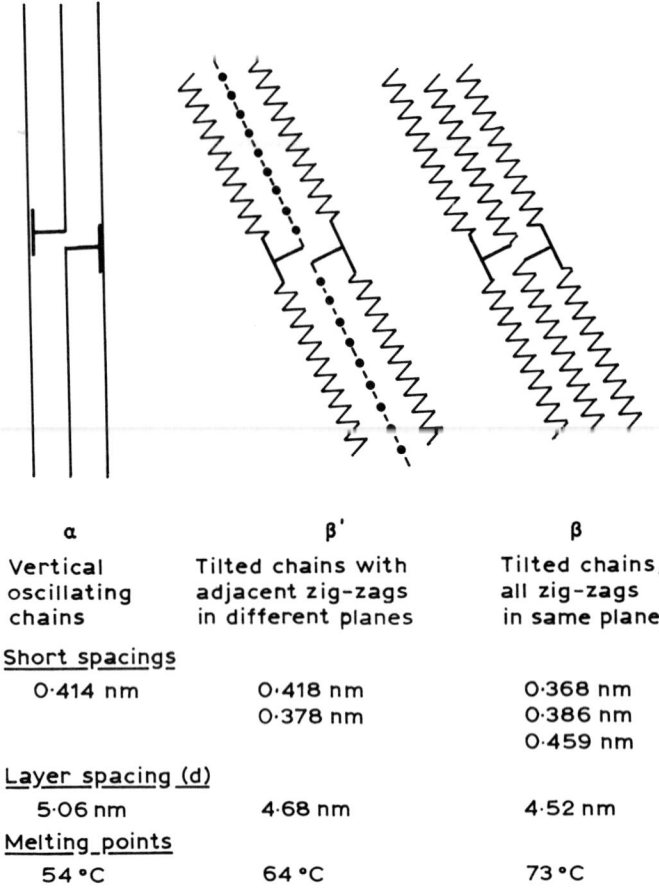

Fig. 3. Schematic diagrams comparing the polymorphic forms α, β' and β as exemplified by tristearin (StStSt).

From the foregoing comments it can be seen that each triglyceride has its own polymorphic and melting behaviour. However, in a mixture of triglycerides they do not behave independently but take on a totally new character in terms of crystallisation behaviour. The systems are so complex that it is easier, in the case of natural fats, to describe them in terms of their different phases; thus the physical properties of a fat can be discussed in terms of its phase behaviour. In a fat or fat blend at a given temperature there will always be a liquid phase and a solid phase and the solid phase can have several components, which can change with temperatures and composition.

A phase diagram (Birker & Padley, 1987) can be used to show how blended fats interact for example to produce minima points (eutectic behaviour) or maxima points (solid solutions). Where the fats are compatible a horizontal line results on the iso-solids line at all different compositions.

Thus, to summarise, the major features defining the firmness, texture and performance of a blended margarine and shortening are (Opfer, 1975):

—the proportion by weight of crystals, which is governed by the solid to liquid ratio;
—the melting point of the crystals;
—the crystal geometry, that is their size, shape and alignment;
—the degree of formation of mixed crystals;
—the ability of the crystals to flocculate into a network which increases firmness.

Normally the greater the quantity of solid triglyceride the greater will be the rigidity of the network because of the increased number of crystals and thus cohesive forces between them. These last prevent flow at stresses below those appropriate for the desired consistency. Changes in temperature will obviously change the product firmness and plastic behaviour by altering the quantity of crystals present, the hardness, and the viscosity of the liquid triglycerides.

Fat crystallisation is initiated by nucleation in a supercooled system. In the manufacture of margarine and shortening the cooling rate, agitation and degree of supercooling control the rate of crystal growth and thus crystal size and crystal agglomeration, which affect the textural and melting properties of the fat product.

PROCESSING

The vast majority of margarine and shortening is now manufactured on scraped surface heat exchangers (Joyner, 1953) and though they vary in design the basic principles apply to them all. The heat exchanger is made of two concentric tubes and in the annular space thus created a compressible refrigerant is circulated. The inner tube has a heated shaft which runs the length of the tube on which are mounted floating scraper blades. As the shaft rotates these blades scrape the internal surface of the tube.

In the process the liquid emulsion or fat blend is pumped along the tube at a fixed speed and the rotating blades remove the chilled product from the walls. This constant renewal of the cooling surface and the turbulence created leads to supercooling and the initiation of crystal nuclei and hence crystallisation. The supercooled and partly crystallised product can then be pumped to a worker unit where crystallisation is completed and the heat of crystallisation released.

It can be seen that it is important that the flow rate of product does not vary, irrespective of the fact that as it is chilled there is a viscosity increase. There is also a temperature differential across the product flow leading to a range of crystalline compositions being created; so in order to minimise product variability tight and continuous control must be maintained at all stages on the evaporation rate of the refrigerant.

The worker unit (or units) is also tubular and can contain a system of beaters, to ensure that the crystal structure is developed in a dynamic environment, hence controlling the size of the crystal aggregates and giving a smooth plastic texture. It is also possible to have no provision for mechanical agitation to induce growth of large crystals from the mass in order to provide a product firm enough to pack into wrapped units.

The system described is the simplest to show the principles. There are now available systems of much greater complexity to improve the texture and plasticity of a widening range of blend types and meet greater specificity in the requirements of the user. Factors such as shaft rotation speed, scraper blade design, size of the annular space and size and location of worker units all can affect the final texture of the product.

BAKERY FATS

Fats and oils in their natural state have been used as a bakery ingredient for years to improve mouthfeel and palatability of the finished foods. The growing sophistication of the bakery industry, both in product range and automation techniques, requires greater control of the ingredients, including the fats.

In designing shortenings and margarine for bakery use it is important to know the application so that the functional properties required can be designed into the product by way of the oil blend used. Once the blend is selected the quality and the process control techniques necessary to maintain the desired properties must be applied.

Comparison of the rheological properties of butter and margarine shows there are considerable differences. Butter is a considerably more complex system than margarine. The fat system is less homogeneous than margarine, as it is made up of liquid and crystalline fat, fat globules and globule fragments interspersed with moisture droplets (Mulder & Walstra, 1974).

The globular fat influences the texture and consistency in that the solid fat inside the globules causes them to go rigid and increase firmness. However, they cannot form solid networks with the crystals outside. Thus when butter and margarine consistency is compared a margarine with less solid fat than butter is equal in firmness to the butter.

The functions of fat in bakery applications are (Hodge, 1986):

—shortening power and lubricity,
—batter aeration,
—emulsifying properties,
—provision of an impervious layer,
—improvement in keeping properties,
—provision of flavour.

The functionality of fat will be discussed in terms of short pastry, cake and puff or flaky pastry in order to demonstrate how fat contributes to the structure and edibility of the product and hence how the fat can be blended and processed to maximise the functions (Pyler, 1973).

Short Pastry

Short pastry is used in a wide range of savoury and fruit products. The major ingredients are simply flour, fat and water. When mixing flour and water the wheat proteins are hydrated to form 'gluten' which develops as a tough cohesive and extensible network which when baked gives a hard brittle structure, often described as having 'flinty' eating qualities. When fat is included in the recipe during mixing it becomes smeared through the dough in irregularly shaped microscopic droplets. The flour particle surfaces are protected from the water and so gluten chain development is interrupted resulting in planes of weakness and so the product becomes 'shorter' eating and more inclined to melt in the mouth. In simplistic terms it can be seen that too little fat will result in tough and harsh eating pastry and too much will so interrupt gluten development that the dough will be loose and soft to handle and too fragile when baked.

The same comments apply when the shortening or margarine is too firm or too soft. A firm fat with solid triglycerides content so high that it will not smear to distribute successfully in the dough will not interrupt the gluten development, thus leading to a flinty product exhibiting shrinkage. Liquid

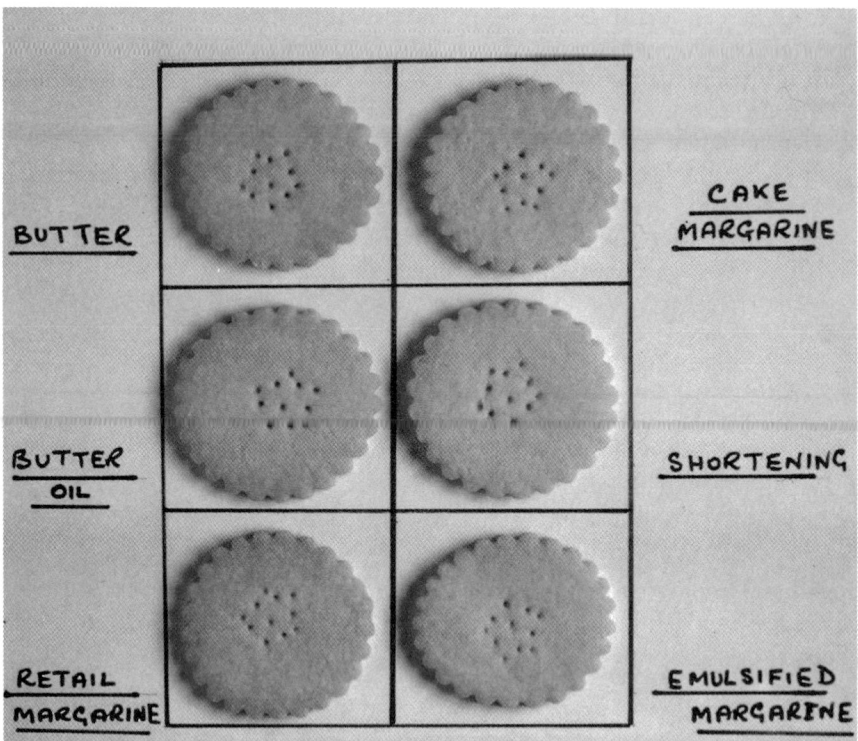

Fig. 4. Comparison of fats in short pastry recipe.

or fluid shortening at the other extreme leads to sloppy and soft and unworkable doughs. Thus a fat for short pastry should be of firm consistency so when being mixed into the dough it retains sufficient body under shear conditions to distribute as protective thin films and droplets through the dough. Figure 4 shows the performance of certain bakery fats.

The traditional fat for short pastry for savoury products is lard which, due to its particular triglyceride structure (Carlin, 1944), crystallises in the β polymorph which has led to the belief that the β form is preferred for short pastry manufacture. However, compounded shortenings containing β and β' polymorphs have been found to perform well in short pastry recipes to give a good 'short' texture and good mouthfeel.

A feature of considerable importance is the ability of the fat to retain its plastic characteristics over a wide temperature range, realistically 15–30°C.

This is a function of the solid/liquid ratio of the fat blend and a relatively high proportion of triglycerides with three saturated fatty acids so that a significant proportion of solid crystalline material is retained at higher temperatures. There is a balance to be achieved in formulating shortening blends in ensuring undue firmness is not achieved whilst giving adequate solids at higher temperature, and also the knowledge that the fat will contribute to the short pastry flavour, thus high residual solid material could detract from the flavour.

Products such as margarine and butter are not extensively used in short pastry as on a strict weight for weight basis more has to be used because these products are only 80% fat, the functional ingredient.

Figure 4 shows the comparison of a number of products made with a typical short pastry recipe. It can be seen that an all fat product performed the best giving little or no shrinkage; with butter oil the result is marginally poorer than the shortening because of its narrower plastic range. Butter and margarine give the poorest performance because of the shortage of fat compared with shortening.

In spite of a rather poorer performance butter has a place in high quality sweetpaste because of the superior flavour of butter based products and the textural 'bloom' they impart. Butter is often blended with lard or shortening in order to improve performance and control cost.

CAKE

The mechanism by which a fat functions in a cake has been the subject of a considerable amount of research work. The process has been described (Shepherd & Yoell, 1976) in terms of the batter preparation and monitoring the changes taking place throughout the period of baking.

The first step in making, say, a madeira cake is blending the ingredients and the method of blending the ingredients has some influence on the fat particle size in the batter.

The traditional methods of batter preparation are the sugar batter method where the sugar and fat are first creamed together or flour batter method where the fat and flour are first blended. The all-in method, where the batter preparation is completed in one stage, has become more popular with the introduction of high speed mixers. Popular in large commercial bakeries are the continuous mixers where a loose slurry of the ingredients is fed to a mixing head where air is injected into the batter.

Examination of batters prepared by these various routes has shown that

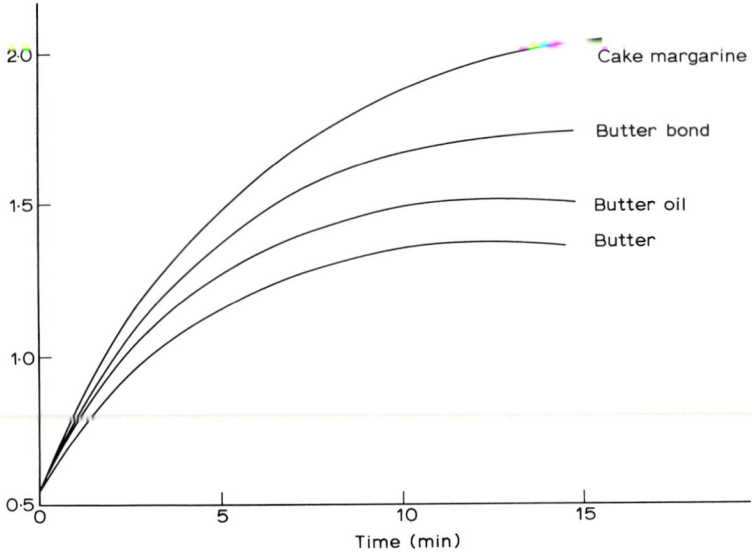

Fig. 5. Rate of air incorporation in a fat/sugar cream.

the air is held initially in the fat phase when plastic fats are used, the method of batter preparation having an influence on the distribution and fineness of the fat particle size in the batter. The finer the distribution of the fat and air the better the final cake volume and crumb structure.

Figure 5 demonstrates that the plastic properties of the fat are important in its ability to incorporate air and retain it. By definition a plastic material contains both solid and liquid portions and this is the case with a plastic shortening or margarine. In the creaming process there must be enough liquid oil available to envelop the air bubble and sufficient crystalline fat to stabilise the system. The small β' crystals are the most effective in stabilising air bubbles. However, crystal aggregates that break up during the process can also stabilise the system. The proportion of crystalline triglyceride at the working temperature must be above a certain minimum which practice has shown to be 5%; however, most commercial bakery fats contain about 20% solid triglyceride.

The test demonstrates the plastic behaviour of the fat and its resistance to work softening, which, if it happens to a significant degree, results in the coalescence of the air cells and loss of volume in the cream. The creaming curves show the expected behaviour of the texturised butter oil as it initially creams very quickly and then fails; butter also demonstrates the limitations of a low solid to liquid ratio leading to a short plastic range. The incorporation of butter oil as a margarine oil blend component significantly improves the performance, though it still does not match the 'tougher' cake margarine.

Following the incorporation of the aqueous ingredients and flour it can be seen that the final batter is a multiphase system where flour particles are suspended in the aqueous phase, but the water continuous phase still has parts which are a water in oil emulsion. The application of heat to this system at the start of the baking process has little effect; however, at about 37°C the irregularly shaped fat particles begin to melt and become droplets of oil and at this point the water in oil emulsion parts of the batter invert to oil in water. As the temperature continues to rise the fat withdraws from the air bubbles which are left in the more viscous aqueous phase, to produce a foam, probably stabilised by the egg protein preventing coalescence of the air bubbles. The flour particles and fat droplets are now distributed through the continuous aqueous phase.

Convection currents in the still fluid batter cause bulk flow such that the air bubbles act as nuclei, for the increase in volume of the total batter, as the carbon dioxide and water vapour moves into them. Studies by Carlin (1944) have shown that no new air cells are created during baking; thus all the air cells which create the cake texture are introduced during the batter preparation.

The continued rise in temperature to 65–70°C results in the start of gelatinisation of the flour and coagulation of the egg protein and the expansion of the air bubbles is very rapid and at 95–100°C the structure becomes fixed.

In practical terms the plasticity of the margarine or shortening must be such that it can be easily whipped into the batter. Figure 6 demonstrates the effects of toughness and wide plastic range on a finished cake. The volumes of cake made with butter and a range of commercial margarines show that butter and butter oil give a lower volume and lighter texture compared to a retail packet margarine. Special cake margarine shows a finer distribution of the air cells in the finished cake and this can be achieved when milkfat is included in the formulation. In this case again butter confers a richness of flavour not achieved by margarine.

Shortenings can be designed to be general purpose in that they can be used in both cake and short pastry applications. However, margarine is preferred in cake manufacture since it not only has the functionality

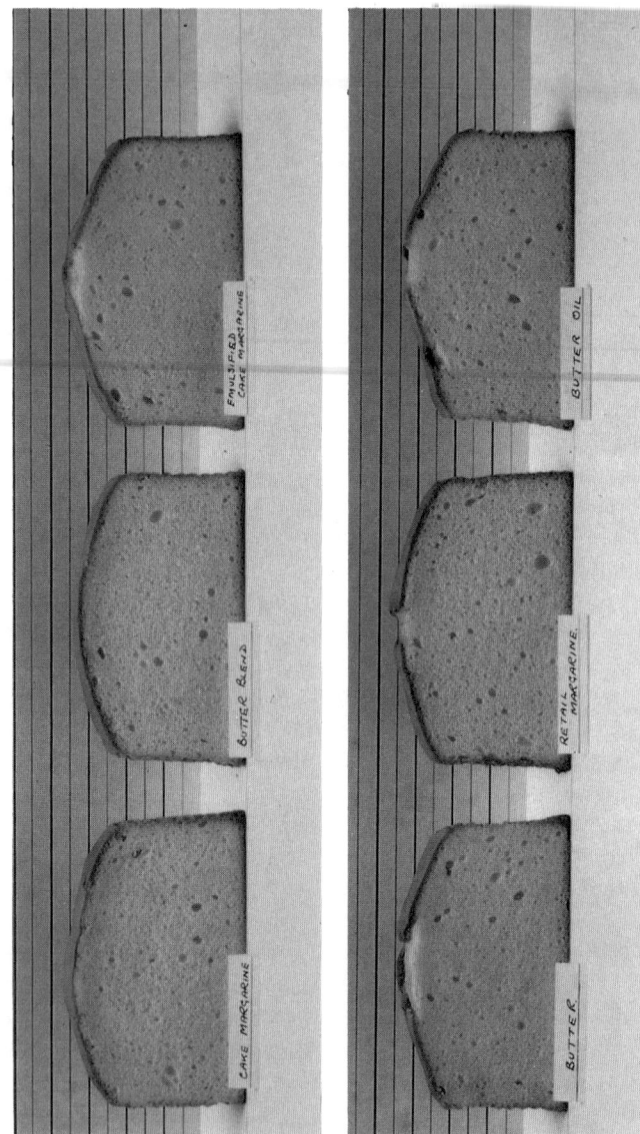

Fig. 6. Commercial madeira cake made with a range of margarines and fats.

described above but can also contribute a richness and flavour to a cake not normally found with a shortening. Butter is of particular value in giving a highly characteristic rich flavour. As well as this margarines are an excellent vehicle for emulsifier systems which can significantly improve performance.

Emulsifiers now have an important position in the manufacture of bakery products. This is best illustrated by a consideration of high ratio cakes. These cakes were developed from a better understanding of the function of all the basic ingredients in a cake recipe—flour, sugar, egg and fat.

A high ratio cake by definition is one that has a higher proportion of liquor in the recipe. Thus the flour used must be milled to a finer powder and treated with chemicals in order to allow it to absorb greater quantities of liquid with the gluten forming proteins largely broken down. It is also necessary that the fat in the recipe should emulsify the increased volume of liquids; thus emulsifying agents were introduced into the fat, the most popular being monodiglycerides of fatty acids. The presence of the emulsifiers not only stabilised the liquids but also provided sites for aeration by reducing batter viscosity allowing the fat to be more finely and evenly distributed through the aqueous phase than in a madeira cake batter to give a cake with a more even texture.

This type of cake altered the rules of recipe balance and the methods of mixing the ingredients, allowing simplification and automation, whilst achieving cakes that were generally moister and more tender.

PUFF PASTRY

Puff pastry illustrates another basic and unique type of bakery product in that it has a light flaky and layered structure which, during baking, increases in volume up to eight times compared to the original dough.

In the preparation of puff pastry, layers of dough, with a well developed gluten network, are arranged so that two layers of dough enclose a layer of fat; then by a complex system of folding and rolling the dough/fat layers a structure of alternate layers of dough and fat are built up.

The layers of fat behave as impervious barriers to the moisture vapour and gases generated during baking. The retained gases expand and so stretch the gluten network to give the well known puffed or flaky texture.

The fat blend for a puff pastry margarine must have a tough and plastic texture as it is required to be rolled and stretched to a thinner and thinner layer and remain continuous. The mechanical stress of the rolling process must not cause the margarine to soften unduly as this would lead to its loss as a layering fat. Further, any brittleness in the texture may cause penetration into the dough during manufacture. The melting point of the fat blend must be such that it keeps the dough layers apart in the initial stages of baking, but without giving the final baked product a 'waxy' mouthfeel.

Butter is often used in the manufacture of puff pastry but it requires handling in such a way that the applied stress is kept below the yield value of the butter which ensures that its plastic behaviour is good and it layers well. The low solid to liquid ratio of butter at ambient temperatures means its layering capability is poor. Figure 7 shows examples of puff pastry made with butter and refrigerated milkfat. Additionally, a processed sample of fractionated butter improved the performance as did the chilling of the butter at the point of usage. Butter and butter oil fractions confer the benefits of a characteristic flavour and being easily digestible as they melt below body temperature.

Danish pastry is a form of puff pastry; however, the dough used is fermented, like a rich bun dough with a short fermentation time. The layering fat is added by the English method and the series of turns is limited to restrict the volume increase. The product is cut to the required shapes, gently proved and then baked. Butter is finding a widening application in the manufacture of Danish pastry and croissants, as it enriches the product but maintains a light eating quality. The volume increase expected is much less than in puff pastry; even so the butter is often refrigerated before use.

Puff pastry margarines are made from oil blends that have a high solid to liquid ratio and often at the expense of the final melting point. Slip melting points of 42°C and higher are not uncommon.

In manufacture the margarine emulsion is shock chilled from a high temperature in order to quickly give a very stable crystal network which is then subjected to a heavy working and kneading routine to prevent establishment of larger crystal networks and obtain a proper balance of reversible and irreversible bonds to prevent the finished margarine becoming to rigid, causing brittleness and flintiness in use.

Puff pastry margarine is made in specially designed tubular chillers to give the shock chilling and plasticising necessary at the high pressures experienced. However, it is still often manufactured on the chilling drum and complector system which is claimed to give a better plasticity. In this case the blend is shock chilled as flakes and crystallisation is largely completed in a static situation. The post-crystallisation working can be adjusted to the final product hardness desired.

Fats in Bakery and Kitchen Products 229

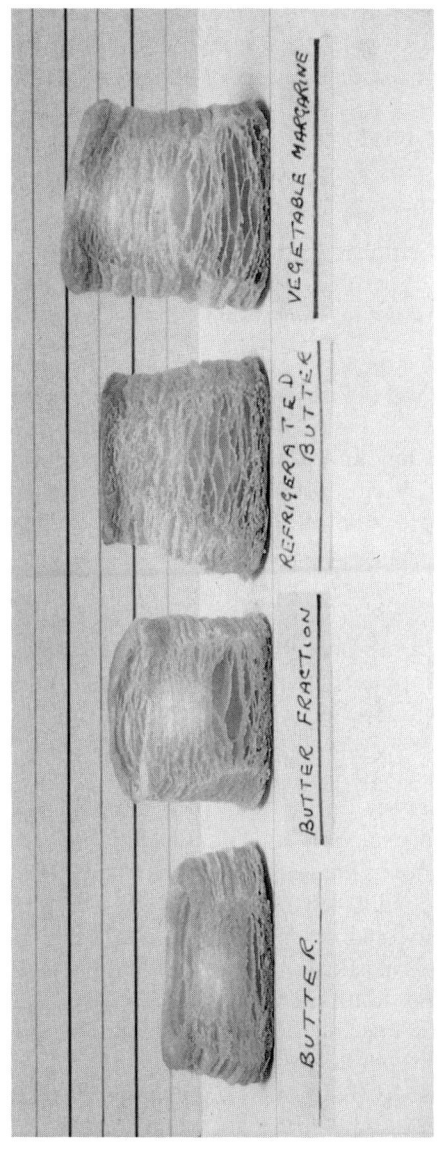

Fig. 7. Comparison of layering fats in puff pastry.

THE INFLUENCE OF EMULSIFIERS IN BAKING

Emulsifying agents, particularly those esters formed from fatty acids and polyvalent alcohols like glycerol, propylene glycol, sorbitol and sucrose and their modifications made by esterification with organic acids like acetic acid, citric acid, lactic acid and diacetyl tartaric acid, have been used extensively in bakery products and other foodstuffs.

The function of emulsifiers (Krog & Laurisden, 1976) in food systems falls into three broad categories:

1. Stabilisation of emulsions and aerated systems.
2. Improvements of texture and shelf-life of starch based products.
3. Dough conditioning by interaction with wheat gluten.

In the stabilisation and aeration of cake batters emulsifiers can play a very important role and the physical state of the emulsifier has a marked influence on the batter. The effect of the emulsifier incorporated in the fat or margarine has been mentioned in high ratio shortenings and in more conventional recipes the emulsifier improves the distribution of the fat and so promotes the distribution of the air. However, hydrates of emulsifiers have been used for many years to improve aeration in cake batter, in particular in fat free sponge cakes.

It has been demonstrated (Krog, 1975) that with distilled monoglyceride dispersions in water at varying concentrations and temperatures a series of liquid crystalline mesophases can be formed. These mesophases are a result of hydration where water penetrates through the layers of the polar groups of the crystalline monoglyceride above the Krafft point. On cooling the hydrocarbon chains crystallise again and the water between the lipid bilayers forms an α crystalline gel structure. Beside this lamellar type of mesophase, cubic and hexagonal structures have been identified. Monoglycerides based on saturated fatty acids form viscous gels where the lamellar structure dominates, whilst unsaturated fatty acid monoglycerides predominate in the cubic structure.

The use of these gels in cake batter has been shown to give a much more uniform air distribution than the shortening containing monoglyceride or monodiglyceride. The finer air distribution increases the viscosity which leads to a better cake volume and texture.

There are several other emulsifiers that show crystalline properties in water dispersions and have been found to improve batter aeration properties such as lactic acid esters of monoglycerides and propylene glycol esters of fatty acids.

Work on starch based products (Krog & Nybo Jensen, 1970) has shown that certain emulsifiers, particularly distilled monoglycerides, have crumb softening and anti-staling properties in wheat bread.

The process of bread staling has been shown to be due to the amylose fraction of the wheat starch. During baking some amylose leaks out of the starch granule and dissolves in the water available to form a gel between the swollen granules of the fresh bread. On cooling of the baked bread the gel contributes to the initial firmness of the bread crumb; however, with time the amylose recrystallises (retrogrades) to its insoluble form and so the bread becomes hard and brittle.

It is now generally accepted that amylose in its helical form has a lipophilic core; hence in this form the amylose can be stabilised by straight chain hydrocarbon molecules, like those found in fatty acids. The saturated distilled monoglycerides have a steric configuration that can easily be enclosed in the amylose helix. The insoluble helical complex raises the gelatinisation temperature of the starch and thus reduces the total gelatinised starch in the bread crumb. The monoglyceride complexed amylose will not retrograde as does the unreacted amylose, thus leading to less amylose being available to be part of the starch gel to give a softer crumb. Additionally the amylose monoglyceride complex does not take part in transporting moisture from the surrounding protein network with the result that this becomes less rigid and hence gives a softer crumb.

Anionic emulsifiers such as the sodium or calcium salts of stearoyl 2-lactylates, diacetyl tartaric esters of monoglyceride and succinylated monoglycerides have been found to impart dough strengthening characteristics in fermented doughs. The effect of dough strengtheners or conditioners is to improve dough processing characteristics and also to give increased volume and a finer texture to the baked product.

During the processing of a dough a gluten network is developed which traps the carbon dioxide produced by the yeast to give the final volume and texture to the finished bread or cake. Any weakening of the gluten during processing will lead to a loss of the gas and a resulting poor volume. The emulsifiers used as dough conditioners interact with the gluten to improve gas retention and dough elasticity to provide tolerance to variations in fermentation times and temperatures and mechanical shock.

There is still some doubt about the mechanism of the interaction of gluten and emulsifiers; however, it has been shown (Larsson, 1980) that the emulsifier can replace some of the flour lipid in association with the gluten suggesting there are lamellar emulsifier structures in the aqueous films at the interface between the gluten strands and starch.

CONTROL OF QUALITY IN MARGARINE AND SHORTENING MANUFACTURE

Since the fat blend is the major component in the manufacture of margarine and shortening most of the analytical control effort is directed towards ensuring not only that the oil blends used are of good edibility and oxidative stability, but that they also have the specified solid/liquid ratio and correct crystal habit.

The tests used to judge the quality and edibility of fats and oils are well documented (Cocks & Van Rede, 1966). Classical tests for free fatty acid content, peroxide value and colour are well known and have been supplemented by tests such as anisidine value used to assess secondary oxidation products and accelerated stability tests such as the Active Oxygen Method (Swift's Test) and the Rancimat test which are both based on bubbling air through the oil or fat at elevated temperature. Finally the flavour of the oil must be judged by an expert panel to ensure it is bland or near bland.

The processor will receive these oils from the refinery with the assurance that qualitative standards have been achieved. It is then important to ensure the blend and solid characteristics of the oils are correct (Zürcher & Hadorn, 1979).

The margarine and shortening manufacturer can either receive refined, deodorised oils which he then blends, or by consultation with the refiner he can receive complete blends. There are arguments for and against both types of operation in terms of quality, efficiency and process control. The system selected usually depends on the way the processor's production organisation has been built up.

Establishment of the correct blend for the duty the margarine and shortening are to perform requires close consultation with the user, and an understanding of his process by the margarine manufacturer. A knowledge of the fatty acid composition and the triglyceride structure of the fats available, using gas chromatographic techniques (Christie, 1973), ensure the crystal habit of the fat is correct.

It is important to know the extent to which a fat or fat mixture crystallises at the temperatures of practical interest and the extent of crystallisation at various temperatures. The modern method for measuring the solids content of fats is based on the difference in molecular mobility in liquid and solid triglycerides (Waddington, 1986), and as the solid to liquid ratio varies with temperature then a temperature profile of the blend solids content can be obtained. The technique in question is wide-line nuclear magnetic resonance (NMR). Pulsed NMR has replaced the much more time consuming technique of dilatometry which measured the solid fat content by the volume contraction during crystallisation.

The phase behaviour of oil blends can be evaluated by plotting pNMR solids content data as a function of temperature and composition. The interaction of two oils or triglyceride types can be shown with these diagrams. These so-called iso-solid diagrams show components are compatible by giving horizontal iso-solid lines. However, where eutectics or compounds are formed the iso-solid lines are not horizontal and so product defects can be forecast, for example a margarine may rapidly develop a grainy or brittle texture due to compound formation (Birker & Padley, 1987).

The analyses discussed above, that is, gas chromatography and solids content determination, are used to establish the oil blend against the user's requirements for performance and eating qualities. Once these parameters are established then blend control can be effected by routine analyses, like solids content of the individual and blended oils and iodine value, supplemented with gas chromatography of the fatty acid methyl esters.

In the production of margarine the manufacturer is faced with additional control problems in that he is making a product which has two phases which, when processed, must be completely stable; also the fat and water levels must reach the statutory levels and any added salt must attain the level specified by the consumer (Andersen & Williams, 1965).

The production of stable water in oil emulsions is facilitated by the addition of emulsifying agents; traditionally monodiglycerides have been used. The influence of emulsifiers on such features as air incorporation, batter stability, etc., has led to a greater sophistication. The emulsifiers also ensure that the water droplets in the emulsion are small (about 5 μ), which leads to a good bacteriological standard and prevents their coalescence. The presence of large droplets would provide a medium with sufficient nutrients where bacteria could grow.

Mixing the phases in margarine manufacture can be performed either in batch process or continuously. The batch process is the more traditional, where the oil phase and aqueous phase, with their soluble ingredients, are premixed in the form of a suspension at a temperature sufficiently high to ensure crystallisation of the highest melting component does not take place. The 'premix' is then fed to the chiller by way of a buffer tank. In many factories continuous metering systems have been installed where proportioning pumps blend the fat and aqueous ingredients immediately prior to the chiller, and the system relies on the agitation and shear characteristics in the chilling tube to give a correctly distributed water globule size.

Both systems can be found in modern factories and can be substantially automated, especially where in-place cleaning is included. However, automation of the chilling process has not been advanced, due to the complexity of the crystallisation process and the limitations on the techniques for measuring product consistency.

Moisture content can be automatically monitored with in-line equipment as well as by evaporation loss. Fat and salt contents are measured by conventional laboratory tests. Most bakery margarines include milk solids, either in the form of whey solids or spray dried skimmed milk powder. These products are added to improve the flavour and the presence of the lactose can improve crust browning.

In modern margarine factories the possibility of microbial contamination has been almost completely eliminated. However, the finished product must be regularly examined for the presence of spoilage organisms, yeasts and moulds and lipolytic bacteria, as well as food poisoning organisms. As well as this, the surfaces and the atmosphere should be monitored. Additionally, close control of the pH value of the aqueous phase of the margarine, combined with maintaining a small water globule size, inhibits the proliferation of spoilage organisms.

As discussed earlier the processing conditions have some influence on the texture of the finished fatty product; hence the chilling and working conditions need to be very closely defined, for example parameters like emulsion or oil feed temperature, throughput speed, refrigerant and evaporation temperature, and product temperature. These controls will lead to a consistent product and the firmness can be confirmed by estimation of the yield value. Final quality testing is done by user tests, for example measuring air incorporation on a standard bakery mixing machine, manufacture of a basic cake which is sensitive to the fat performance and, in the case of puff pastry fats, test vol-au-vents can be made in order to measure the volume increase.

Bakery fats can be processed into other than the plastic fats described so far. These other forms will be discussed.

LIQUID SHORTENINGS

Liquid shortenings by definition are clear and fully liquid at ambient temperature. As discussed earlier batter aeration is dependent on the ability of a plastic fat to retain air; liquid oils do not have this property. However, the advent of high ratio cakes showed that dependency on the fat's plasticity for aeration was reduced by the inclusion of emulsifiers. Thus the developments in continuous methods for cake mixing and the need for bulk storage of ingredients led to the introduction of the fully liquid shortening. In the earlier applications various emulsifier hydrates were added to the cake formulation. However, the development of more complex and oil soluble systems of emulsifiers made it possible for liquid oils to be an effective alternative to plastic fats (Hartnett & Thalheimer, 1979).

Liquid shortenings have been exploited principally in the USA, but have failed to gain popularity in the UK, due, in part, to the high cost of the emulsifier systems, and because the systems have much greater temperature sensitivity than claimed, for example at cooler temperatures (18–20°C), the emulsifiers are often precipitated from solution with the consequence that the product becomes highly variable.

The shortenings are designed for the production of cake, basically of the high ratio variety, and since they are totally liquid they are not appropriate for short pastry.

FLUID SHORTENINGS

These are pourable shortenings, but distinguished from liquid shortenings in that they contain solid suspended particles.

Fluid shortenings were principally developed as frying media to provide a stable but pourable frying oil. Oils like soyabean oil and rapeseed oil, with a relatively high content of linolenic acid, were 'brush' hydrogenated to reduce the linolenic acid level and hence improve the oxidative stability. Instead of winterising the product, to give a clear oil, the addition of a small quantity of a fully hydrogenated fat and a technique for maintaining the solid material in suspension led to a pourable 'slurry'.

A variety of techniques have been patented (US Patents 3,369,909 and 3,528,823) for creating a slurry-like texture fluid over a wide range of temperatures. The critical feature of the suspended particles is their size; if too large they will settle quickly and if too small, though settling slowly, they will pack closely. It has been found that β' crystals are too small and β crystals are better as they are too large to pack closely and treatment to prevent aggregation of the crystals ensures a stable slurry.

The systems described (US Patent 3,395,023) usually require slow crystallisation of the shortening following the addition of a high melting fat component to a liquid vegetable oil and a final homogenisation of the flocculent precipitated crystalline mass. These products can also be made on scraped surface heat exchangers where the control on the rate of cooling is very critical in order to ensure the correct crystal modification is created.

Slurry shortenings still rely heavily on added emulsifiers for their functionality. A number of emulsifier systems have been shown to work well and experience with pilot scale trials has shown that acceptable performance can be achieved with blends containing α monoglycerides, polyglycerol esters, propylene glycol monostearate, lactic acid monoglycerides and sodium steroyl lactylate. Used in high ratio cake recipes these give finished cakes approaching the volume and texture of a cake made with conventional high ratio shortenings. Acceptable quality cakes have been obtained with a reduction of fat in the cake recipe.

Fluid shortenings have temperature limitations in the same way as liquid shortenings. Storage at temperatures below 12–14°C can lead to the product setting up, and temperatures in excess of 35°C cause some melting of the crystals formed, and subsequent cooling will cause the formation of large crystals which will settle.

POWDERED FATS, FLAKED FATS AND FAT POWDERS

The claimed advantages for fat powders, powdered fat and flaked fat are ease of handling in transport, dosing and simplified storage. Blending with the growing number of other dry ingredients is eased.

Before considering the manufacture and application of these forms of fat it is first necessary to define the differences between them. Powdered fats and flaked fats are both similar in that they are entirely made of fat or fat and emulsifier. However, they are manufactured by different methods as their names imply, though there is some overlap in their application.

Fat powders, though they contain substantial amounts of fat, also contain non-fatty material which acts as a carrier. This applies a restriction in their use in that the non-fatty component must be compatible with the final recipe of the user.

Methods of Manufacture

(a) Powdered and flaked fat

In the manufacture of powdered, granulated and flaked fat there are certain common features to be considered. The fat in its final form must be a solid at ambient temperatures, and the flake or particle must be such that the crystallisation must go as quickly as possible to completion so that late crystallisation of the liquid core does not release sufficient heat to cause lumping or caking in the product.

The technique of cooling and crystallising on a cooling drum to create a flaked product has long been used in the margarine industry. The method is similar in creating flaked fat though generally the flakes are thicker so that they can be handled easily in conveying and packaging. Additionally, so that they can crystallise quickly, the fat is usually of a high melting point with high solid fat content at ambient temperature, e.g. pNMR measurement at 20°C of 70% and at 30°C of 30%. Any fat hydrogenated to a high enough melting point can be used. However, a fat with a wide variety of triglyceride types exhibiting little polymorphism is preferred. The inclusion of coconut or palm kernel oils in their hydrogenated form in the fat formulation can assist in achieving the required percentage solids while slightly improving mouthfeel.

Flaked fats can be further pulverised to make them more granular in texture, to improve the flow properties. However, this requires the application of low temperatures.

Powdered fats (Lamb, 1987) are manufactured by the technique of spray chilling, i.e. dosing the fully liquid fat or fat blend into a tower through which cold air is being circulated. The fat (Münch, 1986) must be injected into the upper portion of the tower as a fine spray. Since the globule size is a critical factor in the success of operation a range of systems have been developed such as rotary atomisers, both disc and centrifugal, and high pressure nozzles to control the globule. The system is selected on the requirements of the processor and his products, i.e. flexibility, throughput, product viscosity and formulation complexity.

The major parameter for the successful spray cooling of fat is that the globule size should be such that the total particle is fully solidified before leaving the tower. It can be seen that the holding time in the cold air stream is also important. The difference between the air temperature and the melting point of the fat also plays a part. Thus in principle it can be seen that the smaller the droplet radius and the greater the difference between the melting and air temperatures, the more rapid is the solidification. Other features that have an influence are product temperature and viscosity and the injection pressure. These influence the droplet size and the amount of energy to be removed in the cooling.

The design of the tower is of great importance in that the powder take-off system must not allow outside air or moisture into the tower and the chilled air filtered and cooled for reuse.

It can be seen again that a relatively high melting fat is required in this

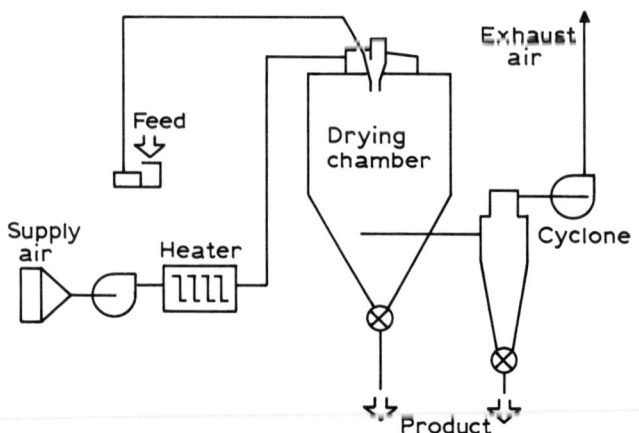

Fig. 8. Diagram of a spray drier system.

system so that rapid hardening is achieved and lumping in store is prevented. The advantage over flaked fat is easier control on dispensing and dispersion with other ingredients. This is particularly important when emulsifiers or fats containing high levels of emulsifiers are being introduced as an ingredient as the quantity is likely to be very small and so its successful distribution is more difficult.

(b) Fat powders

Fat powders are popularly made by the technique of spray drying (Blenford, 1987). The fat is first made into an oil in water emulsion with an aqueous solution of carrier powder, e.g. milk powder, starch, dextrin, etc. The emulsion is then supplied to the spray tower atomiser by a high pressure homogeniser to ensure the feed is homogeneous. The fine spray of emulsion droplets is projected into a hot air stream to evaporate the moisture. The moist air and fine particles are collected in a cyclone whilst the dry fat powder is collected at the base of the drying chamber.

The design of spray driers has advanced significantly from the single stage spray drying system where the drier discharge tended to be at a relatively high temperature to the double and triple stage systems which reduce the energy consumption so that the product could be produced at lower temperatures. Thus more temperature sensitive and higher fat products could be handled. Figure 8 shows the layout for a simple single stage system.

As with powdered fats the design of the atomiser is critical to the success of the plant. The geometry of the spray chambers is also very important where the air flow can be counter- or concurrent and it is important that the spray must not strike the tower wall until dehydrated, otherwise it will stick and burn onto the wall and ultimately interrupt the air stream. The volume of air, its velocity and temperature, must be controlled to be consistent with the heat sensitivity of the product.

Microencapsulation of fats and oils has now been developed to a considerable extent. The method gives a product where the fat is at the core of a non-fatty substance; thus liquid oils can be used without risk of leakage and the temperature sensitivity of the product is reduced.

Microencapsulation utilises a spray drying technique. The fat and oil are thoroughly emulsified with a water solution of the water soluble coating material such as gelatine, gum arabic starch or dextrin. The water is evaporated off as described earlier to leave dry particles in a shell or capsule of dry colloid in which the fatty material is embedded or encapsulated in the form of a minute droplet.

There are several other methods for manufacturing fat powders, for example systems have been developed where molten fat can be sprayed into a stream of dry particles in a spray chiller. This simple expedient of feeding a 'carrier' like flour into the area of atomisation in a spray chiller gives the opportunity for lower melting fats to be used. The system can be used for the manufacture of dry food mixes.

Simple mixing of a powdered fat with a dry component can be used to manufacture fat powders. In this system heating takes place due to the mechanical and shear forces applied; thus a cooling phase prior to packaging of the product is required.

There are also a number of recently patented systems for the manufacture of fat powders. For example European Patent Application 0289069 describes a system of blending a fat or oil with a hydrophilic base substance like starch, caseinate, gelatine, etc., and a small proportion of a polyol like glycerol or propylene glycol to give a free flowing powder which rapidly gives up the oil when dissolved in water, thus being valuable for the manufacture of the seasoning of soups.

(c) Butter powders

Butter powder deserves special consideration when discussing fat powders. A considerable amount of development work has been carried out on this product in order to achieve an 80% butter fat product that performs well in bakery applications and can withstand higher ambient temperatures than butter itself.

The powder is made by way of standard spray drier technology (Frede, Patel & Buchheim, 1987) from an emulsion of milk powder solution and anhydrous milkfat which is homogenised as a 40% total solids solution and then spray dried. The mechanical stability of the powder is improved by the inclusion of tri-sodium citrate, possibly due to its influence on the fat/protein interface. Emulsifier, usually monodiglyceride, is included in the product, even though it has been found to affect adversely the mechanical stability of the powder, as it improves the aeration of the cake batter.

The risk of caking or clumping of the powder can be reduced by using a higher melting fraction of butter fat and by cooling the powder immediately on leaving the spray tower as this creates many crystal nuclei which help to retain the state over a wide range of temperatures. If the cooling is rapid enough the crystalline structure can also give benefits in bakery applications. Butter powder is useful as a bakery ingredient as it is dispersed easily and quickly in the mix and imparts the richness to the product associated with butter.

APPLICATIONS OF FAT POWDERS AND POWDERED FATS

At a time when the demand for convenience foods is increasing the use of powdered products has increased. Powdered products that can be easily reconstituted are now of considerable importance. The best known examples come from the dairy industry where there is a considerable range of milk and milk proteins in powdered form being used as ingredients and nutritional supplements. There is also the opportunity for efficiency improvements in industrial situations where powders give easier handling and storage (Hogenbirk, 1984).

Powdered fat is now being used as a bread improver. This kind of product often includes emulsifiers. Flaked and powdered fats have a limited application because they have a high melting point; however, they have a major role as crystallisation accelerators in products like fondant, and as a stabilising agent for paste products like peanut butter. As well as this lauric hard butters in powder form are used as easy adjusters for the fat content of compounded chocolate. Flaked and powdered lauric fats are used in the preparation of ice cream mixes.

Fat powders can vary considerably in their fat content from as low as 15–20% in microencapsulated powders, through coffee creamers and whippable toppings at 30–40% to butter powder at greater than 80% fat where it is used as a bakery ingredient.

The presence of non-fat dry solids in fat powders means that lower melting fats and fat blends can be used in fat powders than in powdered fats which widens the range of applications compared with powdered fats, though as mentioned earlier some constraint is placed on fat powders in that the carrier must be compatible with the other final recipe ingredients.

One of the earliest examples is coffee whiteners, now used extensively in catering outlets where the carrier is skimmed milk powder. Other convenience foods are powdered sauces and soups that can be easily reconstituted. Bouillon cubes form a special area where liquid oils can be used and in some cases the fat carries the flavouring compounds, e.g. chicken stock. These are 'instantised' products which are readily soluble where the powder has been agglomerated to make it more soluble.

Instant desserts and whipped toppings that can contain up to 40% fat and emulsifier are popular convenience foods relying heavily on the added emulsifier system to ensure good aeration.

Prepared cake mixes for both catering and domestic outlets have been available for many years. The traditional manufacturing method was to distribute a conventional boxed shortening or a pumpable shortening throughout the flour and other dry ingredients or inject a softened shortening into a haze of flour as small droplets. In both cases the crystal structure of the fat is maintained in order to ensure a good bakery performance in the finished mix. The high fat powders which contain between 70 and 85% fat now give the opportunity for bakery mixes to be prepared by mixing all the dry ingredients together. The difficulty that is found is that the fat will not necessarily be in the correct crystalline modification for optimum performance. Developments in emulsifier systems in these products have substantially improved their performance.

It is necessary that consideration is given to the comparison of the advantages and disadvantages of fat powders and powdered fats and then they must be compared with fats as manufactured shortening or liquid form.

In comparing powdered fat and fat powders it can be seen that due to the presence of a carrier in fat powders there is a greater microbiological hazard than in the all fat powdered fat. It has been noted that fat powders deteriorate oxidatively more rapidly than powdered fat because of the contact with the carrier. However, the comparison is confused by the fact that most fat powders are less saturated than powdered fats.

Fat powders and powdered fats are easy to handle and store as are boxed fats. However, liquid oils in bulk require expensive temperature controlled storage installations. Where boxed fats are melted prior to use this gives a significant disadvantage because of equipment, energy consumption and handling the melted fat.

Boxed fats have considerable advantages over fat powders and powdered fats in recipe versatility, though the method of incorporation can be more expensive than with the powders in terms of energy consumption.

The selection of the form a fat product is in depends on the final food product, the ease of handling, incorporation and most importantly the quality of the finished food.

GHEE

Ghee is clarified crystallised butter from buffalo's or cow's milk. It is the most common form in which butterfat is used in India and other countries of the Far East.

Ghee manufacture in India is still a home industry (Ganguli & Jain, 1973). However, increasing industrialisation and the growth of the urban population has led to the establishment of factories using both batch and continuous processes. Anhydrous butterfat is also widely sold as ghee.

The traditional method of ghee manufacture was by way of fermenting whole milk to curd, the curd was churned to butter which was then clarified at high temperature to remove the water. This clarification step takes place at between 110 and 120°C. This is the 'desi' method and it is claimed that it gives the best flavour characteristic because during the moisture removal there is interaction between the fat and the fermented residue of the solids non-fat.

In factory manufacture the process is being simplified where cream or sweet cream butter are used. Heat is applied to either the butter or cream in open jacketed vessels to remove the moisture with agitation. A pre-stratification method can also be used where, after initial heating of the butter, the bottom layer is discarded before the remainder is heated to the desired temperature.

The continuous method is based on pumping cream or butter through a steam heated scraped surface heat exchanger. The superheated cream/butter is passed into a flash evaporator where the moisture is separated from the liquid fat. This process goes through multiple stages to remove the moisture completely.

The processes are selected on the basis of yield and energy efficiency as well as achieving the desired flavour, colour and shelf-life. The quality of the ghee depends on the quality of the milk, cream, the curd or butter, and the temperature of clarification. The lower temperature of 110°C gives a mild flavour, whilst 120°C gives a strong and more 'cooked' flavour.

Ghee quality

The flavour, texture and colour have long served to characterise ghee so they can also be used as indicators of the quality. Where the manufacture of ghee is still a cottage industry, using the desi method, flavour and texture are the only criteria used to judge quality. A nutty, lightly cooked aroma and flavour are generally prized. The texture can vary from granular to smooth and, in some cases, even show a tendency to separate a liquid portion. These textural differences are fixed by local tradition, for example a granular texture is preferred in India with no separation, whilst in Pakistan the product is favoured when it has larger, softer granules which are dispersed in a supernatant liquid.

In the industrial manufacture of ghee considerable effort has gone into providing a product of greater uniformity and extended shelf-life. The quality of ghee depends heavily on the milk which in turn depends on the animal feed, the season and the health of the animals. For example, winter ghee has been shown to have a high acidity, melting point and grain size and milk from animals fed on cottonseed gives ghee which increases in acidity less quickly, but also leads to a product of lower melting point. Where the ghee preparation is by way of cream, 'dahi' or butter the quality of these products also influences the quality of the ghee. Then the method of preparation and temperature of clarification all have an influence.

These factors determine the physiochemical features of ghee; for example the typical analytical characteristics of cow and buffalo ghee are shown in Table 2.

The quality of ghee (Sharma, 1981) is generally measured analytically by parameters such as acid value, peroxide value, flavour and shelf-life. Bacteriological quality is assured by reducing the moisture content to less than 0·3%.

The free fatty acid content of ghee varies with the method of preparation, thus in ghee prepared from ripened cream or butter the free fatty acid is in the range 0·34–0·40% whilst in unripened cream or butter it is 0·23–0·28%. The peroxide value is a less valuable indicator of quality than either free fatty acid content or flavour and, though in the fresh product it should be low, in the stored product it varies considerably, particularly near the point where rancidity is detectable organoleptically.

The flavour of fresh ghee is greatly influenced by the temperature of

TABLE 2

	Buffalo	Cow
% Solid fat °C		
10	51·9	53·4
15	37·5	38·6
20	23·1	22·6
25	16·3	15·7
30	10·8	7·9
35	4·0	3·2
40	NIL	NIL
Slip point (°C)	29·9	33·4
IV	28·4	34·9
Lovibond colour (5·25 in cell)	2·5R 24·0Y	4·3R 44·0Y
% Moisture	0·3 max.	0·3 max.
% Free fatty acid		
Ripened milk	0·34–0·40	
Unripened milk	0·23–0·28	
Unsap. (mg/100)	390	450

clarification. Ghee prepared at 120°C has a distinctive 'cooked' flavour, whilst that prepared at 110°C has a considerably milder flavour. Ghee can be clarified up to 140°C when a 'burnt' flavour is developed. The temperature of clarification also influences off flavour development in that product clarified at 110°C retains its flavour longer than that clarified at 120°C.

The flavour of the ghee is also dependent on the method of preparation. Ripening of the cream or butter is always considered to give an improved flavour and ghee from desi butter is felt to be the best, being nutty with a cooked or caramelised aroma.

It has been found that the keeping quality of desi ghee is better than that of direct cream or creamery butter ghee and longer heating has been found to improve the oxidative stability of ghee, because of the liberation of phospholipids into the fatty matter. Besides organoleptic changes, in storage ghee undergoes textural changes. When filled into containers crystallisation occurs with the formation of solid, semi-solid and liquid layers. Below 20°C ghee has a small grained and compact texture. Storage temperatures of 28–29°C lead to a well defined granular texture though cooling to this too rapidly can give rise to a granular settled portion, a liquid oil portion and a floating hard flake, with each layer having different chemical characteristics, at higher temperatures the texture becomes looser and a liquid oil layer then becomes evident, and at 34°C the product is fully liquid.

The consumer's perception of quality is based on flavour, texture and colour as these three indicate quality and purity. A uniform granular structure, with a white or off white colour is expected. The flavour should be characteristic with desi ghee the most popular.

Anhydrous milkfat, in spite of its more bland flavour, is now being supplied as ghee. The anhydrous milkfat is prepared at temperatures of 80°C so it has greater residual moisture, but less protein.

Uses of Ghee

Ghee is the major culinary fat in the Middle and Far East and is used in a wide range of foods. It is used in the shallow frying of vegetables; in India for example it is used in curries, paratas and dosais. Certain foods like puris and samosas are deep fried in ghee. Basting chicken and pilau with ghee imparts a distinctive and characteristic flavour. Certain dishes demand that the ghee is a solid. Finally, ghee is used as a spread in molten or semi-molten form for chappatis or partially malted and mixed rice. A small proportion of the ghee produced is used in confectionery and to cook sweetmeats based on cereals, milk solids and fruit.

Though ghee is the major culinary fat in the Indian sub-continent, and is used in many foods, not all Indian food is cooked in ghee. In fact many recipes call for liquid vegetable oil. Cooking with ghee has a certain amount of status attached to it, rather like cooking in butter.

VANASPATI

Vanaspati or vegetable ghee is a substitute for natural ghee and is based on hydrogenated vegetable oils. In India the economic situation caused the demand for animal fats to exceed production which resulted in a price increase so that the product was put beyond the reach of the general population. Just after the first world war vanaspati was introduced into the Indian market, where it was quickly accepted because of its similarity to ghee in texture and its relative cheapness. Since that time vanaspati has expanded in use all over the Middle and Far East and in the types of oil used in its manufacture.

In the beginning vanaspati was based on only one hydrogenated oil, for example, groundnut oil or cottonseed oil. However, developments in technology and economic pressure have led to the use of hydrogenated blends utilising oil like soyabean, rapeseed and palm oil. Latterly, there has been the introduction of palm oil fractions (Kheiri & Flingoh, 1982).

Vanaspati is expected to be similar in texture and colour to natural ghee and achieve at least the same shelf-life. It is obviously not possible for the product to have the same flavour as natural ghee; however, there is now a growing acceptance for the product and in some countries it is allowed that flavours are added to make a vanaspati even more characteristic of ghee. In those countries where vanaspati is a major dietary fat its formulation and production are legally specified.

The formulation is now mainly based on soyabean oil, rapeseed oil and palm oil hydrogenated to slip melting points between 35 and 41°C and blended as economic and technical constraints dictate. It has been found, for example, that with levels of palm oil greater than 40% in a blend with hydrogenated soyabean oil there is a greater risk of liquid and solid separation and a poor granularity. It has also been found that oils hydrogenated in such a way that production of *trans* fatty acid is maximised when cooled slowly to manufacture vanaspati give rise to the expected granular texture.

In the manufacture of vanaspati only those oils that have been fully refined, bleached, hydrogenated and deodorised are used and so the quality control requirements that are applied to margarine oils can be applied to vegetable ghee.

The texture requirements for vanaspati can be achieved by slow cooling as in the case of ghee. However, continuous production can be achieved by way of scraped surface heat exchangers with the intensity of chilling and the rate of throughput appropriately adjusted to the oil blend being used. The culinary uses of vegetable ghee are equivalent to those of ghee, where it is used in a range of dishes that are basted and fried.

The improved oxidative stability of vanaspati, compared with ghee, and its higher melting point give it a wider application in confectionery.

SALAD AND COOKING OILS

Salad and cooking oils are major culinary oils. These products are always based on vegetable oils that are liquid at room temperature and so include soyabean oil, rapeseed oil, cottonseed and corn oil and sunflower oil. All of these oils are processed to give blandness in flavour, good oxidative stability and clarity at low temperatures. This can mean, in some cases, as well as refining, bleaching and deodorisation, the treatment of winterisation to remove waxes and the higher melting triglycerides. Soyabean oil is often 'brush' hydrogenated to reduce the linolenic acid content and then winterised to remove the higher melting triglycerides which leads to significant improvement in oxidative stability.

These oils are used in the manufacture of mayonnaise, salad dressings and sauces. Most mayonnaise contains 77–83% oil to give a thick oil in water emulsion.

Salad dressings, like French dressing, contain basically oil, vinegar and spices. The simplest require shaking before use. Salad dressing can also be emulsified and thickened by the use of gums, egg yolk and emulsifiers.

Sauces of the Hollandaise and Bernaise variety, which contain about 30%, must flow and yet adhere to the hot food to give the desired flavour and succulence to the food. The oils described above are often used in these recipes, but clarified butter has a special place as it contributes a special richness and flavour.

A consideration of salad and cooking oils cannot be left without a mention of olive oil. The highest quality olive oil—virgin olive oil—is obtained from the freshest fruit by mechanical pressing and post-pressing clarification. The oil is used with no further purification and so retains a highly characteristic flavour and colour and the presence of tocopherols and certain sterols ensure good oxidative stability. The main contribution in salads and cooking is to the flavour. Clarified butter or butter oil is also used in certain recipes, again to provide a distinctive flavour.

FRYING FATS AND OILS

Deep and shallow frying of food is a rapid method of cooking and the fat or oil serves not only as a heat transfer medium, it also reacts with the protein and carbohydrate in the food to develop colour, flavour and odour. Additionally, the fact that the oil or fat is absorbed by the food means it becomes an ingredient and has a major impact on the palatability of the finished food. The requirements of the fat vary with respect to the type of frying undertaken and the turnover rate of food fried.

In the case of shallow frying (pan frying) the oil or fat is generally only used once, with the consequence that resistance to breakdown during the frying process is of little importance; therefore those oils used extensively as salad and cooking oils can be used and give the food an oily surface and good mouthfeel. Solid fats, based on lard and palm oil, are also used extensively in shallow frying, as well as deep frying.

In the case of shallow frying, butter and margarine can be used to provide distinctive odours and flavours. The presence of milk solids is felt to improve flavour and product colour, though it can lead to the food sticking in the pan. In the case of margarine this is alleviated by the presence of lecithin. There are also available coloured and flavoured oils and oil blends to give the fried food a characteristic flavour.

In deep fat frying the frying medium is exposed to different conditions compared to those of shallow frying. The major differences are that the foodstuff is submerged in the hot oil, the oil is reused and can be held at high temperatures for extended periods, and fatty matter and juices from the food being cooked are transferred to the oil bath.

Deep fat frying is a deceptively simple operation, which can, on occasion, result in abuse of a frying medium (Matz, 1984). The frying operation cooks the food and drives off moisture as steam; thus the frying temperature must always be above the boiling point of water. As water is evolved the void left in the food can take up the frying fat; thus it is necessary that the frying operation is adjusted so that the steam evolution is maintained to give a high internal pressure to prevent undue fat uptake.

It is important that the temperature and correct heat balance in the fryer are maintained in order to give good efficiency. The heat required to cook a given weight of food (which contains a certain amount of moisture) is balanced against the heat input rate. The food entering the fat and steam being driven off cool the fat so that heat input is required in order for the fat to recover to the frying temperature. The heat input should be high enough to maintain a high internal pressure in the food, due to the steam generated, so that the risk of fat absorption is limited. It can be seen from the foregoing remarks that not only is the amount of moisture present important, but so is its availability; thus the size of the food pieces affects the rate of moisture loss. Small food pieces have a larger surface area per unit weight than large pieces of food. Heat penetration into small pieces is rapid, so moisture loss is rapid with the result that a higher heat recovery rate is necessary.

Good fryer design and successful heat balance in the system lead to the preparation of non-greasy fried food. Correct loading of the pan and control of the size of food pieces also have an effect on the quality of the fried food and the life of the frying fat.

Frying fat, as expected, deteriorates in use, leading to off flavours and off odours. Other signs of more serious deterioration are discoloration, smoking and foaming.

Figure 9 is a diagrammatic presentation of the type of reaction taking place in a bath of frying oil, during both heating and frying (Fritsch, 1981). The major chemical changes are:

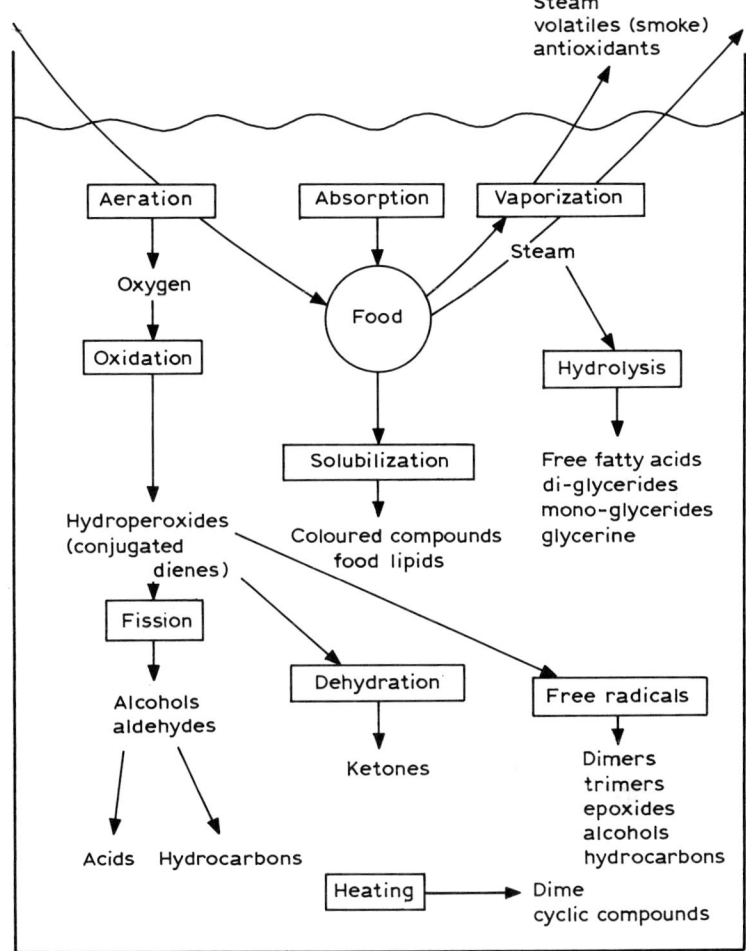

Fig. 9. Diagram of oil changes during deep fat frying.

— hydrolysis which generates free fatty acids, mono- and diglycerides, and glycerol;
— oxidation to give volatile products with strong odours and flavours;
— polymerisation which increases viscosity and leads to fixed foams;
— ring formation leading to cyclic monomers which are nutritionally undesirable.

Controlling these chemical reactions and build up of the by-products is a major part of the skill of the fryer.

Fryer design is of considerable importance since materials of construction, type and control of heating, ease of cleaning, efficient ventilation with no drip back and risk of overloading can all play a part. However, probably the most important factor is the 'turnover'. The 'turnover' is the rate at which the fryer has to be topped up with fresh fat in order to hold it at the full mark, to balance losses from the fryer due to absorption. In practice a make-up rate of about 15–25% of the fryer capacity can hold the free fatty acid content, caused by hydrolysis, at 0·4–0·8% in normal use.

Particles of food can accumulate in the fat during frying. These tend to burn in the fat causing black particles on the fried food. If these particles are not removed on a regular basis by filtration, they cause the development of undesirable flavours and accelerate the deterioration of the fat.

Monitoring the quality of a frying fat during use can be done with simple tests like colour, free fatty acid content and smoke point (Matz, 1984).

The small scale caterer does not have the facilities to carry out laboratory tests of this kind. However, a number of rapid tests have been developed to judge used frying fat quality, for example measurement of dielectric constant which changes as the number of polar by-products change. Colorimetric tests are available which utilise indicators to show the presence of fatty acids and oxidised products. These tests are used to show the residual frying life of the fat. However, they have not been found to correlate well with laboratory techniques (Sulthara & Sen, 1979).

There are tests of greater sophistication for evaluating used frying fat. Estimation of the level of oxidised fatty acid has achieved official status in West Germany for the evaluation of used frying fats. Others include the determination of polymeric triglycerides, cyclic monomers and polar compounds which require techniques of increasing complexity (Croon *et al.*, 1986).

Tests to predict the performance potential of a frying fat fall back on the classical tests that indicate acid value and the oxidation state which indicate the efficiency of the processing and handling; tests for residual phosphatides, tocopherols and sterols which can have an effect on frying performance, the sterol and tocopherol being anti-oxidants, whilst the phosphatides cause darkening.

The introduction of methyl polysiloxane (silicones) not only controls foaming but also has a slight anti-oxidant action which is improved when the silicones are added with phenolic anti-oxidants, citric acid or ascorbyl palmitate. Though these additives improve the performance of the fat, good frying practice with the correct turnover ensures good frying life.

The type of deep fat frying operation dictates the type of fat to be selected, which can vary from meat fats to liquid vegetable oils. Fats with lower unsaturation, like palm oil or hydrogenated vegetable oils, tend to be favoured as giving greater resistance to breakdown in use. However, it has been shown that turnover and the type of frying operation, i.e. continuous or intermittent operation, can mean highly unsaturated oils can be used without risk to the fried product.

Frying fats can be supplied as solid blocks, though the use of a pourable frying medium has advantages when being added to the fryer when cold and heated without danger of burning and handling after use in the clean up operation since they can be filtered when cool. Hydrogenated and fractionated vegetable oils are popular in this case and there are also available pourable frying fats, with a small quantity of suspended solids based on lightly hydrogenated oils. The oils have improved oxidative stability compared with the unhydrogenated oil and the presence of the suspended crystals reduces the risk of migration of added silicones.

The different oils and fats on heating to frying temperature develop characteristic odours, e.g. animal fats becomes 'tallowy', soyabean oil develops a fishy odour and rapeseed oil an acrid, rubbery odour. These odours affect only the frying area and are not transmitted to the fried food. However, frying odours should not be allowed to reach the serving area no matter how bland they may seem.

CONCLUDING REMARKS

Changing nutritional attitudes and an ever increasing demand for qualitative excellence mean that there will be continued pressure on the oils and fats processor and food manufacturer to continue to do research and develop new products and systems. The margarine and shortening manufacturer is well placed to take advantage of those fats created by current and novel methods of modification. Changes in basic fats are also being achieved by plant biology. These techniques, combined with developments in processing equipment to achieve specific rheological characteristics, mean the processor has the opportunity to tailor products to the customer's requirements.

The position of milkfat is less flexible, though butter has a special place in high class foods because of its flavour and mouthfeel. Changes in eating habits with an increased demand for snack foods, combined with a demand for higher quality, give milkfat increased opportunities. For example, croissants and Danish pastry, with sweet and savoury fillings, are a snack of increasing popularity and those made with butter are prized because of their flavour and light eating qualities.

Developments in the introduction of butter fractions have led to some exciting opportunities, e.g. technology is already available which utilises specific butter oil fractions recombined with cream to form low calorie 'butters' which are spreadable from the refrigerator. The use of milkfat as a blend ingredient gives butterfat greater robustness, though control of cooling rates and agitation is needed to develop the correct crystal forms.

The publicity given to the relationship between the so-called saturated fats and high blood cholesterol contents has led to an increased use of margarines and spreads high in polyunsaturated fatty acids. Recipes are now available for the use of these high liquid oil blends in bakery products. Successful products can only be made when the fat or margarine and other ingredients, like the egg, have been chilled to about 5°C.

The continuing demand for the use of fats high in polyunsaturated fatty acids could popularise liquid and slurry shortenings where the baked products' quality will rely heavily on the emulsifiers used as mentioned earlier (Hartnett & Thalheimer, 1979).

The problem has also been approached by way of re-examination of recipes to see if fat levels can be reduced (Hodge, 1986) and some success has been reported.

Reduced fat emulsions with suitable emulsifier systems have been used to make good quality cakes and puff pastry leading to an overall reduction of fat in the recipe.

The development of fat substitutes could herald significant changes in the quantities of fat in culinary recipes to be replaced by, for example, various forms of starch and sucrose polyesters (Gillis, 1988).

The demands for higher quality, improvements in nutritional standard and greater naturalness are major challenges for all sectors of the oils and fats industry and they can only be met with increases in understanding of the material that is handled and developments in processing techniques.

REFERENCES

ANDERSEN, A. J. C. & WILLIAMS, P. N. (1965) *Margarine*, 2nd edn, Pergamon Press, Oxford.

BAILEY, A. E. (1950) *Melting and Solidification of Fats and Fatty Acids*, Interscience, New York.

BIRKER, P. J. M. W. L. & PADLEY, F. B. (1987) *Recent Advances in the Chemistry and Technology of Fats and Oils*, Hamilton, R. J. & Bhati, A. (Eds), Elsevier, Amsterdam.
BLENFORD, D. (1987) *Food* (May), 19–23.
CARLIN, G. T. (1944) *Cereal Chem.*, **21**, 189–99.
CHRISTIE, W. W. (1973) *Lipid Analysis*, Pergamon Press, Oxford.
COCKS, L. V. & VAN REDE, C. (1966) *Laboratory Handbook for Oil and Fat Analysts*, Academic Press, New York.
CROON, L. B., ROGSTAD, A., LETH, T. & KIEUTAMO, T. (1986) *Fette Seifen Anstric*, **88** (3), 87–91.
FREDE, E., PATEL, A. A. & BUCHHEIM, W. (1987) *Molkerei Zeitung Welt der Milch*, **41**, (51–2), 1567–73.
FRITSCH, C. W. (1981) *J. Am. Oil Chem.*, **58**, 272.
GANGULI, N. C. & JAIN, N. K. (1973) *J. Dairy Sci.*, **56** (1), 19–25.
GILLIS, A. (1988) *J. Am. Oil Chem. Soc.*, **65**, 1708–11.
HARTNETT, D. I. & THALHEIMER, W. G. (1979) *J. Am. Oil Chem. Soc.*, **56** (12), 948–52.
HODGE, D. G. (1986) *BNF Nutrit. Bull.*, **11** (3), 153–61.
HOGENBIRK, G. (1984) *S.A. Food Rev.* (Feb/March), 26–7.
JOYNER, N. T. (1953) *J. Am. Oil Chem. Soc.*, **30** (11), 526–33.
KHEIRI, M. S. A. & FLINGOH, C. H. O. H. (1982) *Palm Oil Product Technology in the Eighties*, Palm Oil Research Institute of Malaysia, Kuala Lumpur.
KROG, N. (1975) *Water Relations of Food*, Duckworth, R. B. (Ed.), Academic Press, New York, pp. 587–611.
KROG, N. & LAURISDEN, B. J. (1976) *Food Emulsions*, Striberg, S. (Ed.), Marcel Dekker, New York, pp. 67–139.
KROG, N. & NYBO JENSEN, B. (1970) *J. Food Technol.*, **5**, 77–87.
LAMB, R. (1987) *Food* (Dec.), 39–43.
LARSSON, K. (1980) *Cereals for Food and Beverages: Progress in Cereal Chemistry and Technology*, Inglett, G. E. & Munck, L. (Eds), Academic Press, New York.
MATZ, S. A. (1984) *Snack Food Technology*, 2nd edn, AVI Publishing, Westport, CN.
MULDER, H. & WALSTRA, P. (1974) *The Milk Fat Globule*, Purdoc, Wageningen.
MÜNCH, E. W. (1986) *Deutsche Milchwirtschaft*, **37** (9), 567–9.
OPFER, W. B. (1975) *Chem. Ind.* (18), 681–7.
PYLER, E. J. (1973) *Baking Science and Technology*, Siebel Publishing Co., Chicago.
SHARMA, R. S. (1981) *J. Food Sci. Technol.*, **18** (March–April) 70–7.
SHEPHERD, I. S. & YOELL, R. W. (1976) *Food Emulsions*, Friberg, S. (Ed.), Marcel Dekker, New York, pp. 216–75.
SULTHARA, S. N. & SEN, D. P. (1979) *J. Food Sci. Technol.*, **16**, 208–13.
TIMMS, R. E. (1984) Phase behaviour of fats and their mixtures. *Prog. Lip. Res.*, **23**, 1–38.
WADDINGTON, D. (1986) *Analysis of Oils and Fats*, Hamilton, R. J. & Rossell, J. B. (Eds), Elsevier, Amsterdam.
ZÜRCHER, K. & HADORN, H. (1979) *Gordian*, **79**, 182.

7

Milkfat in Sugar and Chocolate Confectionery

V. K. S. Shukla
International Food Science Centre A/S,
Lystrup, Denmark

SUMMARY

Chocolate can be improved in terms of gloss, shelf-life and flavour by use of milkfat and milk chocolate has been manufactured for well over 100 years. When blending milkfat, the palmitic acid content can be used as an indication of firmness and solids content.

Cocoa butter melts between 32 and 35°C and is predominantly composed of symmetrical triglycerides. A relationship exists between the triglyceride composition of cocoa butter and its solid fat content. About 80% of the worldwide consumption of chocolate consists of milk chocolate in which up to 30% of the fat is milkfat. Milkfat causes a softening effect in chocolate, and in its various commercial states can form eutectic mixtures with cocoa butter while hardly altering polymorphic behaviour.

Milkfat is used in chocolate as either milk powder or milk crumb. Roller dried powder has a more 'cooked' flavour than spray dried powder, and is coarser but easier to disperse. The use of milk crumb can have flavour/taste benefits over milk powder.

Fractionated milkfat may be regarded either as milkfat or a cocoa butter replacer, and assists in inhibiting bloom. Hydrogenated milkfat can also be of benefit. Interesterified milkfat appears to be more compatible with cocoa butter than unmodified milkfat, but lacks flavour. Other confectionery products such as toffee can be improved in flavour/taste terms by milkfat.

Analytical methods extend understanding of the mechanisms associated with fat incorporation in confectionery products and assist in defining safe levels of addition.

INTRODUCTION

The use of milk from animals to supplement the human diet has been acknowledged since the beginning of history. It is the most important food for a mammal and has always been the first food of the newborn offspring. Milk in its various forms has been used in the confectionery industry for a very long time (Cook, 1984) and as milkfat is used largely in chocolate manufacture. The first recorded use of milk in the UK was in 1727 when it was incorporated into chocolate by Nicholas Sanders for Sir Hans Sloane, the first surgeon to King George II. The first milk chocolate recipe was, however, developed by Daniel Peters of Vevey near Geneva in 1876. Milk besides providing nutrition also contributes to gloss and shelf-life and is essential to flavour, colour and texture. Milk chocolate is more popular than plain chocolate, especially when made by the crumb process.

COMPOSITION OF MILK

Milk is a highly nutritious part of the human diet. It is mainly composed of water, but is rich in protein, fat and sugar (Arbuckle, 1986). The approximate composition of milk is shown in Table 1. Milk is the only nutrient which contains the carbohydrate lactose and the protein casein in major amounts. Casein adds important amino acids to the human diet; some of the most important are glutamine/glutamic acid, lysine and tyrosine. Glutamine/glutamic acid plays an important role as an energy source for many metabolic processes that occur within the brain, a role it shares only with glucose. Lysine is widely distributed in body proteins and enzymes and plays an important role in growth and repair mechanisms in living organisms. It is especially important to vegetarians because of the low content of lysine in vegetarian diets. Tyrosine plays an important role in neurotransmission. There are approximately 19 vitamins normally present in fresh milk, including vitamins A, B_{12}, D, E, K, ascorbic acid, nicotinic acid, riboflavin and thiamine. Milk minerals are also of considerable importance as nutrients.

A large number of trace elements are present in milk from pasture fed

TABLE 1
Approximate composition of milk
(Arbuckle, 1986)

Constituent	Amount (%)
Water	87·1
Fat	3·9
Protein	3·3
Lactose	5·0
Ash	0·7

cows, of which selenium, silicon and zinc are the most abundant. Thus milk in its various forms contributes significantly towards the nutritional properties of confectionery products.

MILKFAT

The functional properties of milkfat in confectionery products are mainly governed by the lipid composition, and the effect seasonal and geographical variation has on it.

Fatty Acid Composition

Milkfat is extremely complex, with more than 500 different fatty acids reported in its numerous permutations of triglycerides. Its fatty acid profile is dominated by the saturated acids containing 4–18 carbon atoms. The *cis* C18:1 fatty acid resulting from enzymatic desaturation of C18:0 in both the intestinal epithelium and the mammary gland is the only unsaturated acid present in appreciable amounts. The other dominating fatty acids are butyric acid (C4:0), myristic acid (C14:0), palmitic acid (C16:0) and stearic acid (C18:0) (Taylor & Hawke, 1975; Banks *et al.*, 1987). A typical distribution of fatty acids in milkfat from New Zealand cows (Taylor & Hawke, 1975) is shown in Table 2.

TABLE 2
Fatty acid content in milkfat from New Zealand cows (Taylor & Hawke, 1975)

Fatty acid	Summer (%)	Winter (%)
C4:0	9·6	12·0
C6:0	4·5	4·5
C8:0	2·2	2·3
C10:0	4·2	4·2
C12:0	4·1	4·0
C14:0	11·5	10·8
C14:1/C15:0	2·9	2·2
C16:0	27·6	22·0
C18:0	10·1	13·1
C18:1	17·8	21·5
C18:2	1·4	0·7
C18:3	0·8	0·3

Seasonal and Geographical Variation

The fatty acid composition of milkfat exhibits a great seasonal and geographical variation and therefore the fat melting point and crystallization vary over a wide temperature range. The impact of this effect on confectionery formulations will be dealt with in detail during the later part of this chapter.

In the Northern Hemisphere it has been established (Parodi & Dunstan, 1971; Steen & Andersen, 1974) that the C18 acid content reaches a maximum in the spring and summer months with minimum values in the winter periods. These higher levels of stearic and oleic acid during the spring and summer periods emerge indirectly due to microbial hydrogenation of increased dietary intakes of linoleic and linolenic acids from fresh grass diets. In the countries of the Southern Hemisphere such as New Zealand and Australia, particularly in Queensland (Parodi, 1970), where cows are at pasture throughout the year, the milkfat C18 monoethenoic acid content reaches a maximum during the winter. During this time pastures are generally dry and non-palatable due to lack of rain and frosts. Thus during the winter months cows receive a sub-optimum plane of nutrition with a resulting dependence upon body reserves for synthesis of milkfat. The oleic acid is derived from plasma triglycerides which are rich in stearic acid which in turn is dehydrogenated to oleic acid by an enzyme system in the mammary gland.

An extended study conducted in Ireland (Cullinane et al., 1984a,b) concluded that the fatty acid composition exhibited an obvious seasonal variation with no regional variation being evident. Palmitic acid (C16:0) and oleic acid (C18:1) accounted for 50–60% of the total fatty acid composition. The palmitic acid content was highest in winter and was negatively correlated with oleic acid but not correlated with other C18 acids.

These findings were further confirmed in the author's laboratories (Shukla, unpublished results). Thus high C18:1 coincided with high iodine value, low solid fat contents and softer milkfat. These results are consistent with the findings of other workers in the Northern Hemisphere (Wood & Haab, 1957; Hutton et al., 1969; Steen & Andersen, 1974) and reflect the significant contribution of oleic acid to iodine value and butter firmness (Dixon, 1964). The level of C16:0 is responsible for the major proportion of variation in butter firmness (Cullinane et al., 1984a) and solid fat contents, respectively, and would be a better indication than iodine value for monitoring the change in the physical properties of the milkfat.

TABLE 3
Percentage of *trans* unsaturation in milkfat from different countries (Parodi & Dunstan, 1971)

Country	Summer (%)	Winter (%)
Australia	7·6	4·3
Italy	9·6	8·1
France	7·0	2·3
Canada	6·6	4·9

Trans Fatty Acids in Milkfat

Knowledge of *trans* unsaturation content in milkfat is desirable in order to understand its functionality in confectionery fat model systems. Normally, natural vegetable oils do not contain appreciable amounts of unsaturated fatty acids comprising a *trans* double bond. However, due to microbial attack in the rumen, the ingested feed in the cow undergoes extensive hydrogenation resulting in a mixture of geometrical and postional isomers. The amount and type of feed can influence the microbial population of the rumen which can influence hydrogenation and the formation of *trans* isomer in milkfat (Posati *et al.*, 1975). The results presented in Table 3 show the variation of *trans* isomer contents in milkfat from different countries. Seasonal feed variation results in higher unsaturated fatty acid levels in summer than in winter. The total isolated *trans* fatty acids in milkfat reported in the literature range from 2 to 11% with maximal values in summer and minimal values in winter. These *trans* contents do have a profound influence on the crystallization characteristics of the final fatmix in chocolate production systems.

Flavour Contribution

Milkfat has a great advantage in its desirable flavour. In all products prepared from milk the ultimate flavour achieved may arise from several factors which have a remarkable effect on the properties of the milk chocolate. The short chain free fatty acids (FFA) in milk provide a unique contribution to the flavour of milk chocolate candies.

APPLICATIONS OF MILKFAT IN CONFECTIONERY

The chocolate industry is the largest user of milkfat in the confectionery industry. Cocoa butter is an important major constituent of the chocolate

TABLE 4
Solid fat contents of different types of cocoa butter (Pulse-NMR—tempering mode: BS 684 Method 2)

Type of cocoa butter	20°C	25°C	30°C	35°C
Brazilian	62·5	53·2	30·8	0·2
Ghanaian	77·8	71·1	47·3	0·7
Malaysian	82·6	77·1	57·7	2·6

formulations. Cocoa butter is composed of predominantly (>75%) symmetrical triglycerides (Shukla et al., 1983; Shukla, 1988) with oleic acid in the 2-position. Its unique chemical and physical properties are of great importance for defining the physical properties of the chocolate. It contains approximately 20% triglycerides that are liquid at room temperature and has a melting range of 32–35°C and softens around 30–32°C. This is an essential requirement.

The solid fat contents (Shukla, 1983) of three different cocoa butters normally consumed in the chocolate industries are shown in Table 4. There is a good correlation between the triglyceride compositions and solid fat contents (SFC) of these cocoa butters (Shukla & Nielsen, 1989). Malaysian cocoa butter is the hardest and Brazilian is the softest, whereas Ghanaian lies in between the two.

Milkfat is an important ingredient of milk chocolate together with cocoa butter. Milk chocolate forms more than 80% of all the chocolate produced worldwide. The milkfat contents constitute up to 30% of the fat phase. The use of greater amounts of milkfat is constrained by the fact that milkfat softens cocoa butter very significantly.

The results presented in Table 5 show the typical solid fat contents (Shukla & Nielsen, 1989) of a cocoa butter, Danish butter concentrate

TABLE 5
Solid fat contents by pulse-NMR (tempering mode: BS 684 Method 2)

Sample	20°C	25°C	30°C	35°C
Cocoa butter	79·3	73·1	49·2	1·1
Danish butter concentrate	33·3	29·9	24·8	17·4
Anhydrous milkfat	11·1	9·0	5·0	0

(DBC) and anhydrous milkfat (AMF). These results show that butter concentrate is harder than AMF but far softer than cocoa butter. The compatibility of these fats was examined by preparing several blends of AMF and DBS with all three types of cocoa butters and the solid fat contents of these blends were measured as described earlier using two different tempering modes. One tempering mode is identical to that of the British Standards Institution system (BS 684 Method 2) which is technically similar to that of the IUPAC dilatation procedure. The only difference was that we melted samples to 100°C in order to be absolutely sure that no crystal structures subsisted. In the other extended tempering mode samples were stabilized at 20°C for 64 h.

A critical evaluation of the ISO-NMR curves presented in Figs 1–12 leads us to the following conclusions:

(a) These curves demonstrate the elimination of liquid fractions from DBC. The repeat quality of concentrates was acceptable.
(b) The softening effect (eutectic) of Brazilian cocoa butter is quite evident in these curves for both fractions as compared to the cocoa butters from Malaysia and Ghana.
(c) This eutectic is greatly improved in DBC due to the removal of low melting triglycerides. Thus DBC can be used in modifying the properties of Brazilian cocoa butter.
(d) The eutectic is deeper in the cocoa butter from Ghana.

Thus the addition of milkfat to cocoa butter results in marked lowering of the melting point, adversely affecting the crystallization behaviour and the hardness as shown above.

The softening of chocolate by milkfat is due to a combination of two effects (Timms, 1980). The low melting components of the milkfat soften the chocolate on account of their fluidity, while the solid fat components of dairy butter form eutectics with cocoa butter. However, 30% addition of milkfat to cocoa butter did not alter the polymorphism of the cocoa butter, although the rate of transformation between the polymorphs was reduced (Chapman et al., 1971). These authors concluded that 'the ability of cocoa butter to dissolve up to 30% milkfat without significantly altering its polymorphism is surprising considering the very complex nature of the triglycerides in the animal fat'.

In order to study the variation in milkfat two different batches of Irish milkfats were blended with the same batch of Malaysian and Ghanaian cocoa butters, and the solid fat contents of these blends were measured as

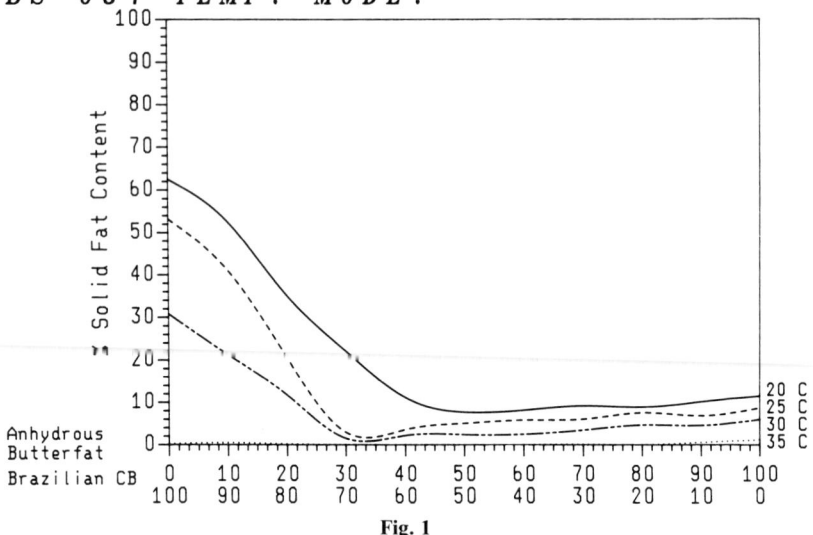

Fig. 1

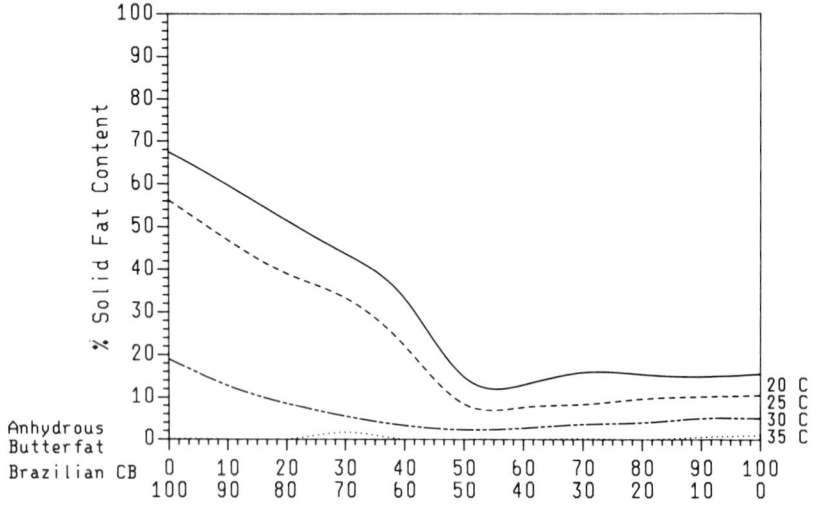

Fig. 2

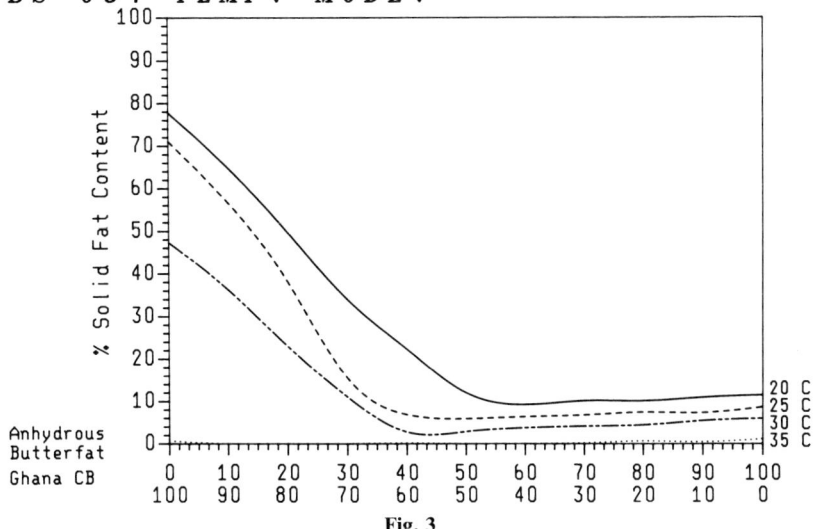

Fig. 3

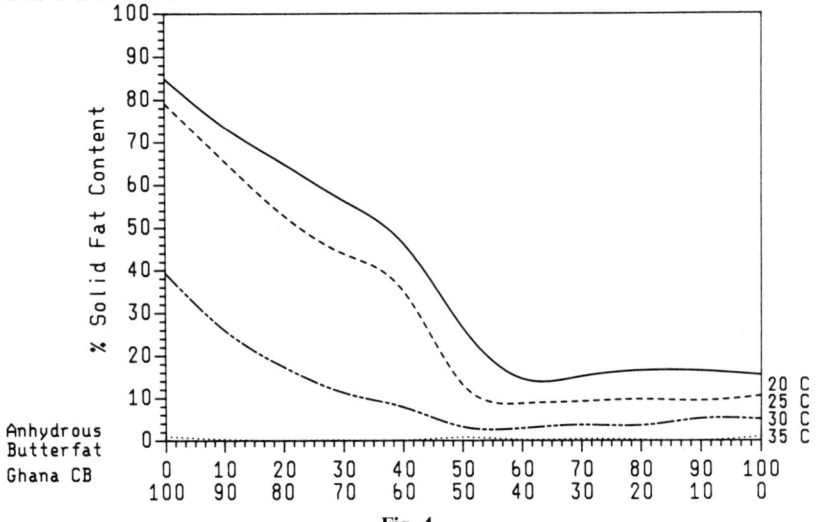

Fig. 4

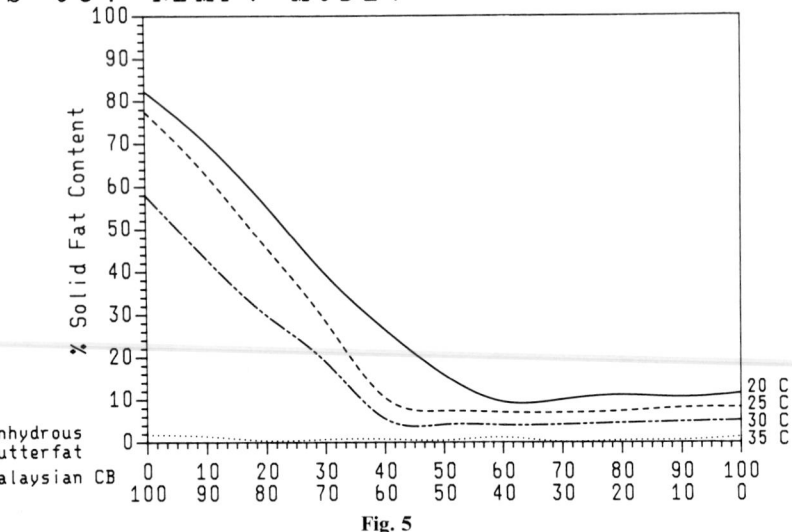

Fig. 5

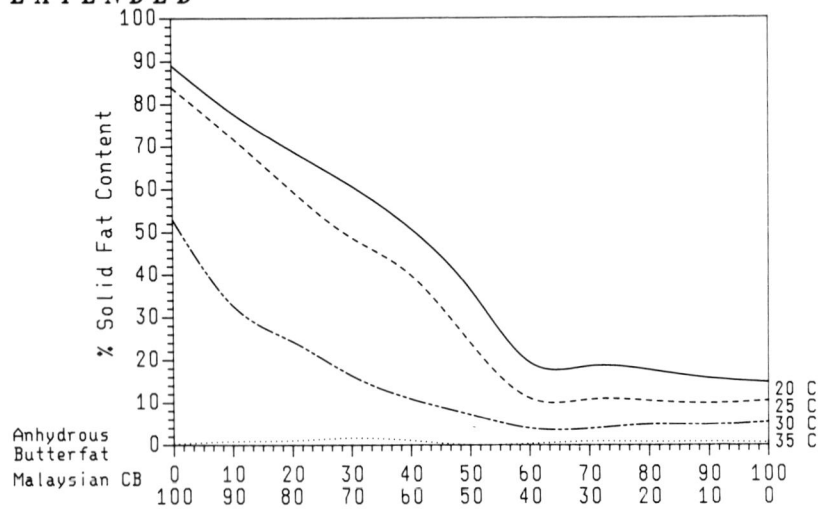

Fig. 6

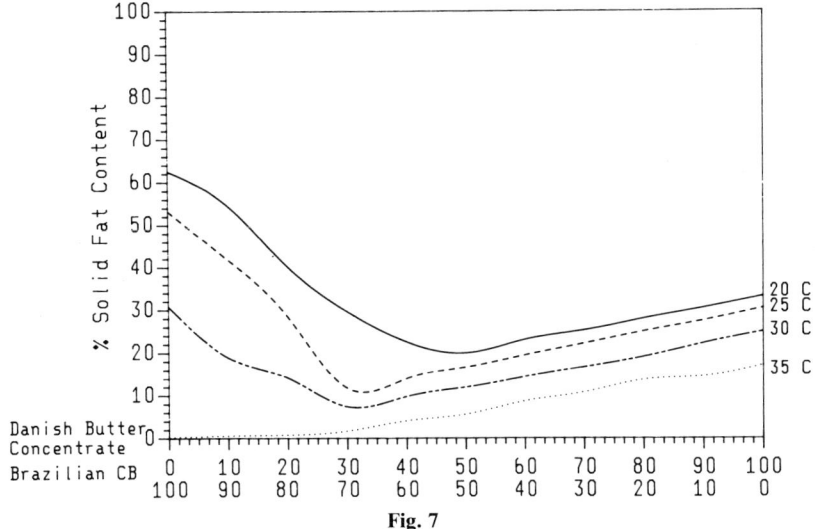

Fig. 7

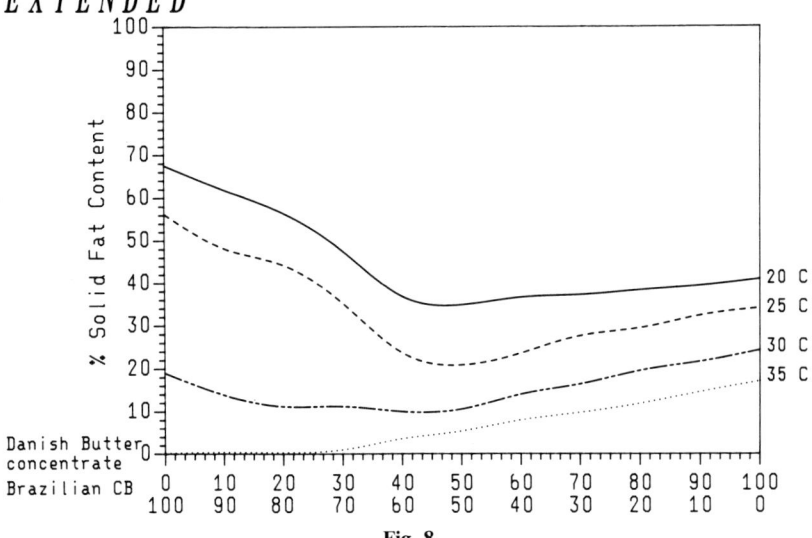

Fig. 8

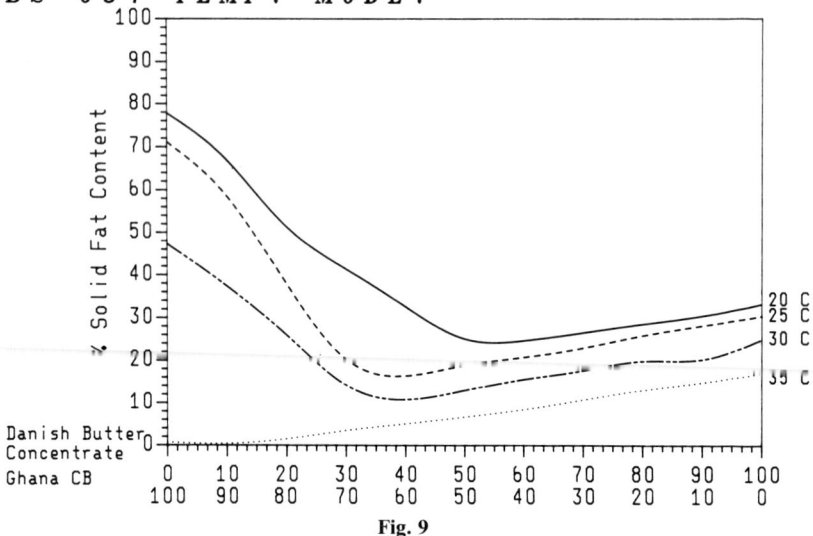

Fig. 9

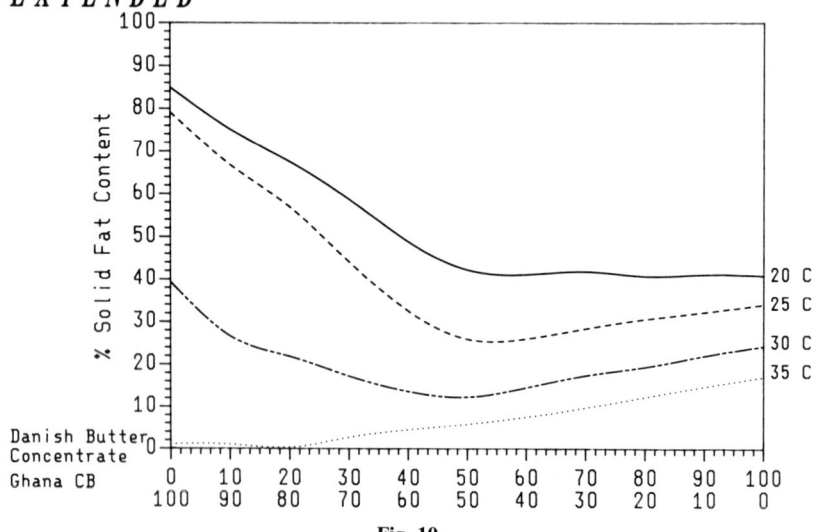

Fig. 10

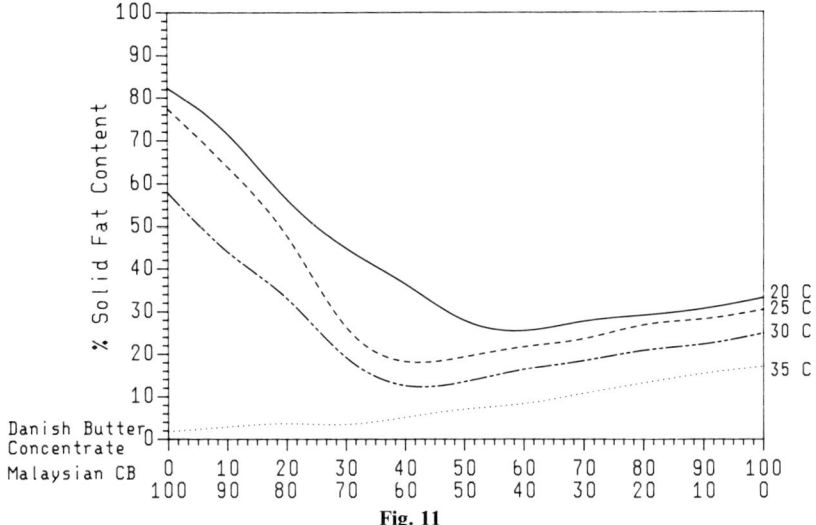

Fig. 11

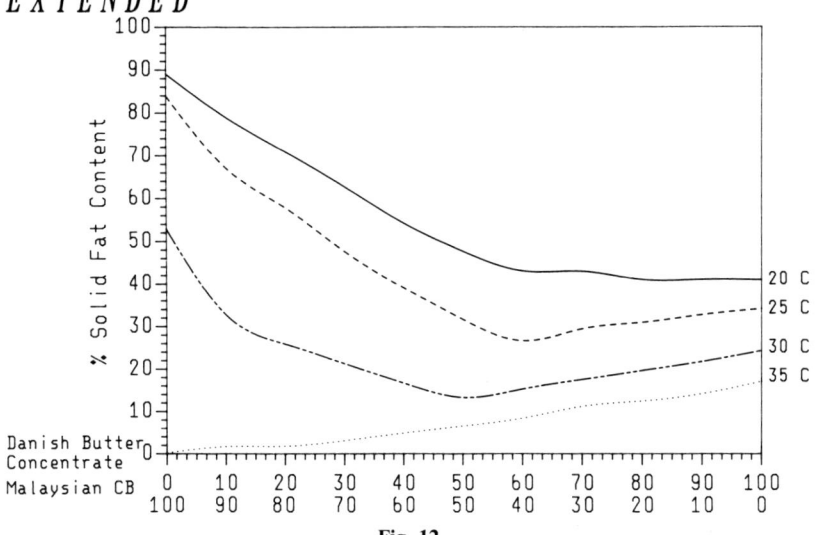

Fig. 12

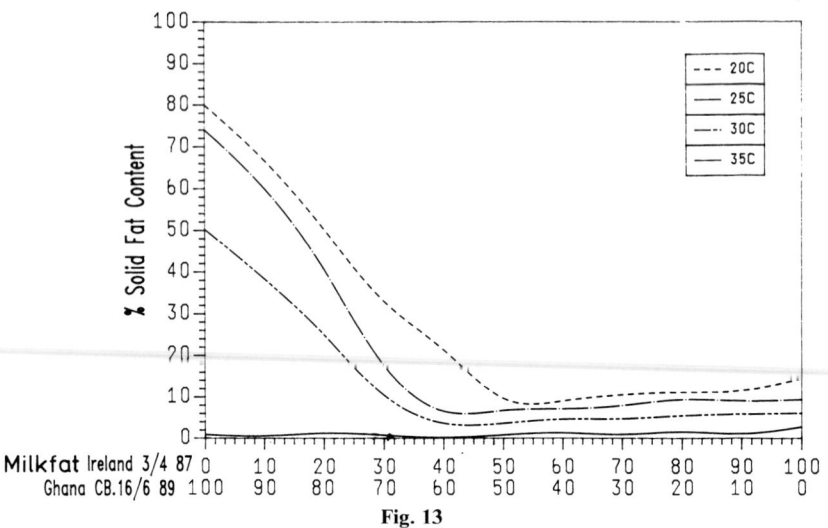

Fig. 13

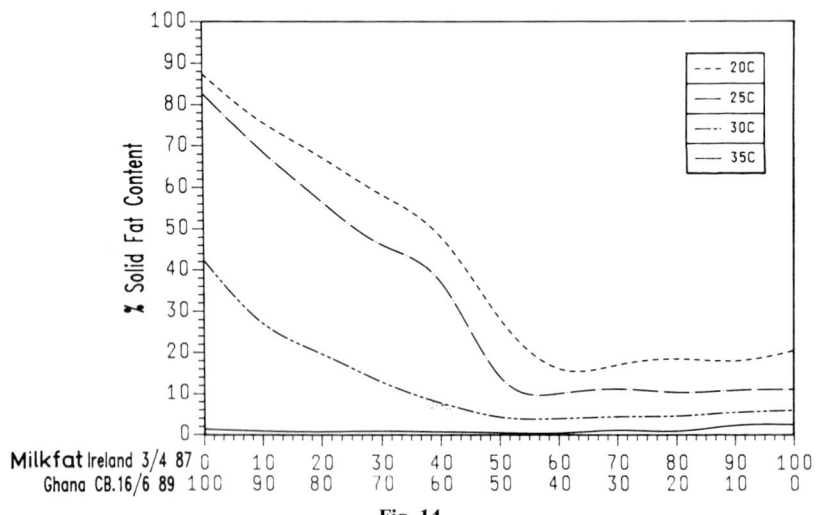

Fig. 14

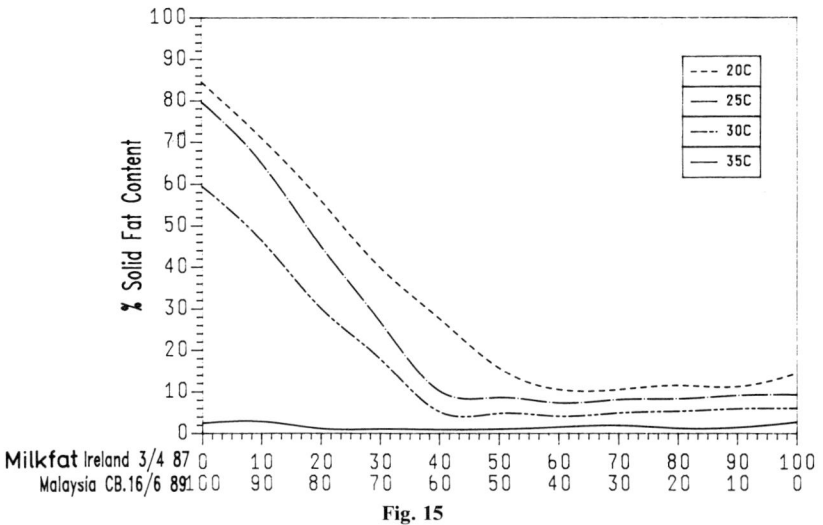

Fig. 15

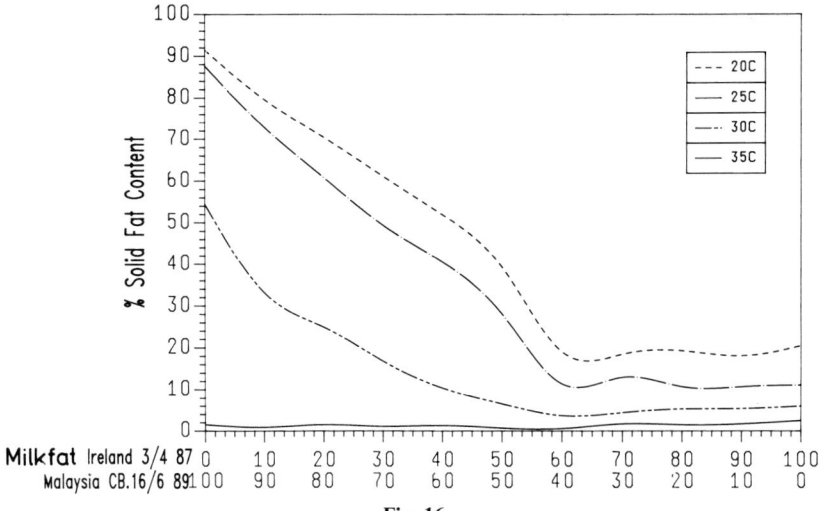

Fig. 16

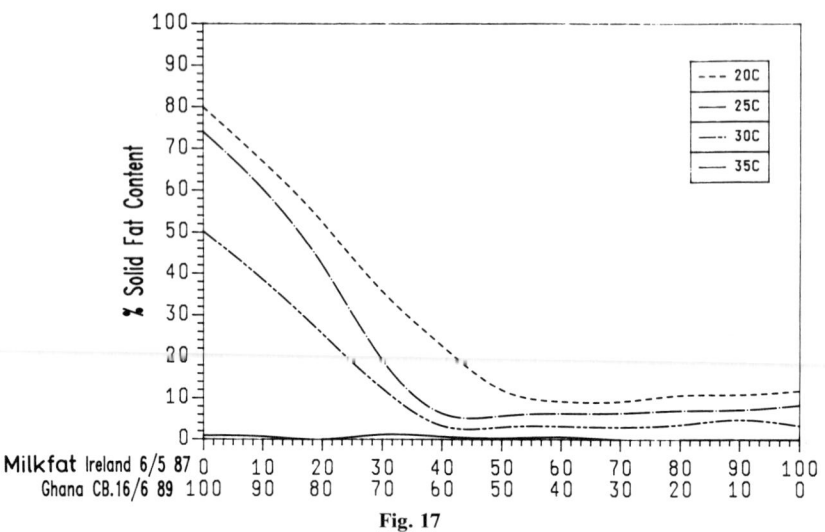

Fig. 17

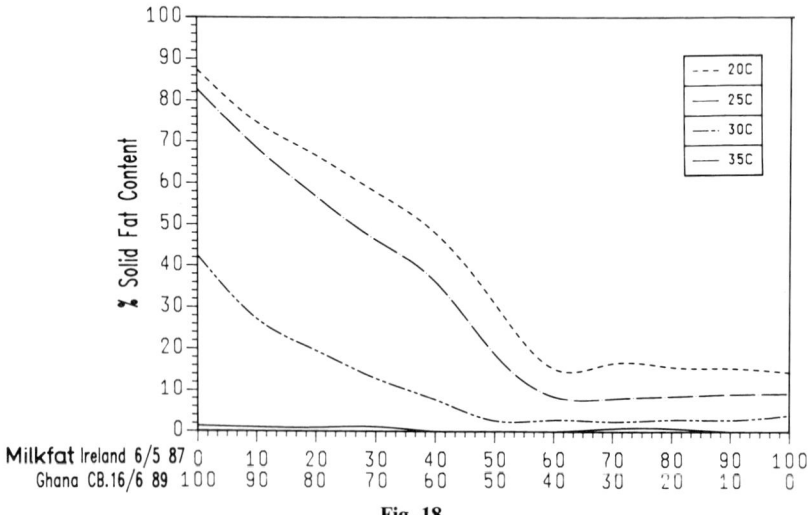

Fig. 18

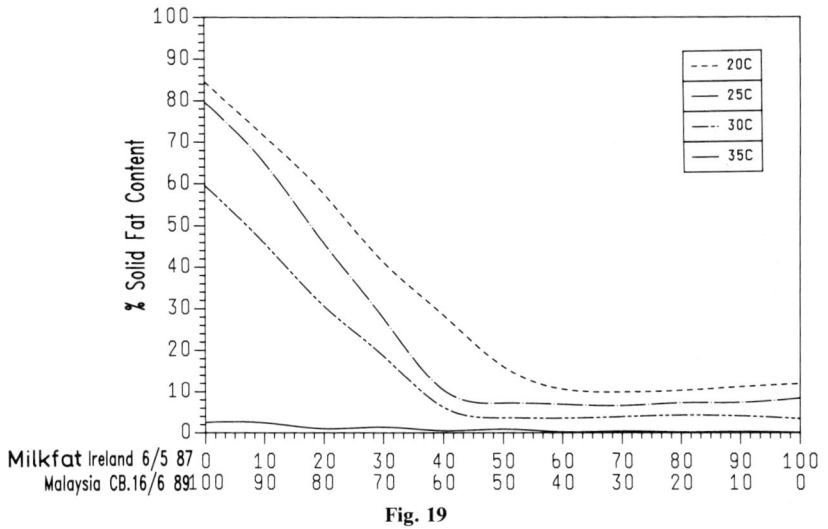

Fig. 19

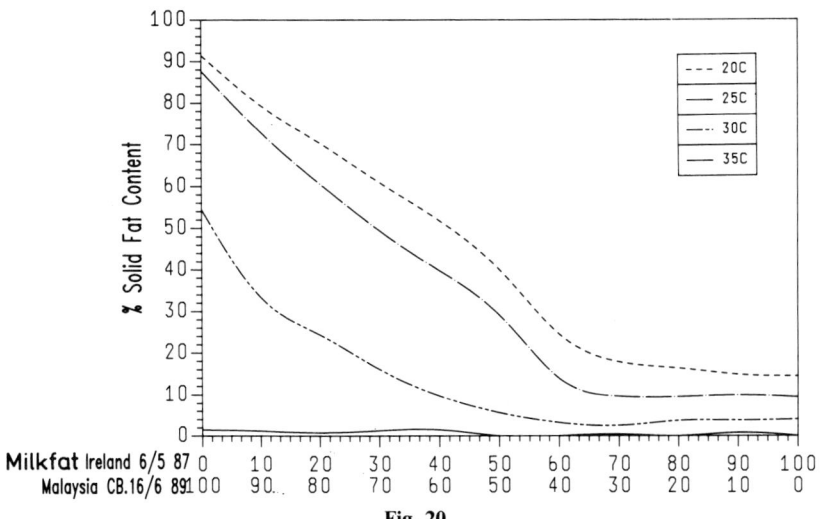

Fig. 20

described earlier. The compatibility curves presented in Figs 13–20 reveal the following:

(a) The eutectic effect is more pronounced in Malaysian cocoa butter and milkfat 3/4–87.
(b) The compatibility curves show less eutectic effect using milkfat 6/5–87 as compared to the milkfat 3/4–87.

There are several other factors contributing towards the quality of milkfat. Therefore stringent quality control should be applied to minimize the seasonal and geographic variation of milkfat to be used in the confectionery industry.

Fractionation and recombination of milkfat have been used for a long time in the production of modified butters for the bakery and dairy industries (Deffense, 1987). The hard, high melting triglyceride fractions are of special interest in the bakery industry for the production of special pastry, e.g. puff pastry, Danish pastries and croissants, and for reconstitution of hard butters for warm climate countries. The softer low and medium melting triglycerides fractions are used especially for biscuit and cheese production (Deffense, 1987).

Recombination of the high and low melting fractions in different amounts has given good results in the production of butter which is more spreadable at refrigerator temperatures than normally prepared butter (Deffense, 1987).

Fractionation of milkfat can also be of interest in connection with the manufacture of chocolate and ice cream.

MILK CHOCOLATE

Milkfat is one of the major ingredients in milk chocolate where it is incorporated as milk powder or milk crumbs. Milk adds to the nutritional value of chocolate by providing protein, calcium, phosphorus and vitamin A. A typical composition of milk chocolate is (Campbell & Pavlasek, 1987):

12·5% cocoa mass/liquor (55% fat),
19·5% cocoa butter,
22·7% milk solids (28% fat),
45·0% sucrose,
0·3% lecithin.

Milk powder is prepared by spray drying the liquid milk or by drying on rotating drums (Minifie, 1974). In both cases the milk is preconcentrated in evaporators. When the milk is to be spray dried the preconcentrated liquid milk is sprayed into a tower in which turbulent hot air is circulating. The spray particles dry immediately and fall to the bottom of the tower where the powder is discharged.

Similar preconcentrated liquid milk is dried as a thin film on internally heated revolving drums. The dried powder is scraped off as the drums rotate.

Spray dried and roller dried milk powder differ markedly in both flavour and physical properties. Roller dried milk powder is heated to a higher temperature during manufacture and has for this reason a more 'cooked' flavour. Some of the components have become water insoluble during the process. The coarser particle structure of the roller dried milk powder makes it wet more easily, so, even though spray dried milk powder is more water soluble than roller dried milk powder, the latter is easier to disperse.

Milk crumbs are preferred in some countries as it generally makes the chocolate more creamy compared to chocolate prepared from milk powder, the latter tending to have a more waxy, slimy and mouthcoating texture. When milk crumbs are used for chocolate manufacture the milk is distributed homogeneously throughout the product (Heathcock, 1987).

Crumbs are manufactured by adding cocoa mass to sweetened condensed milk or evaporated milk. Following mixing the mixture is dried to a maximum moisture content of 1·5%. A typical composition of milk crumbs is as follows:

26–35% milk solids,
50–65% sucrose,
13–18% cocoa mass/liquor.

PLAIN CHOCOLATE

Due to the economic advantage gained by partly substituting milkfat for cocoa butter in chocolate manufacture some research has been done to improve the functionality and performance of milkfat in chocolate: hence the hardening of milkfat by fractionation, hydrogenation or interesterification (Timms & Parekh, 1980).

It is legal in all countries to add milkfat to chocolate, but if the milkfat is modified by fractionation, hydrogenation or interesterification or any combination of these it becomes a legal rather than a technical question as to whether the modified fat is still considered to be milkfat or whether it is to be regarded as a cocoa butter replacer (CBR). If the modified milkfat is regarded as a CBR there are clear maximum limits on how much can be used in 'real' chocolate. In the latter case the modified milkfat will have to compete with a wide range of CBRs currently available on both price and performance. A typical recipe for plain chocolate is:

40% cocoa mass/liquor (full fat),
50% sucrose,
10% cocoa butter.

In contrast to plain chocolate milk chocolate already contains 20–25% milkfat (of the total fat content) from milk powder or crumbs.

Modified milkfat has been evaluated as a component in plain chocolate and in milk chocolate (Timms & Parekh, 1980).

When cocoa butter is partly substituted with fractionated milkfat (hard fraction) the mixtures are not particularly harder (measured as SFC by pulsed-NMR) at ambient temperature than is cocoa butter mixed with unmodified milkfat (Timms & Parekh, 1980). Yet the addition of fractionated milkfat to cocoa butter does stabilize the chocolate against bloom.

The substitution of cocoa butter with hydrogenated fat has to be carefully limited, otherwise it will affect the quality of the chocolate. It is possible to replace 10% cocoa butter with hydrogenated milkfat in plain chocolate (Hendrickx *et al.*, 1971) and replacement of 5–6% in milk chocolate also gives good results. Hydrogenated fats also retard or eliminate fat bloom in the chocolate.

Interesterification increases the hardness, melting point and solid fat content of milkfat making it more compatible with cocoa butter than unmodified milkfat (Timms & Parekh, 1980), but this improvement suffers from the resultant loss of flavour. The high costs involved do not justify the successful application of this technology within the confectionery industry.

TOFFEE

Milkfat imparts a rich flavour. The typical flavour and colour of toffees are produced by the 'Maillard' reaction, which occurs between the milk proteins and the sugars, when these are heated together. Flavours are also produced through 'caramelization' which results from the heat treatment of sugar, in this case lactose from milk. The traditional dairy ingredients for toffee are sweetened condensed milk and butter. Other forms of milk could also be utilized. The presence of small amounts of lecithin in condensed milk helps in emulsification of the milkfat in the caramel mix.

There are not any specific compositional restrictions for toffees, which offer wider ingredient substitution flexibilities than chocolate.

CONCLUSION

The modern analytical methodology available to us today permits us to understand many of the mechanisms associated with the incorporation of fats in confectionery and to put constraints on their safe levels of incorporation, thus avoiding problems in production. The confectionery industry has to rely on a supply of high quality dairy ingredients from the dairy industry.

REFERENCES

ARBUCKLE, W. S. (1986) *Ice Cream*, 4th edn, AVI Publishing, Westpark, CN.
BANKS, W., CHRISTIE, W. W., CLAPPERTON, J. L. & GIRDLER, A. K. (1987) *J. Sci. Food Agric.*, **39**, 303–16.
CAMPBELL, L. B. & PAVLASEK, S. J. (1987) *Food Technol.*, Oct. issue, 78–85.
CHAPMAN, G. M., AKEHURST, E. E. & WRIGHT, W. B. (1971) *J. Am. Oil Chem. Soc.*, **48**, 824.
COOK, L. R. revised by MEURSING, E. H. (1984) *Chocolate Production and Use*, Harcourt Brace Jovanovitch, Foots Cray, Kent.
CULLINANE, N., AHERNE, S., CONNOLLY, J. F. & PHELAN, J. A. (1984a) *Ir. J. Food Sci. Technol.*, **8**, 1–12.
CULLINANE, N., CONDON, D., EASAN, D., PHELAN, J. A. & CONNOLLY, J. F. (1984b) *Ir. J. Food Sci. Technol.*, **8**, 13–25.
DEFFENSE, E. (1987) *Fat Sci. Technol.*, **13**, 3–8.
DIXON, B. D. (1964) *Aust. J. Dairy Technol.*, **19**, 22.
HEATHCOCK, J. W. (1987) PMCA research notes XI no. 1.
HENDRICKX, H., DE MOOR, H., HUYGHEBAERT, A. & JANSSON, G. (1971) *Rev. Int. Choc.*, **26**, 190–3.
HUTTON, K., SEELEY, R. C. & ARMSTRONG, D. G. (1969) *J. Dairy Res.*, **36**, 103.
MINIFIE, B. W. (1974) *Mfg Confect.*, **54**, 19–26.
PARODI, P. W. (1970) *Aust. J. Dairy Technol.*, **25**, 200.
PARODI, P. W. & DUNSTAN, R. J. (1971) *Aust. J. Dairy Technol.*, **26**, 60–2.
POSATI, L., KINSELLA, J. D. & WATT, B. K. (1975) *J. Am. Diet. Assoc.*, **66**, 482.
SHUKLA, V. K. S. (1983) *Fette Seifen Anstrichm.*, **85**, 467–71.
SHUKLA, V. K. S. (1988) *Progr. Lipid Res.*, **27** (1), 5–38.
SHUKLA, V. K. S. & NIELSEN, I. C. (1989) In: *Tropical Forests*, Holm Nielsen, L. B., Nielsen, I. C. & Balslev, H. (Eds), Academic Press, New York, pp. 356–63.
SHUKLA, V. K. S., SCHIØTZ-NIELSEN, W. & BATSBERG, W. (1983) *Fette Seifen Anstrichm.*, **85**, 274–8.

Steen, K. & Andersen, J. O. (1974) *Mælkeritidende*, May 24, 569–611.
Taylor, M. N. & Hawke, J. C. (1975) *New Zealand J. Dairy Sci. Technol.*, **10**, 40–8.
Timms, R. E. (1980) *Lebensm. Wissensch. Technol.*, **13**, 61–5.
Timms, R. E. & Parekh, J. V. (1980) *Lebensm. Wissensch. Technol.*, **13**, 177–81.
Wood, F. W. & Haab, W. (1957) *Can. J. Anim. Sci.*, **37**, 1.

8

Fat Products Using Fractionation and Hydrogenation

K. K. Rajah
Westcliff-on-Sea, Essex, UK

SUMMARY

Fractionation and hydrogenation are two widely used processes which help to alter the melting profile, physical properties and chemical composition of the feed oil or fat and the products so produced are in effect new ingredients suitable for use in applications in which the original oil/fat could never have been used or would have performed poorly.

Fractionation is a thermomechanical process which eventually leads to two new products, an olein and a stearin. The olein can be used in further fractionations. Three types of fractionation systems are available, dry, solvent and detergent. Stages within each process such as nucleation and crystallisation and the conditions under which these are carried out are critical in determining the quality of the fractions. Equally, the separation efficiency of the filters also decides the yield of each fraction. Vacuum filters (e.g. Florentine, Vacuband, Stockdale) are being challenged by positive pressure membrane filters. Recent innovation with respect to the latter is a membrane filter by Krupp used in dry fractionation of fats for cocoa butter replacers. During filtration the pressure can be increased up to an end pressure of more than 50 bar.

Fractionation has enabled the introduction of many fats, particularly milkfat, into new food applications. These fats are now used to better effect in areas such as in the manufacture of stable creams, buttercreams, sauces, infant feed, ice-cream, bakery products and chocolate. The use of fractionated milkfat in yellow fat spreads is increasing, particularly in the production of spreadable butters. Fractionated fats are also used as stable frying oils and salad oils. The fractionated oils or fats can be incorporated directly into some recipes but in others they are used in blends with other oils. Some are also treated in further processes such as texturisation, in this case to make them suitable for use in bakery products.

The measurement of solid fat content at 20°C by pulse NMR is a useful aid for determining suitability of milkfats for specific food applications.

Hydrogeneration decreases the unsaturation of triacylglycerol oils and thereby increases their melting points. The chemical composition of the feed and the conditions under which the reaction is carried out, together with the type of catalyst used, help to determine the physical properties of the fat and the way in which it melts over a temperature range. The choice of the catalyst for the reaction is critical and re-use of catalyst can also generate products with interesting functional benefits.

In a number of applications such as bakery and confectionery products, it is important to have fats which remain firm at ambient temperature but which melt rapidly and are liquid at body temperature. Careful choice of oils and close control of the hydrogenation reactor are important to produce these oils. Equally, it is also sometimes necessary to lightly saturate a liquid oil to improve its oxidative stability, typically for use as a salad oil. Again, similar guidelines apply.

A number of novel processes have been described for the manufacture of stable fats with good taste and colour for use in cooking, as table oils and as base stocks for margarines. Other techniques are available to produce hard butters for bakery coatings and confectionery products. Also, non-conventional oils such as jojoba oil, fish oil, rice bran oil and cottonseed/coconut oils have been hydrogenated under specified conditions to produce chocolate and confectionery fats. Palm oil and palm kernel oil have also been described and used in several novel processes as feed oils for the manufacture of cocoa butter substitutes.

Hydrogenated milk fat recombined with skimmed milk powder has been shown to result in a whole milk which has a more stable flavour. These fats can also partially substitute for cocoa butter in dark chocolate without loss in product quality. They have also been shown to retard fat bloom in chocolate.

PART 1: FRACTIONATION

INTRODUCTION

The functionality and application of oils and fats are limited by their physical properties, particularly melting characteristics. Since oils and fats are mixtures of high, medium and low melting triacylglycerols, the fractionation process can be used to separate oils into two or more acylglycerol mixtures, which can have significantly different physical properties from the feed material. During fractionation the high melting acylglycerols can be crystallised from the melt (Taylor, 1976) and separated. The crystals will form a higher melting fraction, stearin, relative to the original fat, while the liquid phase will be lower melting, olein (Coenen, 1974a, b). The fractionation process therefore offers an important fat modification route for a number of lipids thereby increasing the opportunities for their use in food applications.

FRACTIONATION PROCESSES

There are three well established processes:

(i) detergent fractionation,
(ii) solvent fractionation,
(iii) dry fractionation.

(i) Detergent Fractionation
Fratelli Lanza is generally acknowledged to be the first to describe the principles of detergent fractionation in 1905, hence its other name, the Lanza process. The process commences with dry fractional crystallisation until the crystals are well formed. An aqueous detergent is then added and mixed into the slurry. The crystals are finally recovered through centrifugal separation. One of the best known examples of detergent fractionation is the Lipofrac process (Fjaervoll, 1970) developed and successfully commercialised by the Alfa Laval Company in Sweden.

Norris et al. (1971) adapted the detergent process in New Zealand for milkfat fractionation. Although their pilot scale work was successful it was never developed further by the dairy industry.

(ii) Solvent Fractionation
At a given temperature the solubilities of component triacylglycerols differ markedly from each other. The *cis*-unsaturated triacylglycerols are the most soluble, the *trans*-isomers are less soluble while the saturated triacylglycerols are by comparison least soluble due to their higher melting character. This therefore underlines the principle behind solvent fractionation. Thomas & Paulicka (1976) have described crystallisation from solvents such as acetone and hexane. Solvent fractionation is generally considered to be an expensive process due to the high costs involved in flame proofing the plant, and of the solvents themselves as well as their recovery during operation. Nevertheless, the process is used in the manufacture of high quality fats, e.g. cocoa butter replacers (Deffense, 1985; Rossell, 1985).

Jebsen et al. (1975) reported on the use of solvent fractionation for milkfat. Although there might be advantages in using solvents for the fractional crystallisation of milkfat, flavour problems associated with solvent residues rule this out (Cant et al., 1975).

(iii) Dry Fractionation

Winterisation and dewaxing
Winterisation and dewaxing are held to be limited forms of fractionation (Thomas III, 1985). The former began when it became evident that the cold winter temperatures solidified the palmitic acid rich triacylglycerols in cottonseed oil. Separation of the solid material gave an olein which was stable to clouding at refrigerated temperatures and as a result useful as a salad oil (Weiss, 1967).

Soyabean oil and rapeseed oil and blends of these oils are also partially

TABLE 1
Solid fat content and iodine values for milkfat and fractions separated on a Florentine (Tirtiaux) filter (Deffense, 1983)

Fraction temp.	S-29	OS-29-16	Ref.	O-29	OO-29-16
Sample No.	139-6-Flo	139-7	139	139-6-Flo	139-7
Iodine value	27·4	27·2	34·3	35·3	38·4
Solid fat content (Bruker pc 20)					
5°C	74·6	78·6	57·7	55·2	46·8
10°C	69·3	73·1	48·0	44·6	31·6
15°C	60·3	65·0	35·2	30·2	13·7
20°C	48·1	51·8	17·9	13·9	0·2
25°C	38·8	40·0	10·3	6·2	—
30°C	29·0	26·9	5·1	1·6	
35°C	20·5	15·0	0·5	—	
40°C	9·8	2·5	—		

hydrogenated and winterised to provide a good source of cooking and salad oils (Gooding, 1953, 1962, 1972; Weiss, 1967).

Dewaxing is also processed through the same route as winterisation. The waxes referred to here are complex and high melting. For example the melting point range for waxes in sunflower oil is 70–80°C (Sullivan, 1980).

Fractionation

In the dry fractionation process, both crystallisation and crystal separation take place in the absence of additives. Crystal separation is generally achieved through the use of vacuum filters of which the Tirtiaux Florentine filter and the Stockdale rotary drum filter are prime examples. More recently positive pressure membrane filters have become available from manufacturers such as De Smet Rosedowns, and a high pressure variety is available from Krupp (Wilner et al., 1992).

The Tirtiaux process has been used worldwide for the fractionation of edible fats including milkfat. Analytical results on some milkfat oleins and stearins recovered using a Florentine continuous-bed filter are given in Table 1 (Deffense, 1983). The properties of milkfat fractions from semi-commercial scale studies using a stationary bed filter are shown in Table 2 (Rajah, 1988). During these studies multiple fractionations were also carried out. Some of the results are reported in Table 3. The literature on studies using membrane filters for crystal separation is limited. Some results which were available from recent evaluation of the filter in New Zealand using milkfat are given in Table 4 (Illingworth, 1990).

TABLE 2
Analytical results for milkfat (MF) and fractions separated on a stationary bed Vacuband filter (Rajah, 1988) (reference no. MF 159)

	Milkfat/fraction		
	MF	Olein	Stearin
Carotene (ppm)	6·59	7·21	4·29
Solid fat content (%)			
0°C	67·1	51·4	80·0
5°C	53·7	46·8	78·3
10°C	46·6	39·8	74·3
15°C	31·4	23·3	66·2
20°C	15·8	7·6	55·4
25°C	9·3	2·0	46·7
30°C	4·6		37·0
35°C	1·3		25·6
40°C			13·6
45°C			3·4

MAIN FUNCTIONAL PROPERTIES

The behaviour of edible fats in food systems is determined largely by their physical properties, i.e. viscosity, melting behaviour, solubility, density, etc. (Mulder & Walstra, 1974). The process of melting and solidification (Bailey, 1950; Hannewijk et al., 1964), crystallisation (Mortensen, 1983) and polymorphism (Lutton, 1950; Malkin, 1954; Larsson, 1966; Deffense & Tirtiaux, 1982) are probably the more important of the properties encountered in routine food manufacture.

Fats used in food are either liquid oils or plastic fats. The consistency of the fat is therefore important. It is determined by plasticity and shortening value and these are discussed under bakery products below. Some of the major functional properties of milkfat are summarised in Table 5.

Flavour and Taste

Fats contribute to the flavour and taste in food in one of two ways: first, foods with a high fat content are generally more appetising and give a better mouthfeel during eating. A cake baked with a low level of fat has a poor texture relative to one baked with a higher fat inclusion and this affects taste. Secondly, some fats, for example, milkfat and olive oil, contain some flavour components which contribute significantly to the flavour of the

TABLE 3
Analytical data for milkfat oleins and their fractions from multiple fractionations, i.e. (i) second, (ii) third and (iii) fourth fractionations (Rajah, 1988)

(i) *Reference No. MF 229*
   Second fractionation

	Solid fat content (%)		
	Olein (feed)	Olein	Stearin
0°C	55·8	43·8	67·5
5°C	50·7	36·1	64·1
10°C	43·1	26·5	58·3
15°C	26·0	7·1	45·0
20°C	8·3		23·2
25°C	1·9		7·3
30°C	0·2		2·1

(ii) *Reference No. MF 342*
    Third fractionation

	Solid fat content (%)		
	Olein (feed)	Olein	Stearin
0°C	51·5	33·6	62·2
5°C	45·6	25·7	57·2
10°C	36·3	13·6	50·0
15°C	19·3	0·1	36·8
20°C	2·4		12·8
25°C			0·2

(iii) *Reference No. MF 457*
     Fourth fractionation

	Solid fat content (%)		
	Olein (feed)	Olein	Stearin
0°C	25·0	14·7	42·1
5°C	17·5	7·3	34·4
10°C			17·0
15°C			0·7
20°C			
25°C			

TABLE 4
Solid fat content (by NMR) and dropping point figures on stearins from New Zealand milkfat produced on the membrane filter

Temperature (°C)	Dropping point (°C)	N_0 (°C)	N_{10} (°C)	N_{20} (°C)	N_{30} (°C)	N_{40} (°C)
28	46·6	87·7	84·5	69·5	51·3	23·0
25	44·7	85·7	81·2	61·1	41·0	14·1
22	43·4	85·7	80·2	59·0	37·9	10·3
20	39·1	74·7	73·0	44·9	23·0	2·6

TABLE 5
Functional properties of milkfat

Property	Major products
Air incorporation: creaming	Cake batters, cream, buttercream
Anti-staling: moisture retention	Bread, cakes
Flavour	Toffee, spreads, bakery products, cooking
Flavour carrier	Confectionery, bakery products
Gloss	Bakery products
Heat transfer medium: frying	Potato crisps, dough-nuts
Layering	Puff and Danish pastry
Shortening	Biscuits, cakes
Spoonable/pourable emulsion	Dressings, mayonnaise
Spreadable emulsion	Butter, margarine, spreads

food. In the main, some of these are volatile compounds, but others are found bound within the fat as flavour precursors. Milkfat has both types of flavouring matter (Kinsella, 1975; Urbach, 1979; Manning & Nursten, 1985) and these are present in trace amounts, parts per million or even less. The aliphatic compounds are largely responsible for the dairy flavour in milk and milk products. The flavour of various fats is reviewed extensively elsewhere in this book.

Walker (1972) reported that the concentrations of lactone and methyl ketone precursors in low melting milkfat fractions were higher than those in the original anhydrous milkfat before fractionation. Given that up to 10% of the lactone potential and 20% of the methyl ketone potential of freshly secreted milkfat is lost during the commercial preparation of anhydrous milkfat (Walker, 1972) this still implies that the low melting fraction would be at least close in flavour profile to butter if not better. It is suggested that the level of total free lactone in a recombined butter made from fractionated fat is likely to be somewhat higher than in a traditional butter (Walker & Keen, 1976).

Colour

Colour is important in food acceptability. The bakery trade often requires a 'butter-like' colour for some of their products. In response to this requirement, margarine manufacturers now add carotene or annatto to their cake margarines and some shortenings. The natural colour in milkfat is due to the carotenoids and similar compounds. During the fractionation process carotene tends to concentrate in the olein fraction (Norris et al., 1971; Rajah, 1988; see Table 2). This suggests that it is not directly involved during crystallisation but remains in the liquid phase. Yet the levels present in the stearin fractions were still relatively high (Rajah, 1988). This is attributed mainly to the entrainment of the carotene rich liquid phase within the crystalline structure of the higher melting fraction. Also, when crystallisation is rapid the carotene is trapped in the crystal layers that form.

APPLICATION FOR FRACTIONATED FATS

Frying Oils/Salad Oils/Salad Dressing

The main criterion used to distinguish salad oils is their ability to remain largely liquid in a refrigerator (Krishnamurthy, 1982). Soyabean oil and rapeseed oil meet this criterion but are usually partially hydrogenated to reduce the unstable linoleic acid content, and winterised to generate cooking and salad oils. Cottonseed oil is also winterised (Weiss, 1967) to crystallise out a substantial proportion of the palmitic acid rich triacylglycerols to produce a salad oil. However, these oils and others, e.g. sunflower seed and corn oils, may be used without any further treatment as cooking oils or frying oils. The latter two oils may require dewaxing before they can be used as salad oils. With respect to marine oils, the long-chain fatty acids $C20$–$C26$ impart an unpleasant flavour to the oil (Tirtiaux, 1980). However, when the oil is selectively hydrogenated to an iodine value in the range 100–115, and subsequently fractionated, the olein phase can be commercialised as a salad oil.

Milkfat cannot normally be used as a salad oil because it has a very high content of saturated fatty acids. However, some measure of success was achieved when a milkfat olein was used in a salad mayonnaise formulation (Rajah et al., 1982). It may, however, be used as a frying fat but this is not widespread due to its high cost. Some limited deep fat frying trials by the author using potato chips (frying temperature = 185°C, time c. 10 min per day) have indicated that the quality of food remains satisfactory for at least 7 days, the milkfat olein (m.p. = 17·2°C) gave a buttery aroma for 6 days before turning acrid (free fatty acid was 0·26% as oleic). Palm oil is also a highly saturated fat. Oleins of cloud points less than 10°C, produced from single-stage fractionation, can be used to substitute for soft oils in, for example, frying and cooking applications (PORIM, 1981). Palm olein of iodine value 55, from dry fractionation (Paulicka & Jasko, 1978), can be re-fractionated to produce an olein fraction of iodine value 56, which is suitable for frying. The olein fraction iodine value 56·3 from aqueous detergent fractionation of palm oil can be used as the feedstock for solvent fractionation to generate a second olein, iodine value 58·4, which can be used in frying. Stearin of iodine value 38·8 is used as a cocoa butter extender (Tan et al., 1976).

Cream
Creams which are extremely stable at normal temperatures, particularly at room temperatures reached during the summer, can be produced by specific heat treatment of an oil product, followed by emulsification (Meiji Milk Prods. KK, 1983). The method comprises preparation of the fat phase by adding up to 50% of lipophilic sucrose fatty acid ester and/or up to 20% of a triacylglycerol (melting point > 50°C), e.g. milkfat or soyabean oil or their modified (i.e. fractionated, hydrogenated, interesterified) derivatives. The emulsified creams can be used in applications such as confectionery, salad dressings and in coffee.

A method has been described for the production of whipping creams using a high melting milkfat fraction (Bratland, 1982). This fraction is emulsified to give whipping creams of low fat content and good viscosity and is less sensitive to temperature than conventional creams. The same patent also claims that the low melting fraction can be emulsified to produce a cream which can be churned to give a butter product which is spreadable at refrigerator temperatures.

Ice-Cream
Dairy ice-creams are formulated using cream or milkfat (Hamilton, 1983). However, due to the need for economy, vegetable fats which are generally

less expensive are also used. A 100% hydrogenated palm kernel olein $N_{20} = 40\%$ can be used as the fat phase in ice-cream (Laustsen, 1984). Rossell (1985) has discussed the use of a blend containing 70% palm stearin, 30% palm kernel olein ($N_{20} = 30\%$) which is then interesterified to produce an ice-cream fat. Some investigations were carried out by the author into the use of milkfat fractions in ice-cream. Ice-creams were produced to the following recipe using a low melting milkfat fraction (solid fat content at 20°C, $N_{20} = 10.7\%$) and compared with ice-creams made with anhydrous milkfat (on its own and in a blend with 10% fraction) and a blend of the fraction, 40%, with coconut oil, 60%.

Ice-cream recipe

Ingredients	Composition (%)
Castor sugar	14·0
Skimmed milk powder (medium heat)	11·0
Anhydrous milkfat	10·0
Crodacreme (emulsifier/stabiliser)	0·7
Water	64·3
Total	100·0

The taste-panel results did not show a significant preference for ice-creams made with the low melting milkfat fraction. Although the texture and appearance of all the ice-creams were equally satisfactory, the choice was made primarily on taste. It was therefore concluded that, while low melting fractions could be used successfully in commercial ice-cream recipes, there was no significant advantage or difference over ice-creams made with AMF.

Bakery Products

The applications for fats in bakery applications are determined by their plasticity and shortening value.

Plasticity

Plasticity in a fat may be defined as the ability to retain its shape under slight pressure, but to yield to increased pressure, such as that encountered during rolling, mixing or spreading (McWilliams, 1979). A plastic solid owes its peculiar properties to the tendency of its solid particles to form jams or arches which support the material against shearing stresses. In rheological terms, therefore, plastic fats possess a yield stress. There are additional factors that determine the range within which a material will be plastic. The shape, average size and size distribution of the fat crystals play an important part in this. Equally important is the percentage of solids in a fat and the characteristic continuous variation that occurs with change in temperature. Plastic fats are processed and produced in a stable crystalline form using scraped surface heat exchangers and pin workers (Rajah, 1992). The effect of temperature upon the viscosity of the liquid phase is understood to account for as much as 30–50% of the total consistency variation (Soltoft, 1947).

Shortening value
In addition to plasticity, fats for bakery use should also exhibit shortening power. The ability of a fat to lubricate and tenderise the structure of a baked product is known as its shortening value (Mattil, 1964a).

A good shortening should remain plastic and resistant to significant melt-down of crystalline structure from the range of temperatures reached during mixing. To achieve this, shortenings are formulated using a base oil or fat, and a plasticiser, i.e. a hard fat. Other additions as required are emulsifiers, water, vitamins and nitrogen.

Good shortening ability and plasticity in fats are produced also through re-arrangement of the fatty acids on the glycerol molecule, i.e. through interesterification, to form a fat that stays in the desired crystal form longer. Shortenings can therefore be formulated for a variety of applications. In the bakery sector, shortenings are available, for example, for cake and biscuit manufacture, while special pastry margarines or pastry shortenings or fats are used in Danish and puff pastry manufacture.

Cakes
Milkfats are generally not recognised to be as highly functional as the shortenings in cake making due to their limited creaming ability. The creaming ability is best expressed as the percentage of air incorporated by the fat on the basis of its own volume (Bailey & McKinney, 1941). Cake batters which contain larger amounts of air entrapped within the fat will produce cakes larger in volume than those containing relatively little air (Mattil, 1964b). Different fats may differ markedly in their creaming properties. The main criterion for a good creaming fat is that it should contain a substantial proportion of highly saturated acylglycerols that crystallise in a stable β'-form (Mattil & Norris, 1953). Shortenings that crystallise in or revert to substantial proportions of β-crystals will not cream satisfactorily in cake batters. The creaming quality of butter is variable and generally inferior to shortenings. However, it is possible to overcome this problem by blending together milkfat fractions and

plasticising and texturising them under well-controlled conditions to achieve high creaming power in the finished product (Eyres et al., 1988; Rajah, 1988). This improves performance over butter but still does not match up to those reached with margarines. A low melting fraction, $N_{20} = 11.3\%$, performed well compared to butter and a proprietary cake margarine. In his study Robb (1980) found the low melting milkfat fraction out-performed butter but it remained inferior to the cake margarine. Cake margarines and shortenings, in addition to having the correct blend of liquid oil and solid fat components for a correct solid fat content profile, also have inclusions of the correct type and level of emulsifiers which are normally unsaturated monoglycerides, polyglycerol esters or propylene glycol monoesters. Eyres et al. (1988) have proposed that emulsifiers can be added to cake batters containing the blend of milkfat fractions, producing a result in the finished cake no different from incorporating them in the plasticised fat product. They recommend that these emulsifiers be added in the form of α-crystalline gels normally comprising distilled monoglycerides or polyglycerol esters.

Danish cookies
Low melting milkfat fractions (e.g. $N_{20} = 11.3\%$) are preferred for the preparation of cookies (Rajah, 1988). Tolboe (1984) found that low melting milkfat produces larger cookies which also have a nice crispness. It is also understood that some cookie manufacturers have switched to low melting milkfat with iodine value above 35. This is to avoid problems with fat blooming, which appears as pale strains on the surface of the cookies, formed by crystals of high melting fat (Sanderson, 1985).

Danish pastry
Tolboe (1984) reported that low melting fraction or Danish summer butter (which is low melting) produced inferior pastries because the dough layers did not separate into the desired flaky structure, but remained stuck together in thick layers. However, when high melting fractionated milkfat was used, he found that it produced a Danish pastry of equal flaked structure to that produced when a typical Danish pastry margarine was used. Rajah (1988) confirmed these conclusions using texturised anhydrous milkfat, $N_{20} = 16.4\%$, and high melting fractions, $N_{20} = 30.9\%$ and 32.7%.

Croissant
Butter could be used successfully to produce excellent croissants. However, this is achieved only if the dough is prepared under carefully controlled

temperature conditions, otherwise problems occur. This is because the purpose of the fat in laminated yeast doughs is to assist in the formation of the two dimensional gluten structure in each dough layer, which is necessary for producing the flaky character of the baked roll (Doerry & Meloan, 1986). Here, the fat has to provide the barrier between adjacent dough layers and it must remain as a continuous film throughout the laminating process. Soft and oily fats fail because they tend to be absorbed by the dough. Therefore, when butter from England and Wales, $N_{20} = 13–22\%$, was used at ambient temperatures of 15–18°C, it was found to affect the quality of the product adversely (Rajah, 1988). High melting fractions, e.g. $N_{20} = 32·3\%$, perform extremely well at high ambient temperatures and are preferred to vegetable fat based pastry fats which lack flavour.

Puff pastry

Robb (1980) assessed a high melting fractionated milkfat against a pastry margarine and butter in vol-au-vent rings. He concluded that the high melting fraction had the advantage of not needing to be chilled during the laminating procedure and was capable of producing results comparable with those produced by either butter or pastry fat. Rajah (1988) found that high melting fractions with solid fat content values in the region of $N_{20} = 23\%$ performed well, showing good lift ratios and texture.

Buttercreams

Good flavour, aeration and stand-up properties are the main requirements for fats used in buttercreams. The use of milkfat fractions was investigated in this application.

Recipe

Ingredients	Butter (%)	AMF or Fractions (%)
Butter	50	
AMF or fractions		41
Icing sugar	50	50
Water		9
	100	100

A blend of AMF and an olein (milkfat) ($N_{20} = 13·3\%$) proved most satifactory for this purpose. It creamed well, performed better than the control (butter) or anhydrous milkfat during piping and gave a delicate sweet, good buttery flavour. A low melting fraction, $N_{20} = 7·6\%$ (Rajah, 1988), did not, however, have good stand-up properties, and it became greasy quickly. A blend of high melting milkfat fraction and AMF,

$N_{20} = 25\cdot 6\%$, formed a hard cream and lacked flavour. Robb (1980) found that the aerating proportion of lower melting milkfat fractions during creaming in Madeira cake test trials were comparable with and slightly better than those for butter but slightly less satisfactory than those of cake margarine. This supports the results presented here. Furthermore, the concentration of the lactone and methyl ketone precursors in the olein (Walker, 1972) would contribute towards a flavour profile similar to butter, if not better.

Infant feeding

An infant feed has been developed from sour cream by adding a ferment, acidophilic bacilli and bacteria bifidum in the ratio 0·5:10 (E. Sibe Tech. Inst., 1985). It is claimed that, if 28–20% of the low melting fraction of milkfat is added to the cream before it is warmed, the sour cream is improved in quality and has greater nutritive value. The linoleic acid composition in the end product is also detected to have been increased four-fold to levels found in human milkfat.

Butter Powder

The effect of fractionated milkfat on the quality of butter powder and a spread made from it has been reported by Prasad & Gupta (1984). The authors found that powders manufactured from the hard and soft fat fractions of milkfat, of melting point 38°C and 17°C respectively, contained less free fat and had more expressible fat at 30°C and 45°C than control butter powder made from ripened cream. The powder made from the soft fat fraction had more free fat and expressible fat than that made from hard fat, and it showed a greater tendency to form lumps. Flowability of the experimental butter powders was poorer than that of control butter although bulk densities were similar.

Spreads were made by reconstituting the powders. Products made from the hard fat fractions had a better colour, body and texture, spreadability, flavour and overall sensory quality than that made from the soft fat fraction; it was only slightly less acceptable than the spread made from butter powder from ripened cream.

Sauce Improver

A patented sauce improver has been established which is stable without preservatives or sterilisation (Société des Produits Nestlé SA, 1985). It is based on cream, egg yolk, a liquid milkfat fraction, milk derivatives and flavour enhancers. It has a pH of 5·6–5·8 and a water activity, a_w, of 0·84–

0·9 at 25°C. It contains egg yolk modified with phospholipase. Potato starch or carrageenan may be added as stabiliser or thickener.

Yellow Fat Spreads

A process has been described (Unilever NV, 1983a) for the extraction of short-chain triacylglycerols from milkfat using supercritical carbon dioxide fractionation (Mangold, 1982). The recovered material has a number of important characeristics: it contains a greater proportion of flavour components relative to the feed, and when used in normal margarine formulations as 10–20% additions by weight it confers butter-like properties in terms of plasticity and elasticity.

Unilever NV (1983b) have also described a process for fractionation of hydrogenated milkfat to generate an olein fraction for use in margarine manufacture. Milkfat is hydrogenated to an iodine value up to 10, to convert almost all unsaturated bonds to saturated bonds. Solvent fractionation follows and the olein is claimed to confer improved butter-like properties to the margarine.

Butter which is spreadable from the refrigerator or margarine type products, including low calorie varieties, can be produced (Société des Produits Nestlé SA, 1978) by mixing a fat phase containing amongst others milkfat, particularly milkfat olein, with an aqueous portion containing emulsifiers, stabilisers, salt, etc. Aseptic packaging is required and texture stability of several months is claimed. Low calorie varieties containing 40% fat are also possible.

During the 1970s a number of investigations were carried out in Australia and New Zealand into the use of milkfat fractions for the manufacture of butter with controlled firmness and products which could be spread directly from the refrigerator. Dixon & Black (1974) reported that milkfat was fractionated into 22% high-m.p. fat (HMF), 42% medium-m.p. fat and 36% low-m.p. fat (LMF), with softening points of 40°C, 32°C and 20°C respectively. These were mixed in various proportions, recombined into butter, and vacuum reworked 1 week later. They found that firmness at 5°C and 13°C and organoleptic grade were mainly affected by the proportion of HMF. All butters were soft at 20°C. The results suggested that to achieve good spreadability at refrigerator temperature, together with good thermal stability, an LMF of lower softening point must be used. Berwick (1975) discussed the results of studies carried out on the preparation of mixtures of vegetable oils with whole milkfat and various milkfat fractions, using a Cherry-Burrell shock cooling device or a conventional churn. A blend containing 50% sunflower oil and 50% fractionated milkfat (softening point 40–41°C) was found to show excellent spreadability with good firmness. It also contained added beta-carotene to improve the colour and some propyl gallate to increase the oxidative stability. There is additional information on fractionated milkfat products in spreads in the chapter dealing with anhydrous milkfat.

Chocolate

Rajah (1988) reported on chocolate crystallisation studies where the fat phase used in the chocolates was prepared with mixtures of cocoa butter, summer milkfat and fractionated milkfat (stearin). The objective was to compare the behaviour of these fat mixtures in chocolate and to determine the consequences of using fractionated high melting milkfat in such mixtures. This was done by simultaneously monitoring crystallisation with pulse NMR and the state of temper using the tempermeter.

The pre-treatment of fats when they are solidified from the melt, or tempering as it is otherwise known, determines the final crystalline state of the fat. Melted chocolate is always tempered during manufacture so that this preliminary crystallisation sets the fat phase as homogeneously as possible and with the smallest possible crystals (Kleinert, 1970). This results in a finished product with a hard snap, a good gloss and optimum shelf-life. When chocolate is in the melted state, non-fat cocoa, milk solids, sugar, etc., are suspended in the fat phase which is itself made up of cocoa butter and mixtures of other fats, e.g. milkfat and vegetable fats. Since it is only the fat phase which changes its structure and temperature, tempering therefore affects the fat in the chocolate exclusively. Tempering is therefore used to generate sufficient stable crystals so as to avoid supercooling in the mould. Solidification curves during tempering allow manufacturers to monitor the process. For milk chocolate or plain chocolate containing added vegetable fats besides cocoa butter, tempering requires cooling of the molten chocolate to a temperature below the melting point of β' fat crystals (27·5°C) followed by raising the temperature to a temperature above the β' melting point. However, for plain chocolate containing cocoa butter only, crystallisation into the stable β-form is sufficiently rapid at 28–30°C, that the initial lower temperature crystallisation is not required if the chocolate is seeded initially with stabilised chocolate. Rajah (1988) found that for mixtures containing 25% milkfat stearin and 75% cocoa butter (CB), the time taken after seeding for the N value to reach 4% was found to be almost a quarter of the time taken when 25% summer milkfat was used, and about half the time taken when 15% summer milkfat was used. This implies that the inhibitory influence of milkfat on CB crystallisation is considerably reduced when high melting milkfat fractions are used, and that crystallisation is also excessively fast at 26°C. It was evident that a tempering regime of 70–28°C should be adequate.

The study also showed that although the solid fat content, SFC, of winter milkfat is higher than summer milkfat at 20°C, i.e. N_{20}, blends with cocoa butter containing 25% milkfat showed that the winter milkfat produced a softer blend. This is ascribed to eutectic effects between CB and middle melting triacylglycerols in the winter milkfat. This effect was not apparent at 15% addition.

An important observation was made when mixtures comprising CB/CBE/MF or HMF were assessed (CBE = cocoa butter equivalent, MF = milkfat, HMF = high melting fraction). The results showed that the presence of high melting fractions in the mixtures gave higher SFC values compared to milkfat. This, together with the results which indicated that the high melting fractions accelerated the tempering of chocolate, showed that HMF could displace a greater proportion of cocoa butter or cocoa butter equivalents in commercial chocolate recipes than is current practice with anhydrous milkfat.

PART 2: HYDROGENATION

INTRODUCTION

A number of triacylglycerol oils contain substantial proportions of fatty acids with more than two double bonds, e.g. soyabean oil C18:3 = 5–11% of total fatty acids. Marine oils contain even longer chain fatty acids, C20–C22, with four to six double bonds. This type and degree of unsaturation makes the oils susceptible to autoxidative rancidity and spoilage which can lead to flavour reversion.

Highly unsaturated oils are also liquid at ambient temperature which restricts their range of application in food. Higher melting fats, particularly those melting close to body temperature, and of the appropriate polymorphic form, are necessary for some food applications, for instance in chocolate and most bakery product (deMan et al., 1989).

Unsaturation of triacylglycerol oils can be reduced by hydrogenation. The hydrogenation reaction is the addition of hydrogen to the alkenic linkages in the presence of a metal catalyst. The process is based on the early, 1896, pioneering work of Sabatier (1922) and Senderens. Norman evaluated the process in 1902, and in 1903 patented the process in all major countries. The Procter and Gamble Company acquired the American rights in 1909 and in 1911 successfully launched their first commercial product, Crisco, which was a hydrogenated cottonseed oil shortening. The early products were primarily blends consisting of 15% fully hydrogenated cottonseed oil in refined cottonseed oil or about 10% fully hydrogenated lard added to unhydrogenated lard. At present, hydrogenation is used by the majority, if not all, of the major producers of shortenings, pastry fats, confectionery fats and margarines around the world.

THE HYDROGENATION PROCESS

The hydrogenation reaction mechanisms are discussed in detail by Beckmann (1983), Cecchi (1980), Coenen (1976) and Larsson (1983).

The technology and equipment for hydrogenating fats and oils have been reviewed recently by Grothues (1985) and Carlson (1989). The main objective is to produce hydrogenated products with a desired composition. Therefore close control of the reaction is important. Several factors determine the final outcome of the reaction.

    (i) Catalyst—activity,
              —selectivity,
              —concentration,
              —impurities acting as catalyst poisons;
   (ii) mass transfer;
  (iii) hydrogen pressure;
  (iv) temperature;
   (v) type of oil used.

Catalyst Activity

Hydrogenation is a catalytic process (Mattil, 1964c; Patterson, 1983). The activity of the catalyst is therefore important. Catalysts could be similar in terms of their total metal content but may differ in their individual activity. This is because much of the total metal is inert, with only a small proportion being chemically active. Hence, the need to select catalysts on the basis of their activity and selectivity (Mattil, 1964d).

The activity of a catalyst for the hydrogenation of oils is defined as the iodine value drop in 20 min (IVD20) and can therefore be determined as such when carried out under standard conditions such as those recommended by the American Oil Chemists Society (Recommended Practice Method Ca 17–76).

During the heterogeneous catalysis of oils and fats the catalyst and reactants exist in different physical states, as opposed to homogeneous catalysis in which the catalyst and reactants comprise a single phase. It is a surface phenomenon and, consequently, an active catalyst should have a highly extended surface. Taylor's (1925) 'active spots' theory explains the various phenomena associated with heterogeneous catalysis.

Selectivity

During the hydrogenation process a number of reactions are understood to be in progress concurrently. The addition of a catalyst may accelerate some of these reactions over others, and, equally, specific catalysts may have a more pronounced effect on the different reaction rates relative to other catalysts, e.g. hydrogenation of linoleic acid leading to oleic acid or to its isomers dependent on the catalyst being used (Feuge et al., 1951; Sims & Hilfman, 1953; Allen & Kiess, 1955). These reactions can be non-selective, selective, or highly selective and catalysts which promote these various reactions are termed accordingly.

The early definition (Richardson et al., 1924) for the term selectivity described the main reaction as the conversion of linoleic acid to a monoene, compared to the conversion of the monoene to stearic acid. About 25 years later, Bailey (1949, 1951) dealt with selectivity ratios (SR) as being the ratio of the rate constants for the hydrogenation of linoleic and oleic acids. He also developed kinetic equations to represent the concentration of each acid group as a function of time. Although Bailey (1949, 1951) was the first to report that on a kinetic basis linoleic acid sometimes hydrogenates directly to oleic acid, the chemical model he proposed did not indicate either the positional or geometrical isomerisation which accompanies hydrogenation. In 1965, Albright published a number of graphs that allowed the quantitative determination of the degree of selectivity for the hydrogenation of cottonseed, peanut, corn, soyabean and linseed oils. This selectivity number as defined in Albright's publication is simply the ratio of the hydrogenation of linoleic acid compared to the hydrogenation of oleic acid. Allen (1978) later made some modifications to Albright's graphs. However, Coenen (1976) has introduced additional selectivity concepts, selectivity in *trans*-formation, and selectivity in 18:3 hydrogenation compared to 18:2.

The main factors and conditions necessary for selective hydrogenation are high temperature, low pressure, high degree of agitation and high catalyst concentration.

Mass Transfer

During heterogeneous hydrogenation the actual concentrations of the reactants near the catalyst surface determine the rate, i.e. catalyst activity and selectivity of the reaction, and hence the chemical composition of the final product. Mass transport between the phases is therefore of prime importance during hydrogenation.

Several mass transfer mechanisms are involved in hydrogenation (Eldib & Albright, 1957): transfer of hydrogen from the gas to liquid phase, adsorption onto the catalyst surface of hydrogen dissolved in the liquid phase and release of a fatty acid radical from the catalyst surface to the main body of the liquid. Coenen (1969, 1976) has expanded on these aspects and together with coworkers (Coenen et al., 1964) has discussed the importance of triacylglycerol transport and concentration. Puri & deMan (1977) dealt with the mechanism of hydrogen and unsaturated component transport to the catalyst surface and proposed experimental methods of obtaining gas–liquid and catalyst particle–liquid mass transfer coefficients where both inter- and intra-particle resistances are accounted for.

Re-use of Catalyst

Reduction in catalyst activity may result from catalyst poisoning during hydrogenation. Some of the substances which may act as catalyst poisons are natural components of vegetable oils such as lecithin in soyabean oil (Ottesen & Jensen, 1980) and sulphur compounds in mustard seed oil (Drozdowski & Zajac, 1977). Others which appear, for instance sodium soaps, free fatty acids and oxidation products (Drozdowski & Zajac, 1977, 1980), may be from crude oil processing stages preceding hydrogenation.

Catalysts are frequently re-used in the industrial hydrogenation of oils and fats. It is therefore important to understand some of the underlying chemistry of the reactions when poisoned catalysts are used.

Two methods exist for the use of catalyst, single-use and multi-use (Young, 1986). Fresh catalyst, sufficient for one batch is used in single-use hydrogenation, and discarded after filtration. In multi-use hydrogenation a large amount of catalyst is used, recovered through filtration and retained for subsequent use. Loss of catalyst activity is overcome by replacement of a proportion of the spent catalyst with a similar amount of fresh material (Osinga, 1979). Two of the main reasons for loss of activity in spent catalysts are poisoning of the catalyst by impurities in the oil and oxidation of the recovered catalyst between use.

It has been reported (Urdahl et al., 1972) that marine based oils contain relatively high amounts of sulphur, generated by post-mortal reactions in fish, the levels of sulphur reached in these marine oils being proportional to the storage time of the fish before processing. During hydrogenation sulphur is released due to interaction between hydrogen and sulphur containing compounds at the catalyst surface; the sulphur reacts with nickel at the surface thereby preventing the nickel from readily absorbing and dissociating hydrogen. As a consequence the total activity of the catalyst is reduced (Klostermann & Hobert, 1980). Simultaneously the tendency to promote isomerisation is increased. Both these effects were noted in Krishnarajah's (1982) experiments, using re-used catalysts and sulphur poisoned catalyst.

Others have also reported on the poisoning effect of various sulphur and phosphorus compounds during hydrogenation of rapeseed oil (Babuchowski & Rutkowski, 1969) and fish oil (Magnusson & Notevarp, 1971) while Mork & Norgard (1976) reported that the poisoning effect of halogen containing compounds in fish were in the order $Cl < Br < I$. Ottesen & Jensen (1980) have shown that the presence of phosphorus in the form of phosphatides during soyabean oil hydrogenation mainly affects selectivity.

Effect of Catalyst Type on Selectivity and Product Characteristics

Close control of the process parameters described earlier is essential, but, in addition, the choice of catalyst can be quite important.

A number of hydrogenation catalysts are available, but nickel catalysts remain the most widely used in commercial operations. Three varieties of commercial nickel catalysts are generally available. These are:

(i) selective, highly active nickel catalyst, e.g. Harshaw Harcat, Unichema Pricat 9900,
(ii) highly selective nickel catalyst, e.g. Harshaw Nysel DM3, Unichema Pricat 9906,
(iii) highly selective, *trans*-isomer promoting nickel catalyst, e.g. Harshaw Nysel SP-7, Unichema Pricat 9908.

Highly active nickel catalysts are used in hydrogenations where saturation of double bonds is more important than selectivity considerations, for example hydrogenation of cottonseed oil to the specification of vegetable ghee, melting point 38–42°C at iodine value c. 70 (Harshaw Technical Brochure: Harcat Fat Hydrogenation Catalyst). However, many food products, such as confectionery coatings, cream fillings and table spreads, require fats produced under highly selective conditions, i.e. fats which are solid at ambient temperatures, but which have solid fat content (SFC), as determined by pulse nuclear magnetic resonance spectrometry, close to zero at body temperature. With these characteristics, and particularly if a substantial proportion of the fat melts between ambient and body temperature, a cooling effect in the mouth results as opposed to a waxy sensation when SFC levels remain high at 37°C. Coenen (1976) has suggested that rapid melting of fat below 37°C is caused by trielaidate, stearate–dielaidate, distearate–oleate and stearate–oleate–elaidate while the waxy mouthfeel is contributed by high melting tristearate and distearate–elaidate. High values for S_I, S_T and S_i, i.e. Coenen's (1976) selectivity ratios, promote the formation of these intermediate products and simultaneously limit the formation of the high melting triacylglycerols. However, in some food applications, particularly baking, fats with melting points above 37°C are required, usually up to 40°C. Highly selective hydrogenation is not required for these fats. Typical examples of food use for this range of fat would be puff pastry used for the lids of hot-eating meat pies, vol-au-vent cases, etc., and also for sweet goods such as vanilla slices. Here, a high melting fat is important for layering the dough prior to baking, followed by much folding and rolling. The fat must be sufficiently firm so as not to be absorbed by the dough, but malleable enough to withstand being rolled out into a very thin film during the layering process. High melting triacylglycerols can aid dough structure formation and stability during the early stages of baking. This is most evident in the preparation of short pastry for quiche bases, meat pies, etc., and, in the sweetened form, for a wide variety of tarts, fruit pies and flans.

The commercial manufacture of selectively hydrogenated vegetable fats has been reported by Krishnarajah (1982) who carried out laboratory and factory scale hydrogenation of soyabean oil using a highly selective catalyst, i.e. Pricat 9906, and a *trans*-isomer promoting catalyst, Nysel SP-7, Table 6. He was able to show the difference in levels of *trans*-isomer formation between the two reactions (Table 7). A further series of experiments were carried out using Pricat 9906 which had been used in an earlier hydrogenation on fish oil. The spent Pricat 9906 showed reduced activity which was compensated by using a larger quantity of the catalyst (Table 6). He also reported that *trans*-isomer formation was higher relative to products from fresh catalyst hydrogenation (Table 7). The solid fat content values, at various temperatures for the three different fats produced, are shown in Fig. 1.

TABLE 6
Experimental conditions for the hydrogenation of soyabean oil using Pricat 9906, spent Pricat 9906 and Nysel SP-7 catalysts (Krishnarajah, 1982)

	Pricat 9906	Spent Pricat 9906	Nysel SP-7
Soyabean oil (tonnes)	12	12	12
Temperature (0°C)	195 ± 3	195 ± 3	195 ± 3
Pressure (psi)	5	5	5
Catalyst concentration (kg)	17	34	61

The fats manufactured using Pricat 9906 (highly selective catalyst) and spent Pricat 9906 may be used, typically, as hard base stocks in margarines and as bakery shortenings. The fat with the highest *trans*-level, manufactured using a *trans*-isomer promoting SP-7 catalyst, is used, typically, in confectionery coatings and as a cocoa butter substitute. A number of references are available on sulphur poisoned catalysts (Baltes, 1970, 1972; Osinga & Van Beek, 1974; Poot et al., 1978; Snyder, 1979; Marsch, 1980). Okonek et al. (1980) reviewed the subject in relation to soyabean oil.

Some work carried out by the author has shown that it is quite feasible to hydrogenate rapeseed oil to the melting profile of lauric fats, i.e. coconut oil and palm kernel oil, using sulphur poisoned nickel catalysts. The reaction conditions are listed in Table 8. The results show that the SFC values are quite similar to the lauric fats (Table 9). This type of work enables rapeseed oil to be a direct substitute for lauric fats in ice-cream applications and in confectionery products. Equally, this selectively hydrogenated rapeseed oil can be used to substitute for partially hardened palm kernel oil (melting point 32°C) in typical margarine oil formulations (Table 10).

HYDROGENATION FOR FOOD APPLICATIONS

Salad Oil

Selective hydrogenation is used to produce flavour stability in salad oils. It is well established that off-flavour development is caused by the votalile products, e.g. unsaturated aldehydes, from the oxidative breakdown of oils. The rate of oxidation increases with the degree of unsaturation and so vegetable oils such as soyabean oil, with average values for linoleic acid of 52% and linolenic acid of 8%, are particularly prone to flavour reversion. Selective hydrogenation should preferentially reduce the polyunsaturated

TABLE 7

Isomerisation during soyabean oil hydrogenation. *Trans*-acids at different iodine values for (a) Nysel SP-7, (b) fresh Pricat 9906 and (c) spent Pricat 9906 (Krishnarajah, 1982)

(a) Nysel SP-7

Iodine value	109·1	105·7	97·3	94·3	89·9	86·5	81·3	79·0	78·1	77·4	76·4	75·6
Trans-acids (%)	57·0	60·9	71·3	76·9	72·4	82·0	78·9	82·4	80·0	77·7	75·5	65·7

(b) Fresh Pricat 9906

Iodine value	95·1	88·5	84·5	80·6	77·6	75·3	72·0	69·3	66·3	63·1
Trans-acids (%)	18·6	32·3	33·5	40·1	46·7	44·5	48·1	48·1	43·1	42·3

(c) Spent Pricat 9906

Iodine value	91·1	88·6	85·3	82·2	80·4	78·6	75·9	73·4	69·9	68·6	66·4
Trans-acids (%)	56·6	58·2	62·4	61·5	64·2	62·3	62·3	54·9	64·0	55·6	49·9

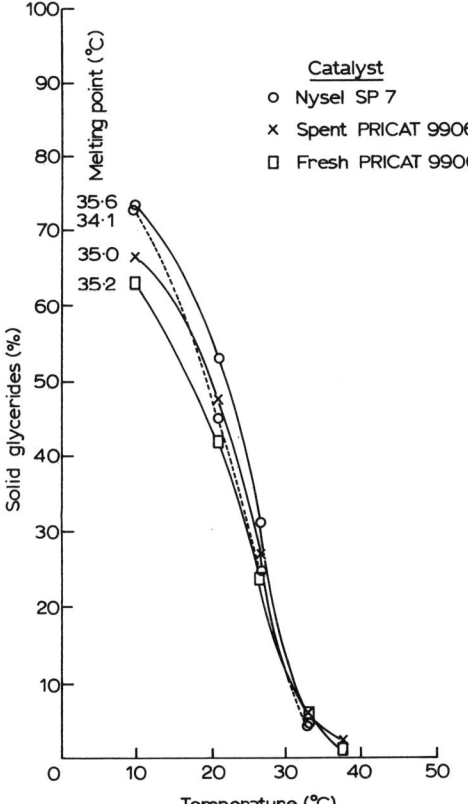

Fig. 1. Solid glycerides–temperature curves for hydrogenated soya oil showing selectivity effects (Krishnarajah, 1982).

acids and lower the oxidation rate and thereby improve the flavour stability of the oil. However, even hydrogenated products can develop characteristic off-flavours produced by specific precursor acids which, on oxidation, release very pungent unsaturated aldehydes (Keppen et al., 1967).

Highly selective nickel catalysts are generally used to lightly hydrogenate oils to improve their oxidative stability. This treatment, however well controlled, tends to produce either *trans*-isomers or stearic acid as the linolenic acid is gradually eliminated. Such oils cannot normally be used as salad oils as they tend to crystallise on standing. To overcome this problem the oil, e.g. soyabean oil, is lightly hydrogenated to an iodine value of 105

TABLE 8
Reaction conditions for rapeseed oil hydrogenation using a sulphur poisoned catalyst

	Experiment	
	1	2
Rapeseed oil quantity	1000 g	1000 g
Catalyst concentration	8 g	5 g
Temperature	$190 \pm 2°C$	$190 \pm 2°C$

Notes
1. Catalyst: Nysel SP-7, a sulphur poisoned nickel catalyst from Harshaw Ltd.
2. In both experiments the reaction was allowed to continue until no further hydrogenation was evident. At the point of completion the iodine values of the oils averaged 86 units.
3. The reactions were carried out under highly selective conditions.
4. RBD rapeseed oil was used for the experiments.

TABLE 9
Comparison of rapeseed oil, hydrogenated using a sulphur poisoned nickel catalyst, with lauric oils

Oil	Experiment		Palm kernel oil	Coconut oil
	1	2		
Iodine value	86·3	86·2	19·0	9·9
Refractive angle	16°07′	16°07′	15°01′	14°44′
Rapid slip point	29·7°C	27·9°C	26·4°C	23·8°C
Melting point	32·8°C	31·4°C	28·1°C	24·9°C
% *Solids*				
10·0°C	66·9	65·8	65·2	64·7
21·1°C	32·2	28·3	36·9	28·2
26·7°C	8·3	5·1	7·2	—
33·3°C	—	—	—	—
27·8°C	—	—	—	—
% *Trans*	71·1	69·8	Nil	Nil

TABLE 10
Substitution of partially hardened palm kernel oil (melting point 32°C) with selectively hydrogenated rapeseed oil in typical margarine oil formulations

Oil blend:	(1) 65% tallow olein 20% HPK 32/4 15% H rape (SP7)	(2) 65% tallow olein 20% H rape (SP7) 15% HPK 32/4	(3) 65% tallow olein 35% H rape (SP7)
% Solids			
10·0°C	52·4	49·7	47·5
21·1°C	21·9	21·9	22·0
26·7°C	11·0	11·3	9·8
33·3°C	3·3	3·5	2·8
37·8°C	0·4	0·9	0·2

TABLE 11
Effect of nickel catalyst versus copper catalyst on the linolenic acid content of selectively hydrogenated soyabean oil (Mounts, 1980)

	Composition of soyabean oils (wt %)		
	Unhydrogenated oil (SBO)	Nickel reduced oil (NiHSBO)	Copper chromite reduced oil (CuHSBO)
16:0	10·2	10·5	10·4
18:0	3·4	4·2	3·7
18:1	22·3	47·6	41·3
18:2	55·5	34·4	44·2
18:3	8·6	3·3	0·4
Calculated IV	137·7	109·3	113·3

and winterised to produce a salad oil. The alternative is to use a copper catalyst. Mounts (1980) has shown (Table 11) that copper chromite catalyst can eliminate linolenic acid almost completely at a calculated iodine value of 113·3 without significantly altering the stearic acid value. By comparison the nickel reduced oil, even at a calculated iodine value of 109·3, had a linolenic acid content of 3·3%. Copper catalysts have been used commercially for the hydrogenation of soyabean oil, both in the USA and in Europe. However, their large scale use has been limited due to their low activity (and hence the need for higher catalyst concentration) relative to nickel catalysts and the susceptibility to poisoning, which affects their selectivity. Special post-treatment of the oil is required to remove traces of copper which is a pro-oxidant.

Mounts (1980) has suggested that citric acid treatment of hydrogenated oils can aid reduction of copper to satisfactory levels. This, however, adds on additional cost. The oil entrained in the filter cake also degrades very quickly and so cannot be recovered. It is estimated that the overall consumption of copper catalysts is about 5–10 times that of nickel. Several references are available on the use of copper catalysts (Nikki & Kabushiki, 1964; De Jonge et al., 1965; Koritala & Dutton, 1965; Lefebvre & Baltes, 1975; Rozendaal, 1983).

Procter & Gamble (1968) have claimed a process for producing saturated materials for use in salad oils and other fat and oil materials from cottonseed, peanut, sesame seed, olive, palm, coconut, corn, sunflower seed, soyabean, rapeseed or safflower oils, lard and tallow, using a suspended particulate catalyst, preferably nickel, platinum or palladium. The USDA (1974) have patented a process for the selective partial hydrogenation of vegetable oil using a nickel–copper chromite catalyst system. It is claimed that the selectivities are only slightly reduced over copper catalysts alone, but the rate of hydrogenation is significantly accelerated. A process is also available for nickel–copper catalysed hydrogenated oils useful in the production of salad cream, margarine, shortenings and dietary fats (Unilever NV, 1975).

Cooking and Frying Products

Rapeseed oil can be hydrogenated at 180–200°C under reduced pressure for greater than 10 min, and then hydrogenated at ordinary or high pressure at 170–220°C, in the presence of copper or nickel catalysts, but preferably reduced nickel catalyst, to give a stable edible oil with good taste and colour (Asahi Electrochem. Ind. KK, 1977).

The cooking and storage stability of soyabean oil can be improved with partial hydrogenation (Hunt-Wesson Foods Inc., 1970). The oil is pre-treated by degassing under vacuum and saturating with carbon dioxide. This is followed by hydrogenation using copper chromite catalyst at 100–310°C to an iodine value greater than 105. The oil is cooled to below 75°C, filtered and deodorised and remains stable with no objectionable flavour even when heated at high temperatures. The nutritional value is maintained and the stearin content remains unchanged. Soyabean oil can be selectively hydrogenated in the presence of nickel catalyst to produce an oil with properties similar to those of groundnut oil, which can be used for frying chips (Holtz & Willemsen GmbH, 1971). A Unilever Plc (1982) process is available, which describes the selective hydrogenation of unsaturated fatty acid derivatives where products containing three or more double bonds are reduced to two double bonds. The process is claimed to be inexpensive, the hydrogenation being carried out in the presence of ammonia and a palladium, platinum, iridium or rhodium catalyst at -20 to $+100°C$. The hydrogenated products can be used as frying oils or table oils as well as in applications such as margarine blends.

Two patents (Yukijirushi Shokuhi, 1984a, b) describe a novel invention where hydrogenated soyabean oil and rapeseed oil are blended and mixed in with food coating crumb. When food is coated with butter, egg and the crumb mixture in this order, it can be roasted on a pan or in an oven toaster, etc., to produce a fried product without once frying in oil.

Fuji Oil KK (1980) have a process for producing a highly stable hydrogenated soft palm oil or its low melting fraction, suitable for application as frying or spraying oils in products such as potato chips, rice cake and instant Chinese vermicelli. Chicken fat can be selectively hydrogenated to an iodine value of 40–65, using nickel or copper chromite catalyst at 140–220°C under normal or elevated pressure (Riken Vitamin Oil KK, 1975). Baked chicken bones are introduced into the hydrogenated fat and the mixture is agitated at temperatures of around 40°C for about 15 min to give the fat a chicken-like taste.

Bakery Coatings

An edible hard butter can be produced for use as a biscuit or cracker coating as well as for enrobing, etc. (SCM Corp., 1980). The product contains selectively hydrogenated glyceride oil and two or more lipoidal emulsifiers. In another patent, a blend of selectively hydrogenated vegetable fatty oil and lauric series fatty oil is used as edible coating oil in, for instance, biscuits (Asahi Electrochem. Ind. KK, 1985). The vegetable oil used should be non-lauric and contain greater than 50% by weight, of monoene acids, e.g. rapeseed oil, olive oil, high oleic safflower oil, almond oil, etc. The fat has slow solidification at room temperature and gives glossy products. A laurine type hydrogenated fat, with a melting point of 30°C or higher, has been developed for icing, to be used with other fat, partial ester of polyhydric alcohol of fatty acid, and sugar (Kanegafuchi Chem. KK, 1981).

Emulsions and Spreads

Asahi Electrochem. (1978a) describe a process where liquid vegetable oil, of iodine value 120–140 and linoleic acid content of greater than 60%, is mixed with palm oil to produce a fat for margarine manufacture. The blend is subjected to random-ester-exchange producing a fat containing 20–35% saturated fatty acid, of which the total proportion of myristic acid and palmitic acid amounts to greater than 80%. The fat is then hydrogenated to produce a *trans*-fatty acid content in excess of 40% and a melting point in the range of 33–37°C. The solid fat index (SFI) measurements are as follows:

SFI >55% at 10°C, >50% at 20°C, >45% at 25°C, >30% at 30°C, <15% at 35°C, <8% at 37°C, <0% at 40°C

A Unilever Plc (1982) process to produce fats suitable for use as intermediates in margarine manufacture has been described earlier.

Sandiness in hydrogenated low-erucic acid rapeseed oil margarine can be prevented by using an oil soluble HLB sugar ester (Grinsted Vaerket A/S, 1982). A typical application would be as in the following formulation:

Water phase: 18%
Fat phase: 0·2% Dimodan PM (saturated distilled monoglyceride)
0·2% soya lecithin
0·2% annatto colour
81·4% fat blend comprising 60 parts rapeseed oil, melting point 32–34°C, 40 parts liquid soyabean oil.

The invention claims that usage levels of up to 1% of an oil soluble low HLB sugar ester (HLB value 1, melting point 62–65°C, 95% tri- and higher esters of mainly myristic, palmitic and stearic fatty acids) prevented β-crystal transformation relative to a control sample without sugar ester which had complete crystal transformation after 1–2 weeks.

Interesterified blends of palm oil and lauric oils can be hydrogenated and/or fractionated to provide fats suitable for use in emulsion and other fat-based food products, such as margarine, artificial creams and ice-creams (Unilever Plc, 1986).

A peanut butter stabiliser has been produced containing an intermediate melting fat fraction and a hydrogenated oil fraction (Procter & Gamble, 1982). The intermediate melting fraction should have an iodine value in the range of 30–40 and preferably made from cottonseed oil hydrogenated to an iodine value of about 36, or a blend of soyabean oil and palm oil such that the final iodine value is 35. It is preferred that the hydrogenated oil fraction is produced from rapeseed oil, and in some instances an amount of hydrogenated cottonseed oil, iodine value of about 4, is also added to the rapeseed oil. Usage levels of stabiliser composition in peanut butter are

1–4%, preferably 1·5–3·5%, especially 2–3% by weight. The stabiliser composition is believed to improve the texture of peanut butter, reduce the stickiness, and provide a fluid consistency over, for example, 10–37·8°C, giving good melt-in-the-mouth properties and stability.

Chocolate and Confectionery Products

A variety of edible oils, e.g. fish, jojoba, palm, palm kernel, rice bran and a mixture of cotton and coconut oils, have been hydrogenated to produce hard butters for use in chocolate confectionery products.

Fish oil

The feed-stock is required to contain between 10 and 30%, by weight, of saturated fatty acid and a similar amount of highly unsaturated fatty acids having more than three unsaturated bonds (Asahi Denka Kogyo, 1980). The oil is selectively hydrogenated to produce a fat containing the following solid fat index:

$>50\%$ at 10°C, $>45\%$ at 20°C, 20–40% at 30°C, $<2·5\%$ at 40°C

The hardened fat is claimed to be highly compatible with cocoa butter and, as such, suitable as raw material for chocolate. The fat is very palatable with satisfactory melt-in-the-mouth properties, and end products are easily released from the mould and are also resistant to blooming.

Jojoba oil

Fully or partially hydrogenated jojoba oil can be used as coating agents for food (Unilever Ltd, 1979). Full hydrogenation is carried out to a melting point greater than 70°C and the product is substituted for beeswax or carnuba wax for imparting gloss to chocolate, etc. The partially hydrogenated product is prepared using a sulphur free metal catalyst, preferably reducing the iodine value of the oil by only about 2. This substantially improves the stability without significantly affecting the melting point or promoting the formation of *trans*-isomers too much. A major application for the partially hydrogenated oil is the coating of dried fruit. The coatings are very stable and are applied in solution in a volatile solvent such as acetone or alcohol. They also preferably contain anti-oxidant, especially tocopherol from vegetable lecithin.

Palm oil

Cocoa butter substitutes can be produced from palm oil through a number of routes. A Russian patent (Fats Res. Inst., 1974) describes a process where hydrogenated palm oil is used as the feed material for fractional crystallisation, the hard fraction obtained being suitable for use as a cocoa butter substitute. A Japanese patent (Fuji Oil KK, 1978) claims a method whereby a hard palatable butter, of good texture, can be prepared inexpensively without the need to fractionate the hydrogenated fat. To achieve this, a mixture containing 20–100 parts by weight of soft palm oil of iodine value greater than 55, and made up to 100 parts by weight with a liquid oil of iodine value 80–140, is hydrogenated under conditions which aid *trans*-isomer formation (elaidinisation). It results in a fat with the following properties: melting point 32–36°C, *trans*-fatty acids of greater than 60% expressed as the rate of trielaidin to total triacylglycerol and SFI values of >65% at 10°C, >60% at 20°C, 20–40% at 30°C and <1% at 40°C. The hydrogenation is carried out in the presence of a nickel catalyst and sulphur containing compound, e.g. methionine, thiamine, etc. The hard butter is suitable for use in chocolate bars and particularly coatings.

Another process is available (Lester *et al.*, 1980) where palm olein is fractionated to produce a second liquid fraction and an intermediate fraction. The intermediate fraction is then catalytically hydrogenated to give an iodine value of 38–45, a linoleic acid content of <2% and a melting point of 33–36°C, or, alternatively, 0·01–1·5%, by weight, of an edible structural modifier is incorporated in the separated intermediate fraction. This method allows the isolation of an intermediate palm oil fraction for the production of a cocoa butter substitute without resorting to a solvent or other chemical additive. An American patent (Kanegae, 1977) describes a similar process where palm olein, iodine value >55, is hydrogenated in the presence of methionine, or, alternatively, cystine or cysteine, and conventional nickel catalyst to give a product having a minimum level of 40% *trans*-isomers. It is claimed that the hard butter may substitute for cocoa butter in chocolate and prevent blooming, even when the product is used without tempering.

Palm kernel oil (Fats Res. Inst., 1978)
A mixture of palm kernel oil and longer-chain fatty acid triacylglycerols is hydrogenated and blended in with a hydrogenated lauryl triacylglycerol free oil. The blend is *trans*-esterified, refined and deodorised and used as a confectionery fat.

Rice bran oil
The manufacture of hard butters from rice bran is described (Asahi Electrochem. Ind. KK, 1978*b*), the end product being a cheap substitute for

cocoa butter, of good quality and satisfactory for use in chocolate manufacture. Amongst its functional attributes are its sharp melting properties, heat stability and resistance to fat blooming.

Cottonseed oil/coconut oil
A mixture of oils, e.g. 40:60 mixture of cottonseed oil and coconut oil, is *trans*-esterified and then hydrogenated (Kharkov Poly, 1977). The hydrogenated fat closely resembles cocoa butter and can be mixed with cocoa powder to produce chocolate.

Amongst other applications described for hydrogenated fats in confectionery products, a high quality grease has been developed for moulds and sheets used in the food industry (Mosc. Fats Res. Inst. & Chermushkin Confection, 1980). The product contains (in weight percentage) 37–41% vegetable oil, 40–44% hydrogenated fat, 10–12% acetone derivatives of monoglycerides and 7–9% of a phosphate concentrate which is added to improve the stability of the composition and to reduce manufacturing cost. The mould coating prevents sticking of the product to the mould surface, thereby producing glossy surface products.

HYDROGENATED MILKFAT PRODUCTS

Edible oil hydrogenation techniques and technology can be applied to modify milkfat. However, the practice is virtually non-existent in the dairy industry and it is unclear if it has been adopted by any of the edible oil processors. Nevertheless a number of studies have been carried out to ascertain the effect of hydrogenation on the properties of milkfat and their application in food.

Rajah *et al.* (1982) have reported on the hydrogenation of normal milkfat and milkfat olein. The results are shown in Tables 12(a), 12(b) and 13(a), 13(b). Some work carried out at Ghent University during the 1970s and early 1980s concluded that the use of bleaching earth on the feed removes most of the pigments but leaves the UV-spectrum of the fat unaltered; the hydrogenation generally follows a first order reaction while an increase in temperature, catalyst concentration, hydrogen pressure and agitation all have a positive influence on the reaction rate. Using differential scanning calorimetry the researchers found that upon solidification milkfat gave a major peak and a smaller peak of high melting triacylglycerols. The researchers found that during hydrogenation there is a shift towards higher temperatures. Timms (1980) found similar results

TABLE 12(a)
Analytical results on hydrogenation experiment using anhydrous milk fat (Rajah et al., 1982)

Sample	Original	1	2	3	4
Melting point (°C)	33·1	33·2	33·4	35·3	35·8
Iodine value	29·1	28·7	28·3	27·7	27·2
NMF % solids					
0°C	69·1	73·3	76·2	78·9	83·5
5°C	65·2	69·4	72·0	75·0	79·9
10°C	57·2	61·3	63·7	66·7	71·7
15°C	41·2	44·4	47·4	50·0	55·3
20°C	21·9	23·9	25·2	27·0	31·0
25°C	13·1	14·3	15·3	16·8	20·0
30°C	6·3	7·0	7·6	8·4	10·3
35°C	0·5	0·3	0·6	1·2	2·2
40°C	—	—	—	—	—

TABLE 12(b)
Fatty acid profiles of hydrogenated anhydrous milkfat from Table 12(a) (Rajah et al., 1982)

Sample	Original	1	2	3	4
C4	3·08	2·91	2·81	2·76	2·92
C6	2·23	2·13	2·11	2·09	2·15
C8	1·45	1·40	1·40	1·38	1·40
C10	3·44	3·33	3·38	3·32	3·38
C12	3·97	3·93	3·85	3·81	3·91
C14	12·33	12·11	12·10	12·11	12·06
C14:1	1·40	1·34	1·35	1·30	0·75
C15	1·17	1·16	1·15	1·11	1·15
C16	30·93	30·99	30·94	31·00	30·84
C16:1	3·42	3·14	3·20	3·12	3·12
C17	1·20	1·18	1·17	1·17	1·13
C18	10·16	10·62	10·75	10·95	11·16
C18.1	22·74	23·12	23·16	23·32	23·42
C18:2	0·87	—	—	—	—

TABLE 13(a)
Analytical results on hydrogenation experiment using a low melting fraction (15·3°C) (Rajah et al., 1982)

Sample	Original	1	2	3	4
Melting point (°C)	15·3	18·0	20·1	22·0	23·5
Iodine value	38·9	37·9	36·6	35·5	34·8
NMR % solids					
0°C	35·4	41·3	47·7	55·2	64·2
5°C	27·2	33·3	39·9	47·2	56·2
10°C	13·6	18·6	24·2	30·9	39·5
15°C	1·5	3·8	6·9	11·0	17·2
20°C	0·2	0·2	1·9	3·7	7·0
25°C	—	—	0·2	0·2	1·4
30°C	—	—	0·2	—	—
35°C	—	—	—	—	—

TABLE 13(b)
Fatty acid profiles of hydrogenated fractions from Table 13(a) (Rajah et al., 1982)

Sample	Original	1	2	3	4
C4	3·44	3·36	3·41	3·26	3·21
C6	2·48	2·44	2·47	2·41	2·40
C8	1·65	1·61	1·66	1·61	1·60
C10	3·64	3·64	3·74	3·73	3·72
C12	4·96	4·84	4·98	4·92	4·91
C14	12·17	12·34	12·54	12·37	12·37
C14:1	1·70	1·88	1·88	1·81	1·00
C15	0·89	0·99	0·99	0·97	0·96
C16	24·97	24·90	25·15	24·96	24·98
C16:1	4·98	4·71	3·96	3·80	3·75
C17	1·01	0·99	1·03	1·01	1·01
C18	6·48	6·73	6·97	7·26	7·70
C18:1	28·23	29·03	29·31	29·44	29·92
C18:2	1·60	1·04	0·39	—	—

during his work on the phase behaviour and polymorphism of milkfat. He also noted that the increase in the heat of melting on hydrogenation was close to the increase calculated from the increase in the amount of stearic acid which occurred during the hydrogenation. Timms (1978) has also reported on the solubility of fully hardened milkfat in liquid oils. He observed that, compared with unhardened milkfat, fully hydrogenated milkfat with sunflower oil showed a marked deviation from the ideal solubility law equation. He suggested that the high melting triacylglycerols of the hydrogenated fat were of widely ranging fatty acid composition, compared with the rather narrow C14/C16/C18 fatty acids comprising the high melting fractions of normal milkfat. Since fully saturated triacylglycerols containing a shorter chain will occur following the hydrogenation of oleic acid, it is possible that this wide range of saturated triacylglycerols leads to more imperfect crystals of variable composition. He maintained that the study showed that an increase in the crystalline imperfection of the solid phase would lead to increased solubility. This observation is particularly useful in the development of milkfat based spreads, where vegetable oils are added to aid spreadability.

The effect of varying conditions of temperature, hydrogen pressure and catalyst concentration was also studied by Yoncoskie et al. (1969). The hydrogenation of milkfat with palladium and nickel as catalysts was researched by Smith & Vansconcellos (1974). They found that palladium was several times more active than nickel.

A number of researchers have reported on the effect of hydrogenation on the flavour stability of milkfat. Selective hydrogenation, using a Raney nickel catalyst, to an iodine value of 20·6 resulted in a milkfat which, on recombination with skimmed milk, gave a more flavour stable whole milk (Weihe, 1956). The strong hydrogenation flavour has always been a problem. Steam distillation or deodorisation also eliminates the desirable flavour components of milkfat. Further investigations have shown that it is possible to trace hydrogenate milkfat without the development of the hydrogenation flavour, the ideal process conditions being a temperature of 120°C, 1% nickel catalyst, and a hydrogen pressure of 20 psig. The iodine value drop was 4 units (Vasishtha et al., 1969). Further improvements were reported (Vasishtha et al., 1970) using a palladium catalyst, 0·5%, at 25°C and 20 psig hydrogen pressure. The reaction time was 85 min, and the iodine value reduction was 4·1 units. The flavour and flavour stability were much improved and the colour remained largely intact.

Candies made with trace hydrogenated milkfats showed that an iodine value drop of 2 units was insufficient to compete with anti-oxidant treated milkfat (Leeder et al., 1974, 1975). However, an iodine value drop of 4 units was more effective in retarding the development of oxidative rancidity in buttercream candies than when anti-oxidants were added.

Hydrogenated milkfat can be used in chocolate manufacture (Huyghebaert et al., 1986). A 10% substitution for cocoa butter in dark chocolate or 5–6% in milk chocolate will not cause significant variation to product quality. Partial substitution of the cocoa butter with hydrogenated milkfat helps retard or eliminate fat bloom in chocolate. Campbell et al. (1969) found that hydrogenation markedly improved the milkfat's bloom inhibiting properties. Chocolate containing 2·5% by weight of fully hydrogenated milkfat (iodine value = 1) took four times as long to show evidence of bloom compared to the control. Partially hydrogenated milkfat was slightly less effective and took three times as long.

REFERENCES

ALBRIGHT, L. F. (1965) *J. Am. Oil Chem. Soc.*, **42**, 250.
ALLEN, R. R. (1978) Paper presented at Am. Oil Chem. Soc. Short Course, June.
ALLEN, R. R. & KIESS, A. A. (1955) *J. Am. Oil Chem. Soc.*, **32**, 400.
ASAHI DENKA KOGYO (1980) Japanese Patent 55, 164, 296-A.
ASAHI ELECTROCHEM. IND. KK (1977) Japanese Patent J52, 085, 206-A.
ASAHI ELECTROCHEM. IND. KK (1978*a*) Japanese Patent 53, 094, 066-A.
ASAHI ELECTROCHEM. IND. KK (1978*b*) British Patent 1, 534, 123-A.
ASAHI ELECTROCHEM. IND. KK (1985) Japanese Patent 85, 050, 420-B.
BABUCHOWSKI, K. & RUTKOWSKI, A. (1969) *Seifen Ole Fette Wachse*, **95**, 27.
BAILEY, A. E. (1949) *J. Am. Oil Chem. Soc.*, **26**, 596.
BAILEY, A. E. (1950) *Melting and Solidification of Fats*, Interscience, New York.
BAILEY, A. E. (1951) *Industrial Oil and Fat Products*, 2nd edn, Interscience, New York.
BAILEY, A. E. & MCKINNEY, R. H. (1941) *Oil and Soap*, **18**, 120.
BALTES, J. (1970) *Fette Seifen Anstrichmittel*, **72**, 425–32.
BALTES, J. (1972). US Patent 3, 687, 989.
BECKMANN, H. J. (1983) *J. Am. Oil Chem. Soc.*, **60**, 282.
BERWICK, B. J. (1975) *Dairy Technol.* **6** (2), 50–3.
BRATLAND, A. (1982) British Patent No. 2, 087, 211 A.
CAMPBELL, L. B., ANDERSEN, D. A. & KEENY, P. G. (1969) *J. Dairy Sci.* **52**, 7.
CANT, P. A. E., MCDOWELL, A. K. R. & MUNRO, D. S. (1975) New Zealand Dairy Research Institute, Annual Report, p. 35.
CARLSON, K. (1989) *J. Am. Oil Chem. Soc.*, **66**, 1547.
CECCHI, G. (1980) *Proceedings of Symposium on Hydrogenation of Oils*, Rimini, Italy, 29–30 September, American Soybean Association, Brussels, pp. 12–40.

COENEN, J. W. E. (1969) *J. Oil Technol. Ass. India*, **16**.
COENEN, J. W. E. (1974a) *Rev. France Corps Gras.*, **21**, 403.
COENEN, J. W. E. (1974b) *Rev. France Corps Gras.*, **21**, 343.
COENEN, J. W. E. (1976) *J. Am. Oil Chem. Soc.*, **53**, 382.
COENEN, J. W. E., BOERMA, H., LINSEN, B. G. & DE VRIES, B. (1964) *Proc. 3rd. Int. Congress on Catalysis*, Amsterdam, Vol. 2, p. 1387.
DEFFENSE, E. (1983) International Conference on Oils, Fats and Waxes, New Zealand, S. A. Fractionnement Tirtiaux, Fleurus, Belgium.
DEFFENSE, E. (1985) *J. Am. Oil Chem. Soc.*, **62** (2), 376.
DEFFENSE, E. & TIRTIAUX, A. (1982), Paper presented at the AOCS Conference, The Hague, Netherlands, October (summary: *J. Am. Oil Chem. Soc.*, **60**, 473).
DE JONGE, A., COENEN, J. W. E. & OKKENSE, C. (1965) *Nature, London*, **206**, 573–4.
DEMAN, L., DEMAN, J. M. & BLACKMAN, B. (1989) *J. Am. Oil Chem. Soc.*, **66**, 1777.
DIXON, B. D. & BLACK, R. G. (1974) *Proceedings of XIX International Dairy Congress*, IE, pp. 650–1.
DOERRY, W. T. & MELOAN, E. (1986) American Institute of Baking, *Technical Bulletin*, VIII, 10, 1.
DROZDOWSKI, B. & ZAJAC, M. (1977) *Am. Oil Chem. Soc.*, **54**, 595.
DROZDOWSKI, B. & ZAJAC, M. (1980) *Am. Oil Chem. Soc.*, **57**, 149.
ELDIB, I. A. & ALBRIGHT, L. F. (1957) *Ind. Engng Chem.*, **49**, 825.
E. SIBE TECH. INST. (1985) Soviet Union Patent 1200875-A.
EYRES, L., BOON, P. M. & ILLINGWORTH, D. (1988) Tailored food ingredients from fractionated milkfat. Paper presented at the 3rd International Exhibition & Conference on Ingredients and Additives, November 15–17, London.
FATS RES. INST. (1974) Russian Patent 273, 987-A.
FATS RES. INST. (1978) Russian Patent 604, 552-A.
FEUGE, R. O., PEPPER, M. B., O'CONNER, R. T. & FIELD, E. T. (1951) *J. Am. Oil Chem. Soc.*, **28**, 420.
FJAERVOLL, A. (1970) *Svenska Mejeritidn.*, **61**, 491.
FUJI OIL KK (1978) Japanese Patent 53, 075, 366-A.
FUJI OIL KK (1980) Japanese Patent 55, 069, 696-A.
GOODING, C. M. (1953) US Patent No. 2, 627, 467.
GOODING, C. M. (1962) US Patent No. 3, 048, 491.
GOODING, C. M. (1972) US Patent No. 3, 674, 821.
GRINSTED VAERKET A/S (1982) RD 213, 001-A.
GROTHUES, B. G. M. (1985) *J. Am. Oil Chem. Soc.*, **62**, 390.
HAMILTON, M. P. (1983) United Kingdom Society of Dairy Technology Ice Cream Symposium, pp. 9–17, paper given 19-20 October 1991.
HANNEWIJK, J., HAIGHTON, A. J. & HENDRIKSE, P. W. (1964) *Analysis and Characterization of Oils, Fats and Fat Products, Vol. 1*, Boekenoogen, H. A. (Ed.), Interscience, London.
HOLTZ & WILLEMSEN GMBH (1971) East German Patent 2157682-A.
HUNT-WESSON FOODS INC. (1970) US Patent 3, 758, 532-A.
HUYGHEBAERT, A., DE MOOR, H. & DECATELLE, J. (1991) *Milk Fat: Production, Technology and Utilisation*, Rajah, K. K. and Burgess, K. J. (Eds), The Society of Dairy Technology, Crossley House, 72 Ermine Street, Huntingdon, Cambridgeshire PE18 6EZ, pp. 44–62.
ILLINGWORTH, D. (1990) Personal Communication, New Zealand Dairy Research Institute, Private Bag, Palmerston North.

JEBSEN, R. S., TAYLOR, M. W., MUNRO, D. S. BISSELL, T. G. et al. (1975) New Zealand Dairy Research Institute, Annual Report, p. 32.
KANEGAE, J. (1977) US Patent 4, 061, 798.
KANEGAFUCHI CHEM. KK (1981) Japanese Patent 81, 036, 897-B.
KEPPEN, J. G., HORIKX, M. M., MEYBOOM, P. W. & FEENSTRA, W. H. (1967) J. Am. Oil Chem. Soc., **44**, 543.
KHARKOV POLY (1977) Russian Patent 246, 307-A.
KINSELLA, J. E. (1975) Food Technol., **29**, 82.
KLEINERT, J. (1970) Rev. Int. Choc. (RIC), **25**, 11 November, 386.
KLOSTERMANN, K. & HOBERT, H. (1980) J. Catal. **63**, 355–63.
KORITALA, S. & DUTTON, H. J. (1965) J. Am. Oil Chem. Soc., **43**, 86–9.
KRISHNAMURTHY, R. G. (1982) Bailey's Industrial Oil and Fat Products, Vol. 2, 4th edn, Chap. 5, Wiley-Interscience, Chichester.
KRISHNARAJAH, K. (1982) Physico-chemical studies on soyabean oil, MSc Thesis, University of Salford, UK.
LARSSON, K. (1966) Acta Chem. Scand. **20**, 2255.
LARSSON, R. (1983) J. Am. Oil Chem. Soc., **60**, 198.
LAUSTSEN, K. (1984) Lipidforum Symposium Proceedings, Milkfat and its Modification, January 26–27, Marcuse, R. (Ed.), c/o SIK, Box 5401, S-402 29, Goteburg, Sweden.
LEEDER, J. G., HWANG, T. M. & CHANG, S. S. (1974) Manufact. Confect., **54**, 70.
LEEDER, J. G., HWANG, P. M. & CHANG, S. S. (1975) Gordian, **75**, 66.
LEFEBVRE, J. & BALTES, J. (1975) Fette Seifen Anstrichm., **77**, 125.
LESTER, C. & CO. LTD, TIRTIAUX, S. A. & PIKE, M. (1980) British Patent 1, 573, 210-A.
LUTTON, E. S. (1950) J. Am. Oil Chem. Soc., **27**, 276.
MAGNUSSON, H. & NOTEVARP, O. (1971) Presented at Sixth Scandinavian Symposium on Fats and Oils, Grena, Denmark, June.
MALKIN, T. (1954) Progress in the Chemistry of Fats and Other Lipids, Vol. 2, Pergamon Press, New York.
MANGOLD, H. K. (1982) J. Am. Oil Chem. Soc., **59** (9), 673A–674A.
MANNING, D. J. & NURSTEN, H. E. (1985) Developments in Dairy Chemistry–3, Fox, P. F. (Ed.), Elsevier Applied Science, London, Chapter 8.
MARSCH, J. T. (1980) US Patent 4, 201, 718-A.
MATTIL, K. F. (1964a) Bailey's Industrial Oil and Fat Products, 3rd edn, Swern, D. (Ed.), Wiley-Interscience, Chichester, Ch. 10.
MATTIL, K. F. (1964b) Bailey's Industrial Oil and Fat Products, 3rd edn, Swern, D. (Ed.), Wiley-Interscience, Chichester, Ch. 8, p. 274.
MATTIL, K. F. (1964c) Bailey's Industrial Oil and Fat Products, Interscience, New York, p. 796.
MATTIL, K. F. (1964d) Bailey's Industrial Oil and Fat Products, Interscience, New York, p. 831.
MATTIL, K. F. & NORRIS, F. A. (to Swift and Co.) (1953) US Patent No. 2, 625, 478.
MCWILLIAMS, M. (1979) Food Fundamentals, 3rd edn, John Wiley, New York, p. 587.
MEIJI MILK PRODS. KK (1983) Japanese Patent 58134961-A.
MORK, P. C. & NORGARD, D. (1976) J. Am. Oil Chem. Soc., **53**, 506.
MORTENSEN, B. K. (1983) Developments in Dairy Chemistry–2, Fox, P. F. (Ed.), Ch. 5, Applied Science Publishers, London, p. 159.

Mosc. Fats Res. Inst. & Chermushkin Confection (1980) Russian Patent 733, 602-B
Mounts, T. L. (1980) *Proceedings of Symposium on Hydrogenation of Oils*, Rimini, Italy, 29–30 September, American Soybean Association, Brussels, pp. 41–55.
Mulder, H. & Walstra, P. (1974) *The Milk Fat Globule*, CAB, Farnham Royal and Pudoc, Wageningen, p. 31.
Nikki, Kagaku & Kabushiki, Kaisha (1964) British Patent 973, 957.
Norman, W. (1903) British Patent 1, 515.
Norris, R., Gray, I. K., McDowell, A. K. R. & Dolby, R. M. (1971) *J. Dairy Res.*, **38**, 179.
Okonek, D. V., Alcorn, W. R. & Cullo, L. A. (1980) Paper presented at the ISF/AOCS World Congress, April, New York.
Osinga, T. J. (1979) *Fette Seifen Anstrichm.*, **81**, 108.
Osinga, T. & Van Beek, W. (to Unilever Ltd) (1974) British Patent 1, 374, 237.
Ottesen, I. & Jensen, B. (1980) Paper presented at the ISF/AOCS World Congress, April 27–May 1, New York.
Patterson, H. B. W. (1983) *Hydrogenation of Fats and Oils*, Applied Science Publishers, London, p. 16.
Paulicka, F. R. & Jasko, J. J. (1978) Palm oil fractionation, Paper presented at American Oil Chemical Society 69th Annual Meeting, St Louis.
Poot, C., Verburg, D., Kirton, D. & MacNeill, A. (to Lever Bros Company) (1978) US Patent 4, 087, 564.
PORIM (1981) *Tech. Palm Oil Res., Malaysia*, No. 4.
Prasad, S. & Gupta, S. K. (1984) *Asian J. Dairy Res.*, **2** (4), 196–200.
Procter & Gamble Co. (1968) US Patent 3, 608, 039-A.
Procter & Gamble Co. (1982) US Patent 4, 341, 814-A.
Puri, P. S. & deMan, J. M. (1977) *J. Inst. Can. Sci. Technol. Aliment.*, **10** (1), 53–5.
Rajah, K. K. (1988) Fractionation of milk fat, PhD Thesis, University of Reading.
Rajah, K. K. (1992) *Lipid Technol.*, November/December, 129–137.
Rajah, K. K., Lane, R. & Middleton, R. (1982) EEC Co-Responsibility Report, Contract 271/82-56.2 entitled Production and Modification of Milkfat Fractions to Alter Their Functional Properties.
Richardson, A. S., Knuth, C. A. & Milligan, C. H. (1924) *Ind. Engng Chem.*, **16**, 519.
Riken Vitamin Oil KK (1975) Japanese Patent 75, 010, 608-B.
Robb, J. (1980) Flour Milling and Bakery Research Association, FMBRA Bulletin, No. 4, August, Chorleywood, UK, p. 179.
Rossell, J. B. (1985) *J. Am. Oil Chem. Soc.*, **62** (2), 385.
Rozendaal, A. (1983) *Proceedings of 3rd ASA Symposium on Soybean Processing—Hydrogenation of Soybean Oil*, June, Antwerp.
Sabatier, P. (1922) *Catalysis in Organic Chemistry* (trans. E. Reid), Van Nostrand, New York.
Sanderson, W. B. (1985) *IDF Newsletter*, June, No. 1, 1.
SCM Corporation (1980) US Patent 4, 209, 547-A.
Sims, R. J. & Hilfman, L. J. (1953) *J. Am. Oil Chem. Soc.*, **30**, 410.
Smith, L. M. & Vasconcellos, A. (1974) *J. Am. Oil Chem. Soc.*, **51**, 31.
Snyder, E. (to Kraft Inc.) (1979) US Patent 4, 169, 843.

Société des Produits Nestlé SA (1978) British Patent No. 1525315.
Société des Produits Nestlé SA (1985) European Patent No. 0166284A2.
Soltoft (1947) *On the Consistency of Mixtures of Hardened Fats* (trans. E. Christensen), Bjarne Kristensen-Bogtrykkerr, Copenhagen.
Sullivan, F. E. (1980) Paper presented at the AOCS Annual Meeting in New York, April 28.
Tan, B. K., Berger, K. G., Hamilton, R. J. & Jacobsberg, B. (1976) Paper presented at Malaysian International Symposium on Palm Oil, Kuala Lumpur.
Taylor, A. M. (1976) *Oléagineux*, **31**, 73.
Taylor, H. S. (1925) *Proc. R. Soc., London*, **A108**, 105.
Thomas, A. E. & Paulicka, F. R. (1976) *Chem. Ind. (London)*, **18**, 774.
Thomas, A. E. III (1985) *Bailey's Industrial Oil and Fat Products, Vol. 3*, Applewhite, T. H. (Ed.), Wiley-Interscience, New York, Ch. 1.
Timms, R. E. (1978) *Austral. J. Dairy Technol.*, December, 130.
Timms, R. E. (1980) *Austral. J. Dairy Technol.* **35** (2), 47.
Tirtiaux, A. (1980) Paper presented at the ISF-AOCS Congress, Tirtiaux Fractionation—The Flexible Way to New Fats, April 27–May 2, New York.
Tolboe, O. (1984) *Lipidforum Symposium Proceedings, Milkfat and its Modification*, Marcuse, R. (Ed.), SIK, Goteborg, Sweden, p. 43.
Unilever NV (1975) Dutch Patent 148, 101-B.
Unilever Ltd (1979) British Patent 2 016 042 A.
Unilever Plc (1982) Dutch Patent 8, 101, 637-A.
Unilever NV (1983*a*) Dutch Patent 8104820-A.
Unilever NV (1983*b*) European Patent 74146-A.
Unilever Plc (1986) EP-170431-A.
Urbach, G. (1979) *Proceedings of Milkfat Symposium, The Flavour of Milkfat*, Timms, R. E. (Ed.), Australian Society of Dairy Technology, p. 18.
Urdahl, H., Saav, H. & Helegrud, A. (1972) Paper presented at the 11th World Congress of J. S. F., Goteborg, Sweden.
USDA (1974) US Patent 3, 856, 719-A.
Vasishtha, A. K., Leeder, J. C. & Chang, S. S. (1969) *Food Technol.*, **23**, 244.
Vasishtha, A. K., Leeder, J. C. & Chang, S. S. (1970) *J. Food Sci.*, **35**, 395.
Walker, N. J. (1972) *N.Z. J. Dairy Sci. Technol.*, **7**, 135.
Walker, N. J. & Keen, A. R. (1976) New Zealand Dairy Research Institute, Annual Report, p. 36.
Weihe, H. D. (1956) *J. Dairy Sci.*, **34**, 910.
Weiss, T. J. (1967) *J. Am. Oil Chem. Soc.*, **44**, 146A, 148A, 186A, 197A.
Wilner, T., Sitzmann, W. and Weber, K. (1992) *Proceedings: Oils and Fats in the Nineties*, Shukla, V. K. S. and Gunstone, F. D. (Eds), International Food Science Centre A/S, PO Box 44, Sønderskovvej 7, DK-8520 Lystrup, Denmark, March 23–26, pp. 162–175.
Yoncoskie, R. A., Holsinger, V. H., Posati, L. P. & Pallansch, M. J. (1969) *J. Am. Oil Chem. Soc.*, **46**, 489.
Young, F. V. K. (1986) *The Lipid Handbook*, Gunstone, F. D., Harwood, J. L. & Padley, F. B. (Eds), Chapman & Hall, New York, pp. 5.5, 209.
Yukijirushi Shokuhi (1984*a*) Japanese Patent 59, 130, 161-A.
Yukijirushi Shokuhi (1984*b*) Japanese Patent 59, 130, 160-A.

9

Fat Products Using Chemical and Enzymatic Interesterification

A. Huyghebaert, D. Verhaeghe and H. De Moor
Laboratory of Food Technology, Chemistry and Microbiology,
State University of Ghent, Belgium

SUMMARY

Interesterification involves the exchange and redistribution of acyl groups among triglycerides and can be divided into three groups according to the reaction type: acidolysis, alcoholysis and ester interchange. Interesterification can be done chemically or enzymatically.

The raw material for chemical interesterification has to satisfy different quality criteria depending upon the catalyst used. Two important types of interesterification are used. By random interesterification with a catalyst the acyl groups are redistributed until a random equilibrium is achieved. The specific glyceride distribution of milkfat (asymmetric) offers an opportunity for modifying milkfat by randomization. A uniform distribution or a rearrangement of fatty acids is pursued. By removing part of the fatty acids or triglycerides from the reaction mixture the equilibrium in the liquid phase is disturbed and by continuing the reaction a new equilibrium is formed in the directed interesterification. Directed interesterification brings about greater changes than random interesterification. Physical properties are affected by the content of side reaction products and must be removed by deodorization by which the typical milkfat flavour is destroyed.

The enzymes used as catalysts are extracellular microbial lipases and catalyse the hydrolysis of fats. The reaction is reversible. One of the advantages of enzymes is their specificity. Lipases show several types of specificity (substrate, positional, fatty acid and stereospecificity). Non-specific lipases produce triglycerides that are similar to those obtained by chemical interesterification. By using specific lipases it is possible to produce triglyceride mixtures which are unobtainable by chemical interesterification. Lipase catalysed interesterification produces side reactions but recent research suggests that in the future side product formation can be controlled by using suitable media. The production costs are higher due to the enzymes.

Chemical interesterification is used industrially to produce fats and oils used in margarines, shortenings and confectionery fats.

INTRODUCTION

Interesterification reactions can be defined as reactions of an ester yielding one or more new ester bonds. They can be divided into three groups according to the reaction type: ester interchange (Fig. 1), alcoholysis (Fig. 2) and acidolysis (Fig. 3). When applied to fats and oils for triglyceride production, the alcoholysis reaction has to be excluded since it yields high amounts of partial glycerides. The other reaction types can be applied, i.e. triglycerides (whether or not mixtures) or triglycerides and fatty acid esters for ester exchange, and triglycerides and fatty acids for acidolysis. Applying fatty acids or their esters modifies the stereospecific structure as well as the fatty acid composition.

Interesterification is carried out in order to alter the melting and crystallization characteristics of fats (melting point, SFI curve, spreadabi-

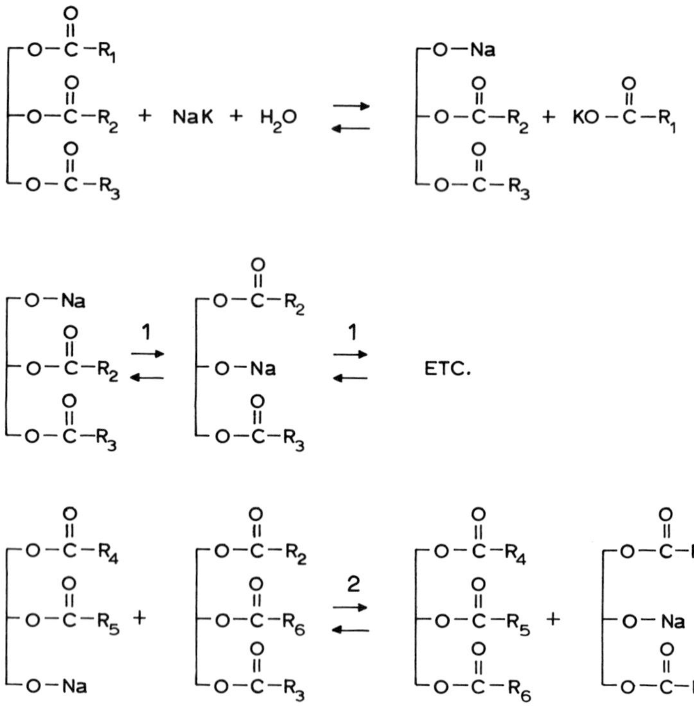

Fig. 1. Ester interchange reaction: (1) intramolecular rearrangement, (2) intermolecular rearrangement (Going, 1967).

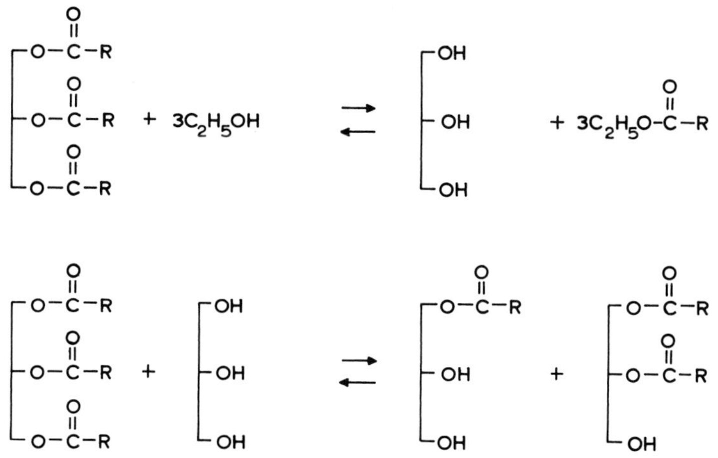

Fig. 2. Alcoholysis reactions (Going, 1967).

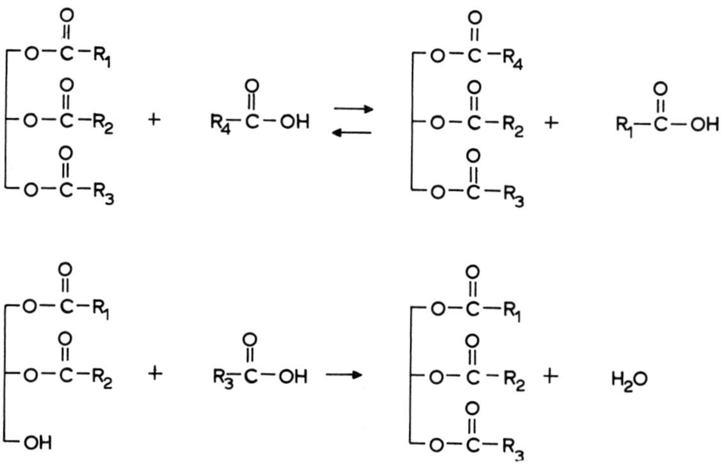

Fig. 3. Acidolysis reactions (Going, 1967).

lity, crystal structure, etc.). It is a much more powerful tool than blending different oils or fats. Thus without hydrogenation, relatively unsaturated fats can be transformed into the equivalent of hardened fats. It avoids elaidinization whilst preserving unsaturated fatty acids. Interesterification may also be combined with fractionation.

Within the general aim of changing the melting properties, two special goals can be identified. First, the application of controlled interesterification results in the production of triglycerides with specific fatty acid composition and stereospecific distribution, i.e. tailor made fats and oils. Due to the market price of cocoa butter, intensive research is focused on the production of CBS (cocoa butter substitutes). Cocoa butter substitutes include the terms cocoa butter replacers (CBR) and equivalents (CBE). The pleasant mouthfeel and melting properties of cocoa butter are associated with:

- the high amount of SOS, POS and POP,
- 85% or more oleic acid residues at the sn-2 position of the glycerol backbone,
- 82% disaturated mono-unsaturated triglycerides on total triglyceride content,
- the suitable crystal structure.

Secondly, it has become possible to use the interesterification process in dietetic margarine manufacturing. Fats are upgraded by the incorporation of polyunsaturated fatty acids (Holemans *et al.*, 1986).

Two important types of interesterification are used (Sreenivasan, 1978; Sonntag, 1982):

—Random interesterification. Oils and fats are a mixture of triacylglycerols in which the acyl groups are distributed in a non-random manner. By random interesterification with a catalyst the acyl groups are redistributed (intra- and intermolecularly) until a random distribution is achieved. The composition of a randomly rearranged oil can be calculated according to the following equations (Sreenivasan, 1978; Sonntag, 1982; Gunstone & Norris, 1983):

% $AAA = a^3/10\,000$
% $AAB = 3a^2b/10\,000$
% $ABC = 6abc/10\,000$

a, b and c are the concentrations of fatty acids A, B and C (mol. %). AAA, AAB and ABC are composed of one, two or three fatty acids respectively. The use of the formulas in the case of 1-stearoyl-2-oleoyl-3-linoleoyl glycerol, results in the following triglyceride composition for the final randomized equilibrium mixture (S = stearic acid, O = oleic acid, L = linoleic acid):

SSS = 3·7%	SOO = 11·1%
OOO = 3·7%	SLL = 11·1%
LLL = 3·7%	OOL = 11·1%
SSO = 11·1%	OLL = 11·1%
SSL = 11·1%	SOL = 22·2%

—Directed interesterification. By removing part of the fatty acids or triglycerides from the reaction mixture the equilibrium in the liquid phase is disturbed and by continuing the reaction a new equilibrium is formed. In directed interesterification, low molecular weight fatty acids or fatty acid esters can be continuously distilled from the fat or else the interesterification can be carried out at low temperature so that higher melting triglycerides crystallize. Thus selective crystallization of a fat or mixed fats may be directed to the effective conversion of all the saturated fatty acids to trisaturated triglycerides. The composition of the final mixture starting with SOL would be:

Solids	SSS	33·33 mol %
Liquids	OOO	8·33 mol %
	LLL	8·33 mol %
	OOL	24·99 mol %
	OLL	24·99 mol %

A third way for directed interesterification uses a selective extraction with dimethylformamide which selectively picks up the more saturated glycerides.

Interesterification can be executed chemically or enzymatically. Enzymatic interesterification is one of the industrial application possibilities for lipases (Macrae, 1983; Posorske, 1984; Chakrabarty, 1985; Macrae & Hammond, 1985; Nielsen, 1985; Schmid, 1986).

CHEMICAL INTERESTERIFICATION

Going (1967), Sonntag (1982) and Sreenivasan (1978) have published surveys concerning chemical interesterification from which some key features may be identified.

Interesterification can occur without the aid of a catalyst, but very high temperatures are required, the reaction proceeds to equilibrium very

TABLE 1
Catalysts for interesterification (Sreenivasan, 1978)

Catalysts	%	Temp. (°C)	Time
Metal salts			
Acetates, carbonates, chlorides, nitrates, oxides of SN, Zn, Fe, Co and Pb	0·1–2	120–260	0·5–6 h under vacuum
Alkali hydroxides			
NaOH, KOH, LiOH	0·5–2	250	1·5 h under vacuum
Alkali hydroxide + glycerol	0·05–0·1 / 0·1–0·2	60–160	30–45 min under vacuum
Metal soaps			
Sodium stearate	0·5–1	250	1 h under vacuum
Glyceride			
Li Al stearate	0·2	250	1 h under vacuum
Na Ti stearate			
Alkali metals			
Na, K, Na/K alloy	0·1–1	25–270	3–120 min
Metal alkylates			
Sodium methylate, ethylate, t-butylate, etc.	0·2–2	50–120	5–120 min
Metal hydrides			
Sodium hydride	0·2–2	170	3–120 min
Metal amide			
Sodium amide	0·1–1·2	80–120	10–60 min

slowly, the triglycerides undergo some decomposition and polymerization and there will be development of free fatty acids.

It has been found more convenient to use a catalyst which speeds up the reaction and lowers the temperature of the reaction. Many catalysts have been proposed (Table 1). The table summarizes information on some important catalysts and conditions for their use. The most popular catalysts can be divided into two classes (Hamilton & Bhati, 1980). The first class works only at high temperature (120–260°C), e.g. zinc chloride, sodium hydroxide with and without glycerol and sodium stearate. The alkali metal alcoholate, sodium methylate, is perhaps the most widely used low temperature interesterification catalyst today.

The raw materials for chemical interesterification have to satisfy quality

TABLE 2
Inactivation of catalysts by poisons

Poison		Catalyst inactivated (kg/1000 kg of oil)		
Type	Level	Sodium	Sodium methoxide	Sodium hydroxide
Water	0·01%	0·13	0·3	—
Fatty acid	AV = 0·1	0·04	0·1	0·07
Peroxide	PV = 1·0	0·023	0·054	0·04
Total catalyst inactivated		0·193	0·454	0·11

criteria, depending upon the catalyst used (Table 2). Interesterification involves three important steps, pretreatment of the oil, reaction with the catalyst and deactivation of the catalyst. The catalyst can be inactivated by moisture, free fatty acids and peroxides. Sodium and sodium methoxide are especially labile towards water and fatty acids: a water content of less than 0·01% and an acid value less than 0·1 are basic requirements for substrates intended for chemical interesterification. By contrast enzymatic interesterification does not impose these standard criteria.

Non-Milkfat Chemical Interesterification

Chemical interesterified fats and oils find applications in shortenings, margarine and confectionary fats. Interesterification improves the physical properties of fats and oils. Similar but different changes in physical properties may, however, also be achieved through the use of blending, fractionation or hydrogenation (partial or complete). Hydrogenation and interesterification are chemical operations, in contrast to fractionation and blending which are physical operations. Production costs, market prices of raw materials and nutritional considerations determine the process used. Whatever the intended use of the products may be, reactions between triglycerides are important and this will be illustrated for the manufacture of shortenings, margarine and cocoa butter substitutes.

Because of its coarse structure lard is the most investigated fat for shortening production. Its original structure consists of 64% palmitic acid on the sn-2 position of the glycerol backbone, whereas oleic acid prefers the sn-1 side. Natural lard tends to crystallize in the β phase, which results in a coarse grainy structure. Esterification reduces a large proportion of the palmitic acid in the 2-position of the S_2U glyceride from 64% to 24%, which in turn produces a smoother textured fat crystallizing in the β' phase.

Overall therefore randomization of lard broadens its plastic range and this makes it a better shortening than natural lard.

The objective in the manufacture of margarines is to produce a fat mixture with a fairly steep SFI curve so that the product is firm in the refrigerator but spreads rather easily on removal and melts rapidly in the mouth. It should also crystallize as β' crystals (Gunstone & Norris, 1983). Depending upon oil costs and availability, treatments such as co-randomization, fractionation, interesterification and direct blending of natural fats may be preferred. Only a few examples are reported.

Margarine oil made with a high content of lauric oils is typically low in melting point and has a short plastic range (Gunstone & Norris, 1983). This produces a margarine that is hard in the refrigerator but melts partly at room temperature. The remedy is to remove or decrease the amount of lauric triglycerides. This is done by interesterifying the coconut oil with an oil such as palm and then blending 60% of the interesterified mixture with 40% of an oil such as sunflower.

Lo & Handel (1983) interesterified refined, unhydrogenated soybean oil and edible beef tallow with sodium methoxide. This was done as an alternative to hydrogenation for the production of plastic fats for use as margarine oils. A blend of 60% soybean oil and 40% edible beef tallow was found to have physical characteristics similar to those of commercial tub margarine oils. The level of polyunsaturated fatty acids was slightly lower and the level of saturated fatty acids slightly higher than the commercial margarine oils.

According to Sonntag (1982) short chain acids (C6–C14) give better melting qualities and long chain acids (C20–C22) on the same triglyceride provide stiffening power in margarines. Both these properties can be combined using blends of randomized oils, for instance 10 parts co-randomized 50% coconut oil–50% hydrogenated rapeseed oil IV4 with 15 parts hydrogenated soybean oil IV95 gives solid content index values of the blend at 10, 20 and 33·3°C of 32, 19 and 1, respectively. Margarine made from this blend has good spreadability, high-temperature stability and a pleasant taste.

List et al. (1977) explored the sodium methoxide-catalysed random interesterification of liquid soybean oil with soy trisaturate blends as a possible route to zero *trans*-margarine oils. For a randomized blend of 80 parts soybean oil and 20 parts of soy stearin, SCI values at 10, 21·1 and 33·3°C were, respectively, 8, 3·4 and 2·2. The polyunsaturated acid content was 51·5% and *trans*-acids were 1·6%. According to List et al. (1977) margarines which contain no *trans*-fatty acids in European markets are formulated, in part, for economic reasons. In these countries fats of high saturated content and oils of low saturated content are available and therefore interesterification rather than hydrogenation may be the processing technology of economic choice to obtain hardened fats, i.e. margarine oils.

Due to the relatively high price of cocoa butter and its occasional unavailability, scientists worked for many years to procure a satisfactory substitute. A number of substitutes can be prepared by blending interesterified lauric acid fats with other fats. According to Sreenivasan (1978) palm kernel oil is a hard butter melting at 46°C and producing a waxy feel in the mouth. On randomization its melting point is reduced to 35°C. By blending together hydrogenated palm kernel oil and its randomized product, a whole series of hard butters with highly desirable melting qualities are obtained. Other possibilities are the blending of non-lauric-containing fats with lauric fats. For instance partially or fully hydrogenated lauric fats can be co-randomized with fully hardened cottonseed oil, or a similar long-chain acid-containing oil such as fully hydrogenated rapeseed or herring oil. Such co-randomized fats have better gloss-producing and quick set-up properties in surface-coating applications.

Milkfat Chemical Interesterification

Using random interesterification and a single oil or fat, changes in melting and crystallization properties are dependent on the original specific triglyceride structure. Data concerning the specificity of milkfat are reported by Walstra & Jenness (1984) (Table 3). Thus, milkfat shows neither a random distribution, nor a 1,3 random 2 random distribution. Butyric and caproic acids show a marked preference for position 3, hence it is often said that milkfat is a highly asymmetric fat. It is striking that 97% of the butyric acid molecules are in the sn-3 position. A similar observation but in the opposite sense can be made for palmitic acid, which has a preference for positions 1 and 2. The specificity is somewhat less pronounced for the other fatty acids, e.g. the intermediate and C18:0 fatty acids. So this specific glyceride distribution offers a possibility of modifying milkfat by random interesterification.

The first paper on chemical interesterification of milkfat appeared in 1958 (Weihe & Greenbank, 1958); more details were reported in 1961 (Weihe, 1961). Random esterification was carried out at 40–45°C for $\frac{1}{3}$–6 h with 0·1–0·3% sodium–potassium alloy catalyst. For directed interesterification, hexane or xylene was usually added before starting the reaction, which was initiated at 25–38°C and then the temperature was reduced in

TABLE 3
Percentage of each fatty acid in designated position of sn-glycerol (approximate average values from various sources) (Walstra & Jenness, 1984)

Notation	1-position	2-position	3-position
4:0	2	1	97
6:0	4	12	84
8:0	13	42	45
10:0	17	50	33
12:0	24	50	26
14:0	27	56	17
14:1	10	45	45
15:0	40	53	7
16:0	46	42	12
16:1	40	37	23
18:0	58	20	22
18:1	43	25	32
18:2	40	40	20
18:3	39	32	29

three to five stages usually to 10–25°C. Directed interesterification brought about greater changes than random interesterification. Increases in melting point were greater in the presence of a solvent than in its absence. The melting point was raised about 15°C by directed and only about half as much by random interesterification. The fraction of interesterified milkfat, after the high melting glycerides (HMG) have been removed, could alone or mixed with normal milkfat be used to make butter of almost any desired hardness.

Riel (1966) studies the effect of various treatments on the physical properties of milkfat. Na–K alloy yielded more high melting glycerides in milkfat in a few minutes than did sodium methylate or ethylate in solution after several hours. Controlled interesterification decreased the proportion of solids at temperatures below 20°C and increased it at temperatures above 20°C.

Interesterification of milkfat resulted in increases in the softening points of milkfat (De Man, 1961a, b). According to the author the large differences in physical properties of milkfat before and after interesterification indicate that the distribution of fatty acids in the glyceride molecule is highly specific. Randomization of the fatty acids resulted in a considerable increase in high melting glycerides (5–7%) and is responsible for a large increase in hardness.

Mickle (1959) demonstrated that a directed interesterification reaction reduced the hardness of the resulting butter. However, the rearranged milkfat had a rancid, metallic flavour. Refining the rearranged fat (neutralization of the free fatty acids and steam injection under vacuum) appeared to remove the flavour (Mickle, 1960). In subsequent research (Mickle *et al.*, 1963), the authors studied the influence of three rearrangement reaction conditions (time 5–55 min, temperature 40–90°C and catalyst $NaOCH_3$ 0·5–5·0% on fat) on the hardness of a semisolid product resembling butter. The hardness of the product was measured at 7–10°C and found to decrease with increased catalyst concentration, but was found to increase with increased time. The largest influence on butter hardness was the linear effect of catalyst concentration. The catalyst effect was related to the formation of diglycerides whose emulsifying properties partially explained the decreased butter hardness. The product resulting from rearrangement had an intensive metallic flavour but this flavour was removed by the refining process applied.

According to Kacherauskis (1966) experimental data on the re-esterification of milkfat are rather scarce and contradictory. Therefore, the author studied the effect of the quantity of catalyst (sodium ethylate, 0·2–5·0%), the duration (10–60 min) and the temperature (30–60°C) of the reaction. Depending on the method of reaction the melting point of the milkfat was reduced by 0·6–20·0°C, mechanical hardness by 2–20 times and the free fatty acid content increased 1·6–40·0 times, depending on the methods of treatment of the milkfat after re-esterification. According to the author it is necessary to continue studying the nature of the re-esterification in order to find optimal conditions for this process.

Chang *et al.* (1969) reported an increase in the softening point and content of HMG by interesterification of milkfat; it did not alter fatty acid composition except for a slight decrease in C4 and C6 fatty acids. The fatty acid composition of HMG from interesterified milkfat differed from that of untreated milkfat, including an increase in C18 acid.

Random interesterification results in a considerable increase in short chain triglycerides (carbon number ± 26) and long chain triglycerides (carbon number ± 50). The mid-fraction decrease is so important that the triglycerides with carbon number 38 are no longer dominating (Timmen, 1978).

Parodi (1979) reported an investigation of the relationship between trisaturated glycerides and the softening points of milkfat. Interesterification of milkfat increased its softening point. The softening points of two milkfats (33·4 and 31·9°C) increased to 36·8 and 36·1°C after interesterification.

In a study of the physical properties of blends of milkfat with beef tallow and beef tallow blends, Timms (1979) reported that interesterified blends would have the disadvantage of a more elaborate configuration and moreover they lacked the flavour of milkfat. In a subsequent paper on the possibilities of using hydrogenated, fractionated or interesterified milkfat in chocolate, Timms & Parekh (1980) reported that interesterification may result in milkfat that is more compatible with cocoa butter than unmodified milkfat, but the improvement is insufficient to offset the cost of interesterification and the resulting loss of flavour in the milkfat.

Using fat analysis it is possible to follow interesterification by the triglycerides carbon number and degree of unsaturation. The analysis of random interesterified milkfat revealed a decrease of 45 and 52% of monounsaturated triglycerides with carbon numbers 36 and 38 and an increase of 26% of trisaturated triglycerides with carbon numbers 44–50. Thus the crystallization range was broadened and the SFI curve moved upwards to higher temperatures (Parviainen *et al.*, 1986).

In experiments at our laboratory on small and pilot-scale level, butter oil was alkali refined to a free fatty acid content of less than 0·05% and vacuum dried to a moisture content of 0·01%. Under a stream of nitrogen, the anhydrous milkfat was interesterified with 0·2% sodium methylate at 90°C. After 1 h, the catalyst was inactivated and washed out by water addition. This product showed a fruity flavour due to short chain fatty acid methylesters and a bitter flavour due to intermediate chain methylesters. A deodorization for 2 h at 170°C and 2 mmHg removed the off-flavours as well as the butter flavour. In order to obtain an interesterified butterfat with a butter flavour, steam deodorization before interesterification could provide a distillate with a heated butter flavour, which could then be put back after interesterification.

Interesterification causes significant changes in the glyceride composition (Table 4). In the untreated milkfat the triglycerides may be divided into two groups: triglycerides with lower molecular weight (lower than 42 C atoms) and the triglycerides with higher molecular weight (C44–C54 triglycerides). Interesterification results in an increase in the low molecular weight triglycerides (C22–C30). This is easily explained by randomization of the asymmetric distribution of lower fatty acids. There is also an increase in the high molecular weight triglycerides where the C50 carbon compounds are predominant.

TABLE 4
The glyceride composition of natural and interesterified milkfat

	Natural	Interesterified
C22	—	0·1
C24	0·3	0·8
C26	0·1	1·4
C28	0·6	1·5
C30	0·9	1·3
C32	1·9	2·0
C34	4·4	3·0
C36	9·5	5·9
C38	13·1	9·1
C40	12·1	9·9
C42	7·7	7·6
C44	6·8	8·3
C46	7·5	10·7
C48	8·8	12·8
C50	11·2	14·2
C52	10·8	10·9
C54	4·6	0·4

During the course of the reaction the C38 content decreases and the C50 content increases: the C38/C50 ratio is a parameter of the interesterification reaction. The theoretical completely randomized milkfat C38/C50 ratio is 0·43. Since the observed ratio is 0·64, the random distribution was not achieved under these interesterification conditions.

Interesterification results in an increase in the melting point of about 5°C and of 2·5°C for the drop point. This can be explained by the greater amount of high melting triglycerides. Changes in the glyceride composition are also reflected in the solid fat index curve (Fig. 4): there is a shift to a higher amount of high melting triglycerides.

The change in the physical properties of milkfat after interesterification offers some technological possibilities. One such possibility is the use of the rearranged fat in recombined butter. To prepare recombined butter the following procedure was adopted.

Milkfat or interesterified milkfat was recombined with water and sweet buttermilk powder by homogenization (40% O/W emulsion). After pasteurization, the cream was cooled and a starter culture added. The biochemical and physical ripening method was as follows: 16 h at 15°C and

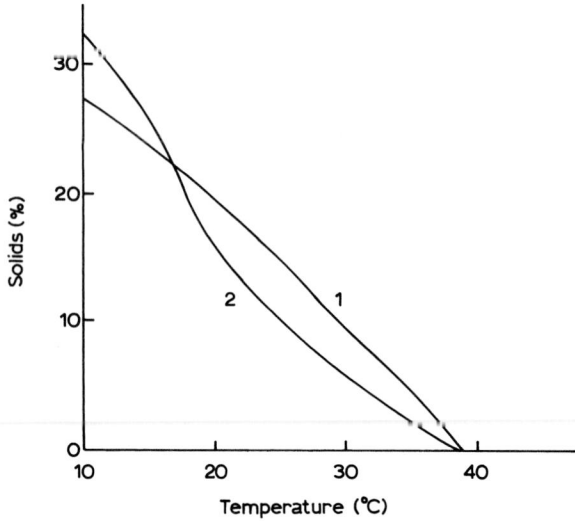

Fig. 4. SFI curves of (1) interesterified milkfat and (2) natural milkfat.

5 h at 5°C. Churning was carried out at 11°C and the butter samples were stored at 5°C.

The texture of both samples was assessed by Instron compression testing at 5°C. The work for a given compression of the interesterified butter is about one-third of that for the recombined natural butter at 5°C. This means interesterification of butterfat results in an improvement of the spreadability of the recombined butter at 5°C. The product, however, is not fully spreadable and can be further improved, mainly by removing the HMG fraction of the interesterified butterfat by fractionation.

Gurr (1983) in a survey on the nutritional significance of lipids and the relationship of blood lipids to diet showed an interesting aspect of milkfat interesterification. Plasma LDL (low density lipoproteins) concentrations are influenced by the amount and nature of dietary fat. But the absolute amounts or proportions of saturated fats in the diet are not the only factors to affect blood LDL levels. It has been shown that, when the fatty acids in the triacylglycerol molecules of butterfat are randomized by chemical interesterification, the fat no longer has the effect of raising blood LDL concentrations when fed to humans. Butter itself has a marked tendency to cause elevated LDL levels (Christophe et al., 1978). The randomized butter was more rapidly hydrolysed by pancreatic lipase *in vitro* (Christophe et al.,

1981) and, in feeding experiments, chylomicrons appeared more rapidly in the plasma than after a similar intake of butter (Christophe et al., 1982).

ENZYMATIC INTERESTERIFICATION

According to Posorske (1984) enzymes have advantages for industrial processing. Three particular benefits offered by enzymes are specificity, mild conditions and reduced waste. Specificity permits control of the products produced. Enzymes usually work at specific sites on a molecule, in contrast to chemical reactions (random reaction). Specificity can also increase yield by reducing side reactions. Enzymes are biodegradable and due to the specificity unwanted side products that normally appear in the waste stream are reduced or eliminated. The specificity is the most salient point.

Lipases show several types of specificity. These specificities, elucidated in hydrolysis reactions, are also applicable to the reverse reactions. One lipase can possess different types of specificities.

Table 5 lists the known and expected types of lipase specificity which may be summarized as:

I. Substrate: (a) different rates of lipolysis of TG, DG and MG by the same enzyme,
(b) separate enzymes from the same source for TG, DG and MG.
II. Positional: (a) primary esters,
(b) secondary esters and
(c) all three esters, non-specific or random hydrolysis.
III. Fatty acid: preference for fatty acids, e.g. short chain, etc.
IV. Stereospecificity: faster hydrolysis of one primary sn ester as compared to the other.
V. Combinations of I–IV.

Eigtved et al. (1988), however, accentuate the difficulties in classifying lipases according to their specificity since the apparent specificity may also depend upon the preparation and the reaction conditions.

Enzyme Catalysed Interesterification of Non-Milkfats

For interesterification processes specificity can be chosen according to the endproduct. 1,3-Positional specific lipases only react at the α position of

TABLE 5
Types of lipase (acylglycerol hydrolase) specificity (Jensen et al., 1983)

Type	Source of lipase	Reaction
I. Substrate		
a. Different acylglycerols: Same lipase	Pancreas	TG > DG > MG 16:0-16:0-4:0 > 16:0-4:0-4:0
b. Different lipases: for TG, DG and MG	Postheparin plasma (lipoprotein lipase)	DG→MG→glycerol
II. Positional		
a. Primary esters	Pancreas	TG→1,2(2,3)DG→2-MG
b. Secondary esters	Geotrichum candidum	Acid must contain cis-9-double
c. All esters	Candida cylindracea	TG→1,2 + 2,3 + 1,3 DG→1 + 2-MG
III. Fatty acid		
a. 4:0–10:0	Pregastric esterase	4:0-10:0 > 12:0-18:0, 18:1
b. cis-9-unsaturation	G. candidum	Acid must contain cis-9-double
c. 8:0–12:0	Rat and human lingual preparations	8:0-12:0 > 16:0, 18:1
IV. Stereospecificity		
a. sn-1-ester	Postheparin plasma, bovine and human milks, adipose tissue, liver (lipoprotein lipase)	sn-1 : sn-3,2 : 1
b. sn-3-ester	Rat and human lingual preparations; termite	sn-3 : sn-1,4 : 1
V. Combinations		
a. Fatty acid and stereospecificities	Rat lingual preparations Human milk lipoprotein lipase	sn-R-R-a-12:0 > sn-R-R-16:0 8:0, 10:0, 12:0 and 18:0 TG > 16:0-16:0-16:0. sn-1:sn-3,2:1

[a] R—long chain fatty acids in human milk.

glycerides and are suitable for tailor made fat manufacture. In altering triglycerides to cocoa butter substitutes, the substrate ought to have monounsaturated fatty acids at the 2-position of glycerol, by preference oleic acid residues. The other sn-places can be modified due to enzymatic action. Thus a hard butter rich in POS and SOS could be prepared from naturally occurring oils with a high content of POP or SOS. Due to market prices, oils containing large amounts of POP are preferred.

A rough selection based on these criteria yields the following triglycerides or the fractionated triglycerides: olive oil, oleic safflower oil, palm oil, low erucic acid type rapeseed oil, shea butter, mango fat and some minor fats (Matsuo *et al.*, 1981; Macrae, 1986). Palm mid-fraction has been widely used as substrate. Hydrogenated palm kernel oil is especially suitable.

Based upon the consideration of the sn-2-oleyl glycerol ester structure and the total fatty acid composition, milkfat has to be rejected as a substrate for CBS products.

Starting with a pool of fatty acids, fatty acid specific lipases such as *Geotrichum candidum* lipase incorporate, by preference, $\Delta 9$ fatty acid residues into the triglycerides.

This is one approach for manufacturing dietetic margarines with increased contents of oleic and linoleic acid (Coleman & Macrae, 1981). An alternative for the production of oils containing increased amounts of $\omega 3$ fatty acids consists of a two-step process of hydrolysis and synthesis.

A number of other publications deal with enzymatic dietetic margarine manufacture. The fat blends which are suitable for margarine consist of a matrix of hardstock crystallized at room temperature in which an oil is occluded which is liquid at room temperature. According to an invention (Holemans *et al.*, 1986) the hardstock comprises H_3, HHU, HUH and H_2M triglycerides (H = saturated fatty acids having 16–24 C atoms, U = mono or *cis*-polyunsaturated fatty acids, M = fatty acids preferably with chain length 12 or 14 C atoms). H_2M triglycerides are very effective as a matrix for emulsions and are produced by a random interesterification and fractionation process, but can also be prepared by an enzymatic process. Thus, a mixture of 1 part soy oil and 0·45 parts lauric acid, dissolved in 9 parts hexane, is percolated through a column filled with *Mucor miehei* (1,3-selective enzyme) on a celite carrier, the liquid leaving the column being extracted with methanol, and hydrogenated to an iodine index of less than 2. Subsequently the saturated product is fractioned in acetone, during which a stearin, a mid- and an olein fraction is obtained. The olein or mid-fraction may be used as a hardstock in the preparation of margarine. The lipases used are 1,3-specific and include: *Rhizopus delemar* (Tanaka *et al.*, 1981*a*; Yokozeki *et al.*, 1982*b*), *Rhizopus arrhizus* (Wisdom *et al.*, 1987), and *Aspergillus* species (Wisdom *et al.*, 1984, 1987).

Besides specificity, activity and stability are important characteristics. Some extremely heat stable lipases have been isolated from *Humicola lanuginosa* and *Thermomyces ibandanensis*. A commercial, immobilized lipase from *Mucor miehei* remains stable at 60°C at which temperature and under selected conditions its half-life is more than 1500 h (Hansen & Eigtved, 1985).

Interesterification activity measurements are always based upon the reaction rates of a mixture of triglycerides and fatty acids in hexane or petroleum ether fractions. The interesterification activity is defined in two ways. The first definition shows analogy with the hydrolytic activity definition in that 1 unit of interesterification activity incorporates 1 μmol guide fatty acid into triglyceride per minute. Wisdom *et al.* (1987) prepared an immobilized lipase with a specific activity of 0·73 units/mg. The second definition starts from the decreasing interesterification rate in batch processes due to the nature of the ester exchange reaction which obeys the laws of probability. The following symbols have been defined (Matsuo *et al.*, 1984):

e = equilibrium percentage of guide fatty acid incorporated into triglycerides,
o = original percentage of guide fatty acid incorporated into triglycerides at time t_o,
a = actual percentage of guide fatty acid incorporated into triglycerides at time t,
x = degree of conversion at time t,
k = proportionality,
S = amount of substrate,
E = amount of enzyme.

The proportionality factor k, is shown by the formula

$$k = \frac{1}{t} \ln \frac{1}{1-x}$$

The degree of conversion of the interesterification is determined by selecting a suitable guide fatty acid (e.g. lauric acid derived from coconut oil) and measuring the distributional change thereof which occurs in a dry reaction system. The term 'completely reacted state' ($x = 1$) means the state in which a constant distribution of the guide fatty acid has been substantially obtained by reaction. When the interesterification is carried out by using a lipase which has specificity toward the 1- and 3-positions of a glyceride, the state in which fatty acids at the 1- and 3-positions of the glycerides are completely at random can be deemed the 'completely reacted state'. The 'interesterification activity' (absolute value) is expressed as

$$K_a = k \ S/E$$

Matsuo *et al.* (1984) showed that *Rhizopus niveus* with high K_a values could be prepared.

Tanaka *et al.* (1981*b*) and Stevenson *et al.* (1979) report enzymatic interesterification without immobilization of the lipase. In order to enhance the stability and economic feasibility, four techniques were proposed to make the lipase recoverable. The lipase of the *Rhizopus arrhizus* mycelium was used in its cell bound form (Patterson *et al.*, 1979; Knox & Cliffe, 1984). Coleman & Macrae (1980) suggest the stabilization of the enzyme in an insoluble form and refer to the enzyme technology used in amino acid manufacture and fructose production from glucose. Covalent binding of *Pseudomonas fragii* lipase to polyethylene glycol (PEG) makes the enzyme soluble in solvents like benzene. The lipase–PEG conjugate can interact with Fe ions and a ferromagnetic lipase was obtained (Tamaura *et al.*, 1986). For *Rhizopus arrhizus* lipase, used for olive oil acidolysis, the improvement of the long term stability largely counterbalances the activity decrease from the immobilization. The lipase is consecutively immobilized on celite and entrapped by a hydrophobic photo-cross-linkable prepolymer instead of by immobilization on celite (Tanaka *et al.*, 1988). Some immobilization procedures are not included in the patent literature (Lavayre & Baratti, 1982; Yokozeki *et al.*, 1982*a*; Wisdom *et al.*, 1984, 1987; Chakrabarty *et al.*, 1988).

The carrier material creates a micro environment for the lipase. Since small amounts of water are essential in obtaining high interesterification and low hydrolytic activity, water has to be coimmobilized (Macrae, 1985). Tanaka *et al.* (1981*b*) recommend the use of glycerol or other dihydric or trihydric alcohols. Thus, the carrier material disperses both enzyme and water (Tanaka *et al.*, 1981*a*). Suitable carriers which show high water retention capacity and low adsorptivity are diatomaceous earth (celite), kaolinite, pearlite, silicate gel, polyvinylalcohol, and cellulose powder

(Matsuo et al., 1984). Chakrabarty et al. (1988) used alumina as an enzyme support. Ion exchange resins, e.g. phenol formaldehyde resin with tertiary amino groups, are suitable for *Mucor miehei* lipase, the moisture content of the immobilized *Mucor miehei* lipase being approximately 10% (w/w) (Eigtved, 1984). Particle characteristics such as porosity, particle size and compressibility are also important (Eigtved, 1984; Wisdom et al., 1984) as they profoundly influence the engineering parameters for fixed bed reactors (Luck et al., 1988).

The specific enzymatic loading and purity of the enzymes also affect the resultant specific activity of the immobilization. Thus controlled vacuum drying from specific starting to end a_w values appears to influence the ratio of lipolytic to interesterification activity (Matsuo et al., 1984; Unilever, 1986). Removing free lipase by water washing from the immobilized enzyme product increases the interesterification rate (Eigtved, 1984). Preliminary contact of catalyst and one substrate further enhances activity (Matsuo et al., 1981).

The importance of water has already been stressed and free water concentration seems to determine hydrolytic and synthesising activity, so influencing the concentration of partial glycerides (Goderis, 1986). In the experiments of Eigtved (1984) and Bhattacharyya & Bhattacharyya (1988), no supplementary amount of water was added to the interesterification system with immobilized *Mucor miehei* lipase. The moisture content of the reaction mixture of Matsuo et al. (1980) was 0·01–0·18% whilst Hirota et al. (1985) employed a reaction mixture with 0·2–1% moisture. Supplementary amounts of carrier material as disperser are considered favourable, e.g. calcium carbonate, glucose or potassium sulphate (Tanaka et al., 1981a). An empirical relationship has been discovered between the original diglyceride content and moisture on the one hand and the final diglyceride content on the other hand (Matsuo et al., 1984). Diglycerides can cause a problem in fat crystallization including the tempering of chocolate. Up to now no economic process for removing diglycerides is available (Kurashige, 1988).

A sequential process consisting of enzymatic interesterification, separation from the reaction mixture by distillation and further treatment of the condensate has been developed. The treatment can be limited to a separation by distillation followed by a partial recycling. In processes with 1,3-specific lipases, fats with short chain fatty acids in the α position of the glycerol backbone facilitate the separation step between triglycerides and fatty acid esters. Another special case deals with fats containing high amounts of the corresponding unsaturated fatty acids of palmitic or stearic acid. In this case, the treatment of the condensate consists of hydrogenation and recycling. These integrated process designs greatly favour economic considerations (Matsuo et al., 1981, 1983; Macrae, 1986).

Several systems to control moisture have been developed such as closed reaction systems, vacuum drying, drying of headspace gases with molecular sieves, condensation, the use of anhydrous salts, concentrated sulphuric acid, the wetting of substrates and hexane (Goderis, 1986). Hydratable silica (40–60 mesh) on zeolite molecular sieves (pore size: 0·3–0·4 mm, particle size > 40 mesh) is very effective (Posorske et al., 1988).

Water-saturated n-hexane has been used as a solvent for interesterification reactions (Yokozeki et al., 1982a; Kalo et al., 1986a, b; Bhattacharyya & Bhattacharyya, 1988; Tanaka et al., 1988). Kwon & Rhee (1985) proved that diisopropylether and isooctane are better suited for interesterification reactions. Solvents lower the viscosity of the substrates and are necessary to promote an increased solubility of fatty acids in triglycerides for acidolysis reactions (Stevenson et al., 1979). The disadvantages of organic solvents (e.g. solvent recovery, the necessity for closed systems in order to reduce explosion risks) can be circumvented by the use of C1–C5 aliphatic alcohol esters of fatty acids (Matsuo et al., 1984). Schuch & Mukherjee (1988) revealed a substrate preference scale for the interesterification of medium chain triacylglycerol esters wtih an immobilized *Mucor miehei* lipase. The reactivity decreased from triacylglycerols through fatty acids to methyl esters (Schuch & Mukherjee, 1987, 1988).

Another possibility for controlling problems with viscosity and solubility is the application of heat stable lipases operating at high temperatures (Hansen & Eigtved, 1985). This strategy is followed by Nakamura et al. (1988). Higher operating temperatures reduce the pressure drop in column systems by altering the viscosity of the substrates. This effect also broadens the substrate choice since some triglycerides have a high melting point, e.g. tristearin.

Enzyme Catalysed Interesterification of Milkfat

Studies on enzyme catalysed interesterification for the modification of milkfat are scarce. Kalo and Antila of the University of Helsinki, Finland, studied the lipase catalysed interesterification of milkfat and milkfat/rapeseed oil mixtures. The enzymes were *Candida cylindracea* (Kalo et al., 1986a, b), *Aspergillus niger* (Kalo et al., 1988) and *Pseudomonas fluorescens* lipases immobilized on celite by adsorption (Kalo et al., 1988) and *Mucor miehei* lipase immobilized on anion exchange resin (Kalo et al., 1988). Kalo & Antila (1989) summarized their results as follows.

—The proportions of saturated and monoene TAGs in the interesterified milk fats were similar whether reactions were catalysed by *Candida cylindracea*, *Pseudomonas fluorescens*, *Aspergillus niger* or *Mucor miehei*, and similar to the proportions calculated according to random distribution.
—Interesterification decreased the proportion of saturated TAGs with 36 and 38 acyl carbons and of monoene TAGs with 38 acyl carbons and increased the proportion of saturated TAGs with 46–50 acyl carbons and of monoene TAGs with 48–52 acyl carbons.
—The degree of hydrolysis depends on the water content of the reaction medium. *P. fluorescens* lipase was active in the medium with a very low content of water.
—The reactions catalysed by *Pseudomonas fluorescens*, *Aspergillus niger* and *Mucor miehei* take place at the sn-2 position, i.e. all three enzymes exhibited apparent non-specificity due to non-enzymatic acyl migration during the long reaction time.
—Determination of solid fat contents (SFC) by pulsed NMR of milkfat interesterified with *Candida cylindracea* lipase in the presence of variable amounts of water showed that between 0 and 20°C the SFCs of the interesterified fats were lower and above 25°C generally higher than those of the untreated fat. An increase of water in the reaction medium decreased the solid fat content at every temperature.

Determination of the TAG composition of mixtures of milkfat solid fraction and rapeseed oil and milkfat solid fraction and hydrogenated rapeseed oil showed interesterification to reduce the proportion of saturated TAGs with 42–52 acyl carbons and increase the proportion of monoene TAGs with 48–52 acyl carbons. The values measured for the proportions of saturated and monoene TAGs were close to the composition calculated according to random distribution. Melting curves determined by differential scanning calorimetry (DSC) showed that, relative to the untreated butterfat, the melting range was narrower, the proportion of fat melting in a high temperature range lower, and the proportion of fat melting at 0–20°C higher.

REFERENCES

BHATTACHARYYA, D. K. & BHATTACHARYYA, S. (1988) Enzymatic acidolysis reaction of some fats. In: *Proceedings World Conference on Biotechnology in Fats and Oils Industry 1987*, Applewhite, T. H. (Ed.), American Oil Chemists' Society, pp. 308–9.

CHAKRABARTY, M. M. (1985) Interesterification reaction of glycerides and their industrial uses (including a discussion of India's edible oil deficits and possible remedies). *Journal of the Indian Chemical Society*, **62**, 1–6.

CHAKRABARTY, M. M., CHAUDHURI, S. G., KHATOON, S. & CHATTERJEE, A. (1988) Comparison of biointeresterification reaction and conventional processes for the preparation of vanaspati and other valuable products. In: *Proceedings World Conference on Biotechnology for the Fats and Oils Industry 1987*, Applewhite, T. H. (Ed.), American Oil Chemists' Society, pp. 288–9.

CHANG, J., YAMAUCHI, K. & TSUGO, T. (1969) Alteration of chemical and physical properties of butterfat by interesterification. *Journal of Food Science and Technology Tokyo*, **16**, 446–52.

CHRISTOPHE, A., MATTHYS, F., GEERS, R. & VERDONK, G. (1978) Nutritional studies with randomized butter. Cholesteremic effects of butter oil and randomized butter oil in man. *Archives Internationales de Physiologie et de Biochimie*, **86**, 413–15.

CHRISTOPHE, A., ILIANO L., VERDONK, G. & LAUWERS, A. (1981) Studies on the hydrolysis by pancreatic lipase of native and randomized butter fat. *Archives Internationales de Physiologie et de Biochimie*, **89** (B), 156–7.

CHRISTOPHE, A., VERDONK, G., DECATELLE, J. & HUYGHEBAERT, A. (1982) Studies on the chylomicronemic response of loading natural or randomized butter fat. *Archives Internationales de Physiologie et de Biochimie*, **90** (B), 100–1.

COLEMAN, M. H. & MACRAE, A. R. (1980) Fat process and composition. GB Patent Specification 1,577,933, Unilever.

COLEMAN, M. H. & MACRAE, A. R. (1981) Fat process and composition. US Patent 4,275,081, Lever Brothers Co., New York.

DE MAN, J. M. (1961a) Physical properties of milk fat. I. Influence of chemical modification. *Journal of Dairy Research*, **28**, 81–6.

DE MAN, J. M. (1961b) Physical properties of milk fat. II. Some factors influencing crystallization. *Journal of Dairy Research*, **28**, 117–22.

EIGTVED, P. (1984) Method for production of an immobilized lipase preparation and use thereof. DK Patent Application, 2510.200, Novo Industri.

EIGTVED, P., HANSEN, T. T. & MILLER, C. A. (1988) Ester synthesis with immobilized lipases. In: *Proceedings World Conference on Biotechnology for the Fats and Oils Industry 1987*, Applewhite, T. H. (Ed.), American Oil Chemists' Society, pp. 134–7.

GODERIS, H. (1986) Immobilized lipase activity in organic reaction media of controlled humidity. PhD Thesis, Leuven, Belgium, Faculty of Agricultural Sciences.

GOING, L. H. (1967) Interesterification products and processes. *Journal of the American Oil Chemists' Society*, **44**, 414–56.

GUNSTONE, F. D. & NORRIS, F. A. (1983) *Lipids in Foods, Chemistry, Biochemistry and Technology*, Maxwell, R. (Ed.), Pergamon Press, Oxford, pp. 144–6.

GURR, M. I. (1983) The nutritional significance of lipids. In: *Developments in Dairy Chemistry-2-Lipids*, Fox, P. F. (Ed.), Applied Science Publishers, London, pp. 365–417.

HAMILTON, R. J. & BHATI, A. (1980) *Fats and Oils: Chemistry and Technology*, Applied Science Publishers Ltd, London, 255 pp.

HANSEN, T. T. & EIGTVED, P. (1985) A new immobilized lipase for interesterification and ester synthesis. Novo Publication A-05930a, Bagsvaerd, Denmark.
HIROTA, Y., TANAKA, Y. & URATA, K. (1985) Verfahren zur Umesterung von Fetten und Ölen und Enzympräparat. DE 3519, 429, Kao Corp.
HOLEMANS, P., SCHIJF, R., VAN PUTTE, K. & DE MAN, T. (1986) Fats and edible emulsions with a high content of cis-polyunsaturated fatty acids. European Patent Application 209 176, Unilever N.V.
JENSEN, R. G., DEJONG, F. A. & CLARK, R. M. (1983) Determination of lipase specificity. Lipids, 18, 239–52.
KACHERAUSKIS, D. (1966) Effect of re-esterification on physical and mechanical properties of manufactured milk fat. XVII International Dairy Congress Section C-2, Heinrichs, E. (Ed.), Hildesheim, pp. 161–6.
KALO, P. & ANTILA, A. (1989) Interesterification of milk fat. B-Doc 164-Addendum. IDF. Annual Sessions in Copenhagen, September 1989, pp. 1–5.
KALO, P., PARVIAINEN, P., VAARA, K., ALI-YRRKÖ, S. & ANTILA, M. (1986a) Changes in the triglyceride composition of butter fat induced by lipase and sodium methoxide catalysed interesterification reactions. Milchwissenschaft, 41, 82–5.
KALO, P., VAARA, K. & ANTILA, M. (1986b) Changes in triglyceride composition and melting properties of butter fat solid fraction/rapeseed oil mixtures induced by lipase catalysed interesterification. Fette, Seifen, Anstrichmittel, 88, 362–5.
KALO, P., PERTTILÄ, M., KEMPPINEN, A. & ANTILA, A. (1988) Modification of butter fat by interesterification catalysed by *Aspergillus niger* and *Mucor miehei* lipases. Meijeritieteellinen Aikakauskirja, 46, 36–47.
KNOX, T. & CLIFFE, K. R. (1984) Synthesis of long-chain esters in a loop reactor system using a fungal cell bound enzyme. Process Biochemistry, 10, 188–92.
KURASHIGE, J. (1988) Enzymatic conversion of diglycerides to triglycerides in palm oils. In: *Proceedings World Conference on Biotechnology for the Fats and Oils Industry 1987*, Applewhite, T. H. (Ed.) American Oil Chemists' Society, pp. 138–40.
KWON, D. Y. & RHEE, J. S. (1985) Effects of organic solvents on lipase for interesterification of fats and oils. Korean Journal of Food Science and Technology, 17, 490–4.
LAVAYRE, J. & BARATTI, J. (1982) Preparation of immobilized lipases. Biotechnology and Bioengineering, 24, 1007–13.
LIST, G. R., EMKEN, E. A., KWOLEK, W. F., SIMPSON, T. D. & DUTTON, H. J. (1977) "Zero *trans*" margarines: preparation, structure, and properties of interesterified soybean oil–soy trisaturate blends. Journal of the American Oil Chemists' Society, 54, 408–13.
LO, Y. C. & HANDEL, A. P. (1983) Physical and chemical properties of randomly interesterified blends of soybean oil and tallow for use as margarine oils. Journal of the American Oil Chemists' Society, 60, 815–18.
LUCK, T., KIESSER, T. & BAUER, W. (1988) Engineering parameters for the application of immobilized lipases in a solvent-free system. In *Proceedings World Conference on Biotechnology for the Fats and Oils Industry 1987*, Applewhite, T. H. (Ed.), American Oil Chemists' Society, pp. 343–5.
MACRAE, A. R. (1983) Extracellular microbial lipases. In: *Microbial Enzymes and Technology*, chap. 5, Fogarty, W. M. (Ed.), Applied Science, London, pp. 225–50.

MACRAE, A. R. (1985) Interesterification of fats and oils. In: *Studies in Organic Chemistry 22: Biocatalysts in Organic Synthesis.* Proceedings of an international symposium held at Noordwijkerhout, The Netherlands, 14–17 April 1985, Tramper, T. (Ed.) Elsevier, Amsterdam, pp. 195–208.

MACRAE, A. R. (1986) Edible fats. European Patent Application 185,524, Unilever N.V.

MACRAE, A. R. & HAMMOND, R. C. (1985) Present and future applications of lipases. *Biotechnology and Genetic Engineering Reviews*, 3, 193–217.

MATSUO, T., SAWAMURA, N., HASHIMOTO, Y. & HASHIDA, W. (1980) Producing a cacao butter substitute by transesterification of fats and oils. UK Patent Application 2,045,359, Fuji Oil Co. Ltd.

MATSUO, T., SAWAMURA, N., HASHIMOTO, Y. & HASHIDA, W. (1981) Method for modification of fats and oils. Australian Patent Application 77,623, Fuji Oil Co. Ltd.

MATSUO, T., SAWAMURA, N., HASHIMOTO, Y. & HASHIDA, W. (1983) Method for modification of fats and oils. US Patent 4,420,560, Fuji Oil Co. Ltd.

MATSUO, T., HASHIMOTO, Y. & HASHIDA, W. (1984) Method for enzymatic interesterification of lipid and enzyme used therein. European Patent Specification 35,883, Fuji Oil Co. Ltd.

MICKLE, J. B. (1959) Factors affecting the spreadability of butter. *Journal of Dairy Science*, 42, 389.

MICKLE, J. B. (1960) Flavour problems in rearranged milk fat. *Journal of Dairy Science*, 43, 436–7.

MICKLE, J. B., VON GUNTEN, R. L. & MORRISON, R. D. (1963) Rearrangement of milk fat as a means for adjusting hardness of butterlike products. *Journal of Dairy Science*, 46, 1357–61.

NAKAMURA, K., YOKOMICHI, H., OKISAKA, K., NISHIDE, T., KAWAHARA, Y. & NOMURA, S. (1988) Process for transesterifying fats. European Patent Application EP 0257,388A2, Kao Corporation.

NIELSEN, T. (1985) Industrial application possibilities for lipase. *Fette, Seifen, Anstrichmittel*, 87, 15–19.

PARODI, P. W. (1979) Relationship between trisaturated glyceride composition and the softening point of milk fat. *Journal of Dairy Research*, 46, 633–9.

PARVIAINEN, P., VAARA, K., ALI-YRRKÖ, S. & ANTILA, M. (1986) Changes in the triglyceride composition of butterfat induced by lipase and sodium methoxide catalysed interesterification reactions. *Milchwissenschaft*, 41, 82–5.

PATTERSON, J. D. E., BLAIN, J. A., SHAW, C. E. L., TODD, R. & BELL, G. (1979) Synthesis of glycerides and esters by fungal cell-bound enzymes in continuous reactor systems. *Biotechnology Letters*, 1, 211–16.

POSORSKE, L. H. (1984) Industrial-scale application of enzymes to the fats and oil industry. *Journal of the American Oil Chemists' Society*, 61, 1758–60.

POSORSKE, L. H., LEFEBVRE, G. K., MILLER, C. A., HANSEN, T. T. & GLENVIG, B. L. (1988) Process considerations of continuous fat modification with an immobilized lipase. *Journal of the American Oil Chemists' Society*, 65, 922–6.

RIEL, R. R. (1966) Etudes de propriétés dilatométriques et meltométriques de la graisse de lait modifiée. In: *XVII International Dairy Congress Section C: 2*, Heinrichs, E. (Ed.), Hildesheim, pp. 295–302.

SCHMID, R. D. (1986) Innovationsfeld Biotechnology-Ansatzpunkte in der Fettchemie. *Fette, Seifen, Anstrichmittel*, **88**, 555–60.
SCHUCH, R. & MUKHERJEE, K. M. (1987) Interesterification of lipids using an immobilized sn-1,3-specific triacylglycerol lipase. *Journal of Agricultural and Food Chemistry*, **35**, 1005–8.
SCHUCH, R. & MUKHERJEE, K. D. (1988) Interesterification of lipids by an sn-1,3-specific triacylglycerol lipase. In: *Proceedings World Conference on Biotechnology for the Fats and Oils Industry 1987*, Applewhite, T. H. (Ed.), American Oil Chemists' Society, pp. 328–9.
SONNTAG, N. O. V. (1982) Fat splitting, esterification and interesterification. In: *Bailey's Industrial Oil and Fat Products, Vol. 2.*, Swern, D. (Ed.), John Wiley, New York, pp. 127–74.
SREENIVASAN, R. (1978) Interesterification of fats. *Journal of the American Oil Chemists' Society*, **55**, 796–805.
STEVENSON, R. W., LUDDY, F. E. & ROTHBART, H. L. (1979) Enzymatic acyl exchange to vary saturation in di- and triglycerides. *Journal of the American Oil Chemists' Society*, **56**, 676–80.
TAMAURA, Y., TAKAHASHI, K., KODERA, Y., SAITO, Y. & INADA, Y. (1986) Chemical modification of lipase with ferromagnetic modifier—a ferromagnetic modified lipase. *Biotechnology Letters*, **8**, 877–80.
TANAKA, A., KAWAMOTO, T., KAWASE, M., NANKO, T. & SONOMOTO, K. (1988) Immobilized lipases in organic solvents. In: *Proceedings World Conference on Biotechnology for the Fats and Oils Industry 1987*, Applewhite, T. H. (Ed.), American Oil Chemists' Society, pp. 123–30.
TANAKA, T., ONO, E., ISHIHARA, M., YAMANAKA, S. & TAKINAMI, K. (1981*a*) Enzymatic acyl exchange of triglyceride in *n*-hexane. *Agricultural and Biological Chemistry*, **45**, 2387–9.
TANAKA, T., ONO, E. & TAKINAMI, K. (1981*b*) Method of producing improved glyceride by lipase. US Patent 4,275,011, Ajinomoto Co.
TIMMEN, H. (1978) Modifizierte Milchfette: Herstellung, Charakterisierung, Verwendung. *Deutsche Milchwirtschaft*, **29**, 1127–34.
TIMMS, R. E. (1979) The physical properties of blends of milk fat with beef tallow and beef tallow fractions. *Australian Journal of Dairy Technology*, **34**, 60–5.
TIMMS, R. E. & PAREKH, J. V. (1980) The possibilities for using hydrogenated, fractionated or interesterified milk fat in chocolate. *Lebensmittel Wissenschaft und Technology*, **13**, 177–81.
UNILEVER (1986) Fat processing. European Patent Specification 64, 855, Unilever NV.
WALSTRA, P. & JENNESS, R. (1984) *Dairy Chemistry and Physics*, John Wiley, New York, 467 pp.
WEIHE, H. D. (1961) Interesterified butter oil. *Journal of Dairy Science*, **46**, 944–7.
WEIHE, H. D. & GREENBANK, G. R. (1958) Properties of interesterified butteroil. *Journal of Dairy Science*, **41**, 703.
WISDOM, R. A., DUNNILL, P., MACRAE, A. & LILLY, M. D. (1984) Enzymic interesterification of fats: factors influencing the choice of support for immobilized lipase. *Enzyme and Microbial Technology*, **6**, 443–6.
WISDOM, R. A., DUNNILL, P. & LILLY, M. D. (1987) Enzymic interesterification of

fats: laboratory and pilot scale studies with immobilized lipase from *Rhizopus arrhizus*. *Biotechnology and Bioengineering*, **29**, 1081–5.

YOKOZEKI, K., TANAKA, T., YAMANAKA, S., TAKINAMI, T., HIRESE, Y., SONOMOTO, K., TANAKA, A. & FUKUI, S. (1982*a*) Ester exchange of triglyceride by entrapped lipase in organic solvent. In: *Enzyme Engineering* 6, Chibata, I., Fukui, S. & Wingard, L. B. (Eds), Plenum Press, New York, pp. 151–2.

YOKOZEKI, K., YAMANAKA, S., TAKINAMI, K., HIROSE, Y., TANAKA, A. & SONOMOTO, K. (1982*b*) Application of immobilized lipase to regio-specific interesterification of triglyceride in organic solvent. *European Journal of Applied Microbiology and Biotechnology*, **14**, 1–5.

10
Flavours Derived from Fats

G. Urbach* and M. H. Gordon†
*CSIRO, Division of Food Science and Technology, Victoria, Australia
and †Department of Food Science and Technology,
University of Reading, UK

SUMMARY

Fats can have a pronounced effect on the flavour of foods of which they are an ingredient. Several fats, notably milkfat, cocoa butter, olive oil, beef fat and lard can contribute positive flavour notes, depending on minor components and their state of oxidation. Chemical changes can induce off-flavours. Other fats already contain degraded components and must be refined to render them palatable before use. Mixing other ingredients with fats can significantly alter the flavour impression by reason of dilution or the interactive effect of secondary components on aroma impression. The physical state of the food on the palate is also a contributory factor; such variables as viscosity, emulsion characteristics, and the relative migration rates of aroma compounds between fat, water and the air headspace on the palate are all capable of modifying the perceived aroma. It has been proposed that certain minor ingredients of a non-volatile nature can influence the perception of flavour volatiles.

The chemistry of the flavour of milkfat is discussed as also is the influence of certain key factors such as enzymatic action, cooking and method of manufacture. The processing of milkfat to closely related products, such as fractionated milkfat, ghee, etc., can influence flavour. The principal factors affecting the flavour of key non-milkfats are highlighted together with the main degradation processes involved in off-flavour development.

With the current increasing tendency towards more natural foods, consumers may in future develop a taste for the indigenous flavours of certain natural oils such as groundnut or soya, and hence methods for extending the shelf-life of such oils could be of commercial value.

PART 1: MILKFAT AS A SOURCE OF FLAVOUR
(G. Urbach)

INTRODUCTION

Flavour consists of two elements, taste and aroma. Taste is usually defined as the sensations of sweet, sour, bitter and salty which are experienced on the taste buds of the tongue; the Japanese define an additional taste which they call umami (Yamaguchi, 1979) and which is the sensation of pleasantness imparted by such potentiators as monosodium glutamate. Taste is mainly caused by involatile molecules such as sucrose, lactic acid, quinine and sodium chloride, while aroma is the sensation which occurs when volatile molecules strike the back of the nose, e.g. fruity aromas produced by various esters.

There are many types of butter with regional preferences for different varieties, but the major divisions are cultured butters which are preferred in Europe and sweet-cream butters which are more common in the USA, Australia and New Zealand.

The flavour of butter immediately on churning is much stronger than that after storage for a few days. This is not due to the evaporation of the flavour compounds but could be due to chemical reactions between flavour compounds, or to a redistribution of the flavouring material between the phases of the butter.

Although most of the flavour of butter appears to be the result of metabolic processes in the animal, the feed does have some effect. Thus, during the spring and summer months, butter made from fresh cream has a delicious aromatic flavour which originates from the spring and summer grasslands. This flavour is lacking from fresh-cream butter made in the winter months, due mainly to the cows being fed chaff and bran and other dry feeds (Wigan, 1951). It is not known what compounds are responsible for this difference.

CULTURED BUTTER

It is likely that the desirable flavour of cultured butter is provided by a background of the compounds contributing to the flavour of sweet-cream (non-cultured) butter combined with the effect of the components produced by the starter bacteria. As early as 1929, van Niel *et al.* recognized the importance of diacetyl to the flavour of cultured butter and, in 1963, Winter *et al.* identified a range of aldehydes, ketones and their condensation products in fresh French butter made from unpasteurized cream. Winter *et al.* (1963) suggest that some of the aldehydes originate from amino acids by enzyme-catalysed oxidative decarboxylation.

Lindsay *et al.* (1967) prepared a synthetic cultured-butter flavour based on their own investigations into the flavour of cultured butter and found that the optimum flavour was obtained from a blend of lactic acid, acetic acid, acetaldehyde, dimethyl sulphide and diacetyl. Their grading panel found no significant difference between butters flavoured with this synthetic mixture and butters cultured in the normal manner. The pH of the butter was particularly important, as off-flavours were evident below pH 4·5 while the best flavour was obtained by adjusting the pH into the range 4·5–4·8 using sodium bicarbonate; the released carbon dioxide appeared to have beneficial effects on the flavour. Dutch workers (Nederlands Instituut voor Zuivelonderzoek, 1975) imparted the flavour of ripened-cream butter to sweet-cream butter by adding a flavouring composition based upon 4-cis-heptenal, phenylacetaldehyde, dimethyl sulphide, diacetyl, acetic acid and citric acid. The method permits the manufacture of ripened-cream-flavoured butter without the need for large quantities of sour buttermilk. An alternative method, developed at the same institute, also avoids the formation of sour buttermilk (Veringa et al., 1976). Aroma compounds, produced in the form of a mixture of starters, and lactic acid, in the form of a culture concentrate, are worked into sweet-cream butter.

Mick et al. (1982) isolated many volatile compounds from cultured butter (Table 1), all of which do not necessarily have flavour significance. The method of isolation favoured lower boiling compounds, hence the results are not quantitative. To understand the flavour significance of all the identified compounds it will be necessary to determine both their flavour thresholds and their recoveries.

PRECURSOR FLAVOURS

Fats and oils consist mainly of triglycerides; some are made up of relatively few fatty acids, others are more complex, the most complex probably being milkfat which is made up of hundreds of fatty acids. The fatty acids which are important from a flavour point of view are:

(1) γ- and δ-hydroxyacids and possibly γ- and δ-ketoacids,
(2) β-ketoacids,
(3) lower fatty acids having ≤ 12 carbon atoms,
(4) unsaturated acids.

Hydroxyacids

Hydroxyacid-triglycerides rearrange spontaneously to give γ- and δ-lactones which have significant flavour characteristics. When butter was stored at $-10°C$ for a year, half the esterified γ- and δ-hydroxyacids were converted to their corresponding lactones. However, at 180°C, conversion was complete within 2 h (Stark & Urbach, 1976).

TABLE 1
Volatiles from sour cream butter (Mick et al., 1982) (with permission of Milchwissenschaft)

Benzene and C_{7-9} homologues
Alkanes C_{10}, C_{12-17}
Alkenes C_{16-18}
Terpenes: β-pinene, α-3-carene, limonene, (Z)-ocimene, γ-terpinene, p-cymene, camphor, 2,3-dihydrofarnesene, α-terpineol
Alkan-2-ones C_{6-13}, C_{15}
Alkan-3-ones $C_{7,8,19}$
Alkanals $C_{8,9}$
Alk-2-enals $C_{6,7,9,10}$
Styrene
Anisole
2-Methyloctan-3-one
1,3-Dichlorobenzene
Ethyl esters: acetate, hexanoate, octanoate, benzoate, undecanoate, lactate
Benzaldehyde, benzonitrile, 4-methylbenzaldehyde
Benzothiazole
2-Methyltetrahydrothiophen-3-one
Methyl esters: benzoate, decanoate, salicylate, hexadecanoate
Acetophenone, 2-hydroxybenzaldehyde
Naphthalene, 2-methylnaphthalene
Propyl benzoate, hexyl benzoate, benzyl benzoate
Dibutyl phthalate
Butan-2,3-dione, pentan-2,3-dione, acetoin
3-Hydroxypentan-2-one
Alkan-2-ols $C_{4,5,7}$
Pentan-3-ol
Alkan-1-ols $C_{4,iso5,7,14-17}$
Hex-2-en-1-ol, hex-3-en-1-ol, hex-4-en-1-ol
Cyclohexanol
Alkanoic acids: $C_{2,4-10,12}$, 2-methylpropanoic acid
Oct-1-en-3-ol
3-Methylbut-2-en-1-ol
Deca-2,4-dienal
2-Phenylethyl acetate, benzyl alcohol, 2-phenylethyl alcohol, phenol
Diphenyl
Cyclopentanone
2,5-Dimethylpyrazine
6-Methylhept-5-en-2-one
Furfural, linalooloxide
2-Ethylhexan-1-ol
δ-Lactones: $C_{6,8-10,12}$

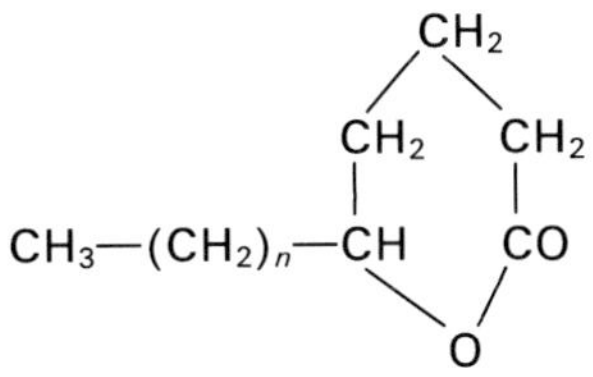

$$CH_3-(CH_2)_n-\underset{\underset{O}{\diagdown\diagup}}{CH}\overset{\overset{CH_2}{\diagup\diagdown}}{\underset{|}{CH_2}\underset{|}{CH_2}}CO$$

Fig. 1. Saturated δ-lactones in heated milk fat ($n = 0, 2, 4, 6, 7, 8, 9, 10, 12$).

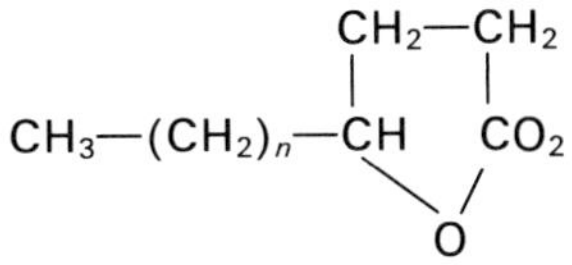

$$CH_3-(CH_2)_n-\underset{\underset{O}{\diagdown\diagup}}{CH}\overset{CH_2-CH_2}{\underset{|}{}\underset{|}{}}CO_2$$

Fig. 2. γ-Lactones ($n = 1-11, 13$).

Freshly secreted milkfat contains practically no free lactones (Boldingh & Taylor, 1962), some lactones are formed during processing and there is further formation of lactones on storage, but most importantly during frying or baking.

Saturated, straight-chain δ-lactones have been isolated from butterfat (Fig. 1). Of these, those containing an even number of carbon atoms are much more abundant than those containing an odd number, as is the case for the corresponding fatty acids (Kinsella, 1969; Stark & Urbach, unpublished data). Honkanen et al. (1968), however, have isolated the branched-chain 7-carbon lactone, *trans*-4-methyl-δ-hexalactone from so-called zero milk, i.e. milk from cows fed on an odourless, purified, protein-free diet consisting of cellulose, starch, sucrose, and urea and ammonium salts as the nitrogen source.

Butterfat also contains varying quantities of the precursors of unsaturated δ-lactones, *cis*- and *trans*-δ-tetradec-8-enolactone, a conjugated *cis,trans*-δ-decadienolactone (Urbach & Stark, 1978a), δ-dodec-9-enolactone and δ-tetradec-9-enolactone (van der Zijden et al., 1966).

The homologous series of saturated γ-lactones from 6–18 carbon atoms, excluding C17, have also been found in butterfat, although only γ-dodecanolactone occurs in amounts which are significant for flavour (Fig. 2) (Jurriens & Oele, 1965).

TABLE 2
Flavour thresholds and optimum levels of lactones in relation to their concentrations in milkfat before and after heating (lactones in ppm)

Lactone	Flavour threshold	Optimum level	Concentration	
			Before heating	After heating
$\delta 6$	5	10		2
$\delta 8$	0·1	0·2	0·05–0·6	0·2–4
$\delta 10$	1·0	2	1·2–7·4	7–23
$\delta 12$	10	10	1·5–13·4	10–58
$\delta 14$	50	50–100		24–96
$\gamma 12:0$	1	1	0·5	0·2–182
$\gamma 12:1$	1			0·05–24
Bovolide				0–3
Dihydrobovolide				0–3

$\delta n = \delta$-lactone having n carbon atoms.
$\gamma 12:0 = \gamma$-dodecalactone.
$\gamma 12:1 = \gamma$-dodec-*cis*-6-enolactone.

There are also some unsaturated γ-lactones in butterfat; γ-dodec-*cis*-6-enolactone is always present, while bovolide (Boldingh & Taylor, 1962), dihydrobovolide (Stark & Urbach, 1979) and a conjugated *cis,trans*-γ-tetradecadienolactone can sometimes occur (Urbach & Stark, 1978a).

Table 2 shows flavour thresholds of some of these lactones (i.e. the minimum amount which can be perceived in the mouth) determined in 'synthetic butter' (i.e. butter oil in which all flavour compounds have been generated and removed by vacuum steam distillation at 180°C is then re-emulsified with 16% distilled water) (Urbach *et al.*, 1972). The table also gives optimum levels of these lactones as well as their concentrations in butterfat before and after heating.

δ-Hexalactone is present in butterfat at a level where it would not be tasted, but, as it is more soluble in water than in fat, it is considered to be an important contributor to milk flavour (Badings & Neeter, 1980). At 5 ppm, in synthetic butter, it is reminiscent of whey, at higher concentrations it tends towards sweet almond.

δ-Octalactone has by far the lowest flavour threshold of all the lactones identified in butterfat and can be present in heated butterfat at up to 40 times its flavour threshold, although the most common value is 10–20 times the flavour threshold, i.e. 1–2 ppm (Urbach, 1979). At 0·2 ppm it imparts a sweet creaminess, tending towards fresh coconut at higher levels. It is an important flavour precursor of butterfat.

The flavour threshold of δ-decalactone is 10 times that of δ-octalactone and its maximum concentration in heated butterfat is approx. 20 times its flavour threshold. On this basis it has about the same importance for butter flavour as δ-octalactone. Its flavour is that of a mixture of peach and coconut, in both of which it occurs (Maga, 1976).

The flavour threshold of δ-dodecalactone is again an order of magnitude higher than the threshold of δ-decalactone and its concentration in heated butterfat is usually only twice its flavour threshold, occasionally its value is 4 or 5 times the threshold. Thus its flavour contribution is less than that of δ-decalactone. At its threshold it does not have a definable flavour but merely imparts fullness to synthetic butter, at higher concentrations it has a definite desiccated-coconut flavour and, interestingly, was described as reminiscent of margarine. δ-Decalactone, δ-dodecalactone and presumably δ-octalactones are considered to be responsible for the typical coconut flavour of condensed milk (Tharp & Patton, 1960).

Even in heated butterfat δ-tetradecalactone is only present in concentrations around the flavour threshold and as such it and the higher straight-chain lactones do not contribute a significant flavour to butterfat.

In some heated butterfats γ-dodecalactone can be the most significant lactone. With a flavour threshold of 1 ppm it is not normally present in unheated butterfat above its threshold. However, on one occasion it was reported to be present at a concentration of 182 ppm (Urbach & Stark, 1978b); in this case it was the single most potent flavour compound in the butterfat. Its unsaturated analogue, γ-dodec-*cis*-6-enolactone, has a very similar flavour and threshold of the same order (Wilson, 1989). Both lactones have a strong, sweet, fruity odour, reminiscent of artificial raspberry flavour. With a level in heated butterfat ranging from practically nothing to 24 ppm, γ-dodec-*cis*-6-enolactone can be a very important flavour contributor (Urbach & Stark, 1978b). Wilson (1989) found that the level of γ-dodecalactone in European butterfat was higher than in New Zealand butterfat and that this difference accounted for the sweeter flavour of European cheeses as compared with New Zealand and Australian cheeses. He (private communication) was able to raise γ-dodecalactones in milkfat by a diet of oats and sunflower seeds plus pasture. The increase was sufficient to influence flavour.

Homologous saturated δ-hydroxyacids are mainly synthesized from acetate in the mammary gland (Walker *et al.*, 1968), but there is evidence that the pathway for the formation of γ-lactone precursors is entirely different (Stark et al., 1978). When ruminants are fed with a supplement of oil or oil seeds whose unsaturated fatty acids have been protected against biohydrogenation in the rumen (by encapsulating the oil in crosslinked casein), the milkfat and body fat of these animals can contain approx. 20% of di- or polyunsaturated acids (Scott et al., 1972). The meat from such animals can have an objectionable, sweet, fruity flavour which is intensified on cooking (Park et al., 1974). The same flavour is also present in dairy products where it is not quite such a problem because dairy products are not usually heated to the same extent as meat and because a sweet, fruity flavour is not necessarily objectionable in dairy products (Stark & Urbach, 1974). This sweet, fruity flavour was shown to be due to γ-dodec-cis-6-enolactone and γ-dodecalactone. The level of these lactones was in no way related to the level of δ-lactones, but appeared to result from the protected oil seeds and unprotected oil (Stark et al., 1978). Anet (1973) suggested that the linoleic acid in the feed was converted by the rumen microflora to 10-hydroxyoctadec-cis-12-enoic acid which was subsequently converted to 4-hydroxydodec-cis-6-enoic acid by three β-oxidations (Fig. 3). This mechanism would account for the position and configuration of the double bond. The saturated γ-dodecalactone can be formed in an exactly analogous manner from the oleic acid in the feed. Other homologues of γ-dodecalactone may be derived from the 9-enoic acids which are always present in milkfat in small quantities. This accounts for the fact that only traces of other saturated γ-lactones are found in butterfat.

For γ-lactones to be produced, the specific microflora and precursor must react, a combination which was obviously present during the production of the meat and dairy products high in di- and polyunsaturated acids. Further work (Urbach, 1982) indicated that, to raise the level of γ-dodecalactone in milkfat, both a readily available source of starch, such as is available in crushed, but not in whole, oats, and an excess of precursor are required.

Higher levels of saturated δ-lactones are produced when cows are fed on rations which contain very little fat. An extreme case of this is zero milk (where a cow received only 37 g oil per day as compared with about 1 kg per day which a cow on pasture can ingest) (Honkanen et al., 1970). Conversely, when cows were given a protected supplement (containing in the order of 1 kg oil per cow per day), the saturated δ-lactones in heated butterfat were reduced to two-thirds or less of what they were before (Urbach & Stark, 1978b). δ-Lactones were also depressed when cows were given a fat-depressing diet (Dimick et al., 1966). The stage of lactation also affects the δ-lactone level, which is low immediately after calving and rises steadily with length of lactation (Dimick & Harner, 1968). It has been

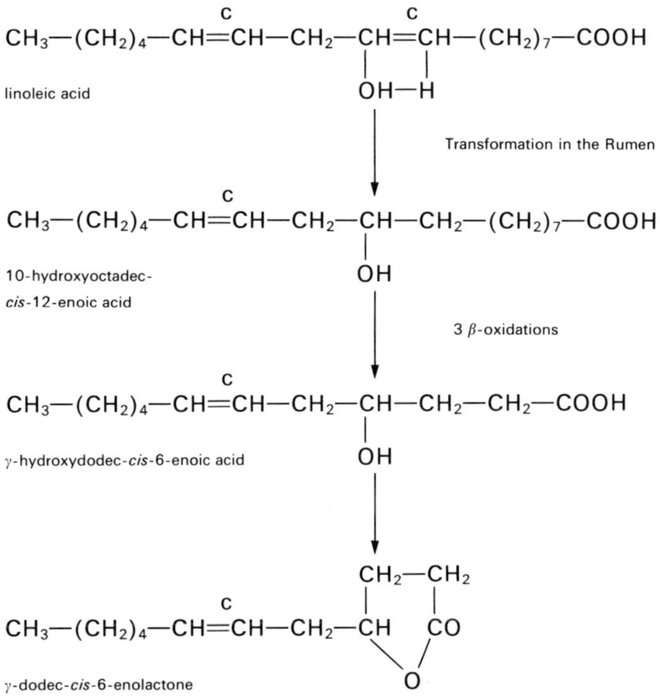

$$\text{CH}_3-(\text{CH}_2)_4-\overset{c}{\text{CH}}=\text{CH}-\text{CH}_2-\overset{c}{\underset{\text{OH}-\text{H}}{\text{CH}}}=\text{CH}-(\text{CH}_2)_7-\text{COOH}$$

linoleic acid

↓ Transformation in the Rumen

$$\text{CH}_3-(\text{CH}_2)_4-\overset{c}{\text{CH}}=\text{CH}-\text{CH}_2-\underset{\text{OH}}{\text{CH}}-\text{CH}_2-(\text{CH}_2)_7-\text{COOH}$$

10-hydroxyoctadec-cis-12-enoic acid

↓ 3 β-oxidations

$$\text{CH}_3-(\text{CH}_2)_4-\overset{c}{\text{CH}}=\text{CH}-\text{CH}_2-\underset{\text{OH}}{\text{CH}}-\text{CH}_2-\text{CH}_2-\text{COOH}$$

γ-hydroxydodec-cis-6-enoic acid

↓

$$\text{CH}_3-(\text{CH}_2)_4-\overset{c}{\text{CH}}=\text{CH}-\text{CH}_2-\text{CH}\begin{array}{c}\text{CH}_2-\text{CH}_2\\|\quad\quad\quad|\\\diagdown\quad\;/\\\text{O}\end{array}\text{CO}$$

γ-dodec-cis-6-enolactone

Fig. 3. Suggested mechanism for the formation of γ-dodec-cis-6-enolactone from linoleic acid (Stark et al., 1978) (with permission of the *Journal of Dairy Research*).

claimed that the non-oxidative flavour deterioration of milk powder is due to lactone formation. Therefore it would be advisable to use low-lactone milk for production of whole-milk powder and high-lactone fat for butter and dairy products destined for baking (Kinsella, 1969).

Van der Ven (1964) reported the presence of esterified δ-keto-octanoic, -decanoic and -dodecanoic acids as well as traces of γ-keto-decanoic, -undecanoic and -dodecanoic acids in milkfat. Theoretically, when butterfat is heated to high temperatures (180–200°C), these ketoacids should form enol lactones and eventually unsaturated lactones. These lactones have not been reported in milkfat but Thompson et al. (1978) found the unsaturated γ-lactones

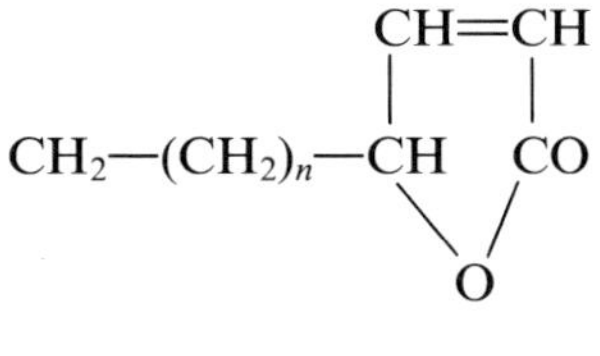

$$\text{CH}_2-(\text{CH}_2)_n-\text{CH}\begin{array}{c}\text{CH}=\text{CH}\\|\quad\quad|\\\diagdown\;/\\\text{O}\end{array}\text{CO}$$

$n = 1-5$

in the volatile decomposition products of fried trilinolein where they imparted a deep-fried flavour. Chang & May (1973) and May et al. (1978) also claim that the isomeric γ-lactones

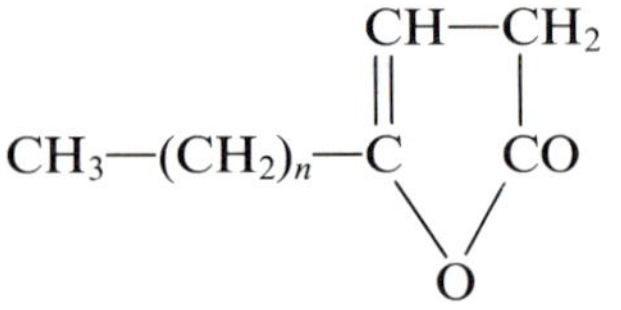

$$CH_3-(CH_2)_n-\underset{\underset{O}{\diagdown\diagup}}{C}\overset{CH=CH_2}{\underset{CO}{|}}$$

$n = 3, 4$

improved the flavour of margarine when added at a level of 2·5 ppm. These products are likely to be the result of oxidation of linoleate rather than simple heat-induced rearrangement.

Stark & Urbach (unpublished data) found three methyl substituted γ-lactones, dihydrobovolide, bovolide and hydroxybovolide in heated butterfat (Fig. 4). These may have arisen from ketoacid triglycerides although there was no evidence to prove this. It is possible that bovolide is an oxidation product of dihydrobovolide because, from the same butterfat, Stark and Urbach sometimes found bovolide and sometimes dihydrobovo-

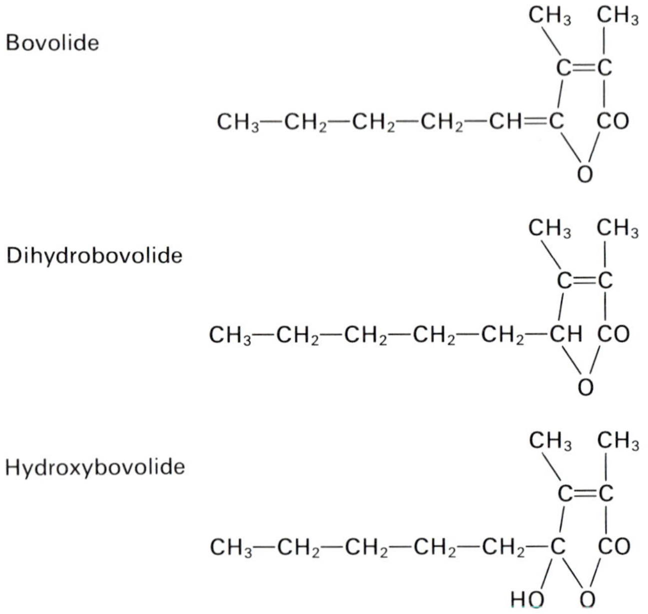

Bovolide

$$CH_3-CH_2-CH_2-CH_2-CH=\underset{\underset{O}{\diagdown\diagup}}{C}\overset{\overset{CH_3\ \ CH_3}{\diagdown\ \ |}}{\underset{CO}{\underset{|}{C=C}}}$$

Dihydrobovolide

$$CH_3-CH_2-CH_2-CH_2-CH_2-\underset{\underset{O}{\diagdown\diagup}}{CH}\overset{\overset{CH_3\ \ CH_3}{\diagdown\ \ |}}{\underset{CO}{\underset{|}{C=C}}}$$

Hydroxybovolide

$$CH_3-CH_2-CH_2-CH_2-CH_2-\underset{\underset{HO}{\diagup}\underset{O}{\diagdown}}{C}\overset{\overset{CH_3\ \ CH_3}{\diagdown\ \ |}}{\underset{CO}{\underset{|}{C=C}}}$$

Fig. 4. Bovolide and related lactones from butterfat.

lide. All these compounds are claimed to have a celery odour (Boldingh & Taylor, 1962; Kaneko & Mita, 1969) and, at a level of 3 ppm (Stark & Urbach, unpublished data), certainly could contribute to the flavour of heated milkfat. Compounds with mass spectra identical to those of dihydrobovolide and bovolide were also found in the body fat of sheep (Suzuki & Bailey, 1985). They were more concentrated in the fat of animals finished on forage compound than that of animals finished on grain (Bailey, 1988, Private communication); this finding could be explained by the presence of dihydrobovolide in alfalfa (Kami, 1983), a widely distributed forage crop.

Although lactone precursors are normally regarded as being present as triglycerides, lactones, fatty acids, indole, skatole, various phenolics, vanillin derivatives and aromatic acids have also been found in milk as conjugated glucuronides and sulphates. They can be liberated by β-glucuronidase and aryl sulphatase but the literature does not mention the action of heat (Brewington et al., 1973, 1974). Re-investigation of the conjugates of milk is overdue.

The δ-lactones isolated from butterfat are dextrorotary but have lower specific rotations than optically pure dextrorotary δ-lactones (van der Ven & de Jong, 1970). Therefore both optical antipodes exist in butterfat with the dextrorotary isomer predominating. Palm et al. (1991) found distinct differences in the ratios of R and S forms of δ-octa-, δ-deca- and δ-dodecalactones in cultered butter, margarine and coconut (Table 3(a)). It must be assumed that the antipodes are formed by different pathways. The odour of dextrorotary and laevorotary δ-decalactones is identical according to Tuynenburg Muys et al. (1962), but Mosandl & Gessner (1988) did find some differences between the organoleptic properties of the laevo- and dextrarotary isomers of δ-octa-, δ-deca- and δ-dodecalactones (Table 3(b)). This could easily contribute to the differences in flavour between butter and coconut.

Unsaturated lactones with the double bond α to the carbonyl group have not been isolated from butter, but several of them have been synthesized and are claimed to have a buttery odour (Nobuhara, 1968, 1969a, b, 1970; Kikoman Syoyu Co. Ltd, 1971).

β-Ketoacids

β-Ketoacids also occur as triglycerides in milkfat (van der Ven et al., 1963). Although there are practically no methyl ketones in fresh milk and they do not form when butter is stored (Stark & Urbach, 1976) the β-ketoacid triglycerides are converted to the corresponding methyl ketones by the

TABLE 3(a)
Ratios of S and R isomers of δ-lactones in cultured butter, margarine and coconut

	δ-Octalactone S:R	δ-Decalactone S:R	δ-Dodecalactone S:R
Cultured butter	85:15	19:81	10:90
Margarine	50:50	50:50	50:50
Coconut	12:88	26:74	61:39

Reproduced with permission from Palm *et al.* (1991) *Z. Lebensm. Unters. Forsch.*, **192**, 209. © Springer-Verlag.

TABLE 3(b)
Structure and properties of 5-alkyl substituted lactones

δ-Lactone	Configuration $[\alpha]_D^{20}$		Odour
δ-Octalactone	5R	$+61°$	Distinct coconut with spicy-sweet aspect
	5S	$-60°$	More fatty, less intense coconut
δ-Decalactone	5R	$+60°$	Fruity, sweet, milky note
	5S	$-61°$	Creamy peach, fatty, buttery, more intense than R
δ-Dodecalactone	5R	$+46°$	Fruity-sweet, apricot, fatty-green
	5S	$-46°$	Similar to R+, more intense

Reproduced with permission from Mosandl and Gessner (1988) *Z. Lebensm. Unters. Forsch.*, **187**, © Springer-Verlag.

action of heat and water (van der Ven *et al.*, 1963). Much more vigorous conditions are required to liberate methyl ketones than to form lactones. When butterfat is distilled at 35°C and less than 7 mPa for 5 h, about half the potential lactones are generated whereas no ketones are formed (Stark *et al.*, 1976*a*). However, when butter is used for baking and even more so when it is used for frying, methyl ketones would be expected to be liberated and to contribute to the flavour imparted by the butter.

The flavour thresholds of methyl ketones are in the range 0·1–1·0 ppm

TABLE 4
Yields of methyl ketones generated from butterfat by moist heat

Ketone	Quantity (ppm of butterfat)
Acetone	1–17
Pentan-2-one	2–9
Heptan-2-one	7–19 (29)
Nonan-2-one	3–12 (18)
Undecan-2-one	6–19 (23)
Tridecan-2-one	9–27 (37)
Pentadecan-2-one	12–58 (86)

(Forss, 1972). Table 4 shows the range of quantities of methyl ketones produced when butterfat is heated for 2 h at 180°C; this value is generally called the methyl ketone potential of the butterfat (Urbach, 1979). The values for acetone and pentan-2-one are low because of the method used for recovery. The values in parentheses were obtained from butterfat from cows which had been fed on barley and pasture hay. When the cows were fed exclusively on pasture hay, the values were still above the normal range but somewhat lower than the values in parentheses.

In general, methyl ketone potentials are high in ruminant milkfat when there is little fat in the diet and are suppressed by feeding oil, either unprotected or protected against biohydrogenation in the rumen (Urbach & Stark, 1978b; Stark et al., 1978). Methyl ketone potentials are low at the beginning of lactation and gradually increase as lactation proceeds (Dimick & Walker, 1968), a trend which also holds for saturated δ-lactone potentials and fatty acids synthesized in the mammary gland. This correlation is to be expected as all three groups are synthesized from acetate in the mammary gland, but Cook & Scott (1975) suggested that the fatty acids ingested with protected supplement could cause an inhibition of fatty acid biosynthesis.

Lipase partially liberates methyl ketones from the β-ketoacid-glycerides when cheese is matured (Lawrence, 1967). The flavour of methyl ketones is reminiscent of blue-vein cheese but the β-ketoacid triglycerides of butterfat are only partially responsible for the flavour of blue-vein cheese. The mould, *Penicillium roqueforti*, used in the manufacture of blue-vein cheese, actually oxidizes the saturated fatty acids to β-ketoacids which then form methyl ketones (Kinsella & Hwang, 1976). The methyl ketones are in turn partially reduced to alkan-2-ols which have a flavour potency close to that of methyl ketones and are important in the flavour of blue-vein cheese (Forss, 1972).

The alkan-2-ols are also formed from the β-ketoacids when butterfat is catalytically hydrogenated (Stark *et al.*, unpublished data). Contrary to other fats, which develop an unpleasant odour on catalytic hydrogenation and have to be deodorized before they can be used, hydrogenated butterfat has a pleasant, fruity odour due to a mixture of saturated lactones and alkan-2-ols.

Lower Fatty Acids

Butterfat differs from most other fats and oils in that it contains lower fatty acids (Hilditch & Williams, 1964; Swern, 1964; Kurtz, 1974), which, when released, by either heat or lipolysis, produce a very strong flavour. Where this flavour is undesirable, as in fresh milk, it is called lipolytic rancidity and, at levels of the order of 100 ppm, butanoic, hexanoic, octanoic, decanoic, dodecanoic and tetradecanoic acids all produce off-flavours in synthetic butter (Urbach *et al.*, 1972); at lower levels they probably contribute to the totality of desirable butter flavour. McNeill *et al.* (1986) observed a rancid flavour in butter when the level of butyric acid reached 35 ppm and the level of hexanoic acid 26 ppm. However, free fatty acids certainly contribute to the desirable flavour of cheese, particularly Provolone types (Harper, 1959). A part of the free fatty acids in cream can be removed by vacreation, the extent of removal ranging from 44% for decanoic acid to 78% for acetic acid (Labuschagne, 1975).

Pregastric esterases, i.e. enzymes originating from the mouth tissue of calves, lambs and kids, have been especially effective in producing cheese flavours, each source of pregastric esterase producing a characteristic cheese flavour (Huang & Dooley, 1976). These enzymes are specific for short-chain acids located in the α-position of triglycerides and it has been assumed that it is these short-chain acids which are largely responsible for certain characteristic cheese flavours. It is quite likely, however, that other trace acids, such as the lactone and methyl ketone precursors and possibly as-yet-unrecognized precursors, are also liberated and form part of the cheese flavour.

It is claimed that, with very low additional levels of pregastric-esterase-modified butterfat, a sensation of richness is imparted without any detectable free-fatty-acid-flavour character. As additions are increased, the flavours imparted resemble cream or butter, and lipolysed butterfat is added to shortbread to increase its buttery flavour. When the amounts added are relatively high, the flavour approaches that of cheese. Pregastric-esterase-lipolysed butterfat is relatively free of soapy or bitter flavour notes which occur in other lipase-modified butterfat, since other lipases presumably liberate higher fatty acids (Paulet et al., 1974). Dairyland Food Laboratories in the USA have built up a thriving business on the supply of pregastric esterases and pregastric-esterase-treated butterfat to the food industry (Nelson, 1972), e.g. Capalase-KL (i.e. a mixture of pregastric esterase from kid and lamb) is sometimes used in the manufacture of feta cheese (Efthymiou & Mattick, 1964). Numerous patents, involving the production of flavours by the action of lipolytic enzymes on butterfat and other fats, have been issued, for example, for the production of Italian cheese flavour, buttery flavour, cultured-cream flavour, blue-cheese flavour, yoghurt flavour, and many companies are involved in the production of a large range of enzymes (Kilara, 1985). However, Foda et al. (1974) found that for true Cheddar flavour pure milkfat was absolutely essential.

Some milks have a much greater tendency to produce lipolytic rancidity than others. It has been found that underfeeding of the cows is one of the factors which contribute to the tendency of the milk to undergo lipolytic rancidity (Astrup, 1980).

4-Ethyloct-2-enoic acid was isolated from goat's milk in Switzerland by the Givaudan Company (Smith et al., 1984). Ha and Linday (1991) examined goat's milk, sheep's milk and cow's milk cheeses for volatile fatty acids using a technique which allowed them to identify branched-chain acids (Ha and Lindsay, 1990). The level of branched-chain fatty acids in goat's and sheep's milk cheese was much higher than in cow's milk cheeses. Qualitative profiles of most free fatty acids were similar in the sheep's, goat's and cow's milk cheeses which they examined, except that 4-ethyloctanoic acid was not found in cow's milk cheese. Milkfat of cows contained low concentrations of 4-methyloctanoic acid, but milkfat of sheep and goats contained significant amounts of both 4-methyloctanoic and 4-ethyloctanoic acids, which contributed mutton-like and goat-like flavours, respectively. Brennand et al. (1989) list the flavour thresholds for 29 branched- and straight-chain fatty acids, and aroma qualities for 24 of these, in the range of 2 to 11 carbons. In many cases the branched-chain fatty acids have much lower thresholds than the straight-chain acids with the same carbon number, e.g. the threshold for 4-methyloctanoic acid is 20 ppb compared to 2·4 ppm for nonanoic acid and the threshold for 4-ethyloctanoic acid is 6 ppb compared to 1·7 ppm for decanoic acid. Traces of branched-chain fatty acids due to slight lipolysis of milkfat can thus make a major contribution to flavour.

When lambs were fed with a supplement of oil, protected against

TABLE 5
Flavour threshold (ppm) of fatty acids in various media

Acid	Water		Milk	Butter oil	Butter		Vegetable oil
Acetic	54[a]	22[b]		10[d] 7[e]	5[g]		
Butyric	6·8[a]	6·2[b]	25[a]	0·66[e]	50·5[f]	3[g]	0·6[h]
Hexanoic	5·4[a]	15[b]	14[a]		93·7[f]	10[g]	2·5[h]
Octanoic	5·8[a]	5·83[b]			454·6[f]	10[g]	350[h]
Decanoic	3·5[a]		7·0[ci]		363·0[f]	5[g]	200[h]
Lauric			8·5[ci]		303·4[f]	50[g]	700[h]
Myristic					50[g]		

[a] Patton (1964a); [b] Siek et al. (1971); [c] Kinsella (1969); [d] Forss & Patton (1966); [e] Siek et al. (1969); [f] McDaniel et al. (1969); [g] Urbach et al. (1972); [h] Feron & Govignon (1961); [i] Skim milk.

biohydrogenation in the rumen, the conventional lamb odour was significantly decreased as might be expected from lamb fat which contained a much lower proportion of low-molecular-weight fatty acids than conventional lamb fat (Park et al., 1976). The proportion of low-molecular-weight fatty acids in goat's and cow's milk fat are also reduced by feeding with oil supplement (Start et al. 1978).

The flavour of deep-fried foods is due, in part, to fatty acids formed during processing. Kawada et al. (1967) and May (1972) isolated over thirty aliphatic acids from oil kept under simulated restaurant conditions.

Acidity and partition affect the flavour impact of fatty acids in heterogeneous media. Due to the greater water solubility of short-chain acids (butyric and hexanoic), they pass in greater amounts into the aqueous phase of milk where their flavour threshold is higher and their flavour is diminished due to salt formation. This accounts for the relatively large amounts of unesterified fatty acids in normal milk with no associated rancid flavour (Kintner & Day, 1965).

Table 5 lists the flavour thresholds of short-chain fatty acids in various media. The medium has a great influence on the value of the threshold, for example, in the absence of the buffering capacity of milk protein (Urbach et al., 1972) professional graders were able to taste the lower fatty acids at about an order of magnitude less than a laboratory panel (McDaniel et al., 1969) could taste them in butter.

Unsaturated Fatty Acids

Butterfat contains significant amounts of unsaturated fatty acids, mainly arachidonic (20:4), linolenic (18:3), linoleic (18:2), and oleic (18:1) acids and their isomers (Kurtz, 1974). These and other mono- and polyunsaturated acids are also widely distributed in plant and fish oils. They oxidize readily to produce various off-flavours known collectively as oxidative rancidity. Butterfat contains some short-chain unsaturated fatty acids (Forss *et al.*, 1967*a*) and it has been claimed that some of these produce a buttery flavour in the presence of higher methyl ketones (Sevenants, 1983, 1986).

Oxidation of unsaturated fatty acids can be initiated by OH· radicals generated by copper and ascorbic acid (Wilkinson, 1964). This is a non-enzymic system and depends on the copper being in an available form. Dairy factories are now very careful to avoid copper contamination. Another mechanism is enzymic (Hill, 1979) and is shown in Fig. 5.

In the presence of xanthine oxidase, the inactive form of oxygen, known as triplet oxygen, reacts with a suitable substrate, such as xanthine or the aldehydes which have been produced by the copper/ascorbate system, to produce H_2O_2 plus the superoxide radical ($O_2^{-\cdot}$). The H_2O_2 is then available as a substrate for the action of lactoperoxidase which can act on ascorbic acid or phenols or thiocyanate. The oxidation product of

(1) Xanthine or aldehydes or other substrate
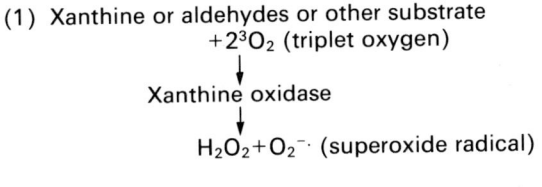

(2) Ascorbate or phenols or
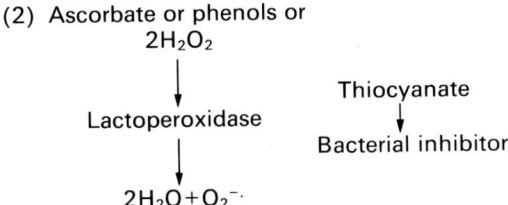

(3) $2O_2^{-\cdot} + 2H^+ \rightarrow H_2O_2 + {}^1O_2^*$ (singlet oxygen active form)

(4) ${}^1O_2^* + \text{fat} \rightarrow \text{Oxidation products} + R\cdot$ (free radicals)

$R\cdot + \text{fat} \rightarrow \text{Oxidation products} + R'\cdot$

Fig. 5. Enzymic oxidation of milkfat.

thiocyanate is inhibitory to the growth of bacteria, a mechanism which appears to have evolved because of the natural advantage which it confers. Both xanthine oxidase and lactoperoxidase produce superoxide radical. This dismutates spontaneously to produce the highly active form of oxygen, $^1O_2^*$, known as singlet oxygen and this can attack unsaturated fatty acids to produce a host of breakdown products and free radicals. These, in turn, can attack the unsaturated fatty acids. Thus, once oxidation is initiated either by copper or by enzymes, it can continue as long as there is substrate, unless the free radicals are destroyed.

The xanthine oxidase/lactoperoxidase system can be largely inactivated by pasteurization at 80°C instead of 72°C (Hill, 1979). This gives the added bonus of binding the copper to the casein, up to a level of about 0·15 ppm, in a form which inactivates the copper as a pro-oxidant. Much of the residual action of the xanthine oxidase/lactoperoxidase system can then be removed by the combined action of superoxide dismutase and catalase. Superoxide dismutase removes the superoxide radical formed by both xanthine oxidase and lactoperoxidase by converting it to hydrogen peroxide and the inactive form of oxygen—triplet oxygen. The catalase then removes the hydrogen peroxide formed by both superoxide dismutase and xanthine oxidase and converts it to water and more of the harmless triplet oxygen.

It has been suggested that slight oxidation contributes to the desirable flavour of dairy products, e.g. hept-cis-4-enal is claimed to impart a creamy flavour to butter at a level of 1 part in 10^9 (Haverkamp Begemann & Koster, 1964; Schogt et al., 1967) whereas at 1 in 10^8 it is reminiscent of the cold-storage defect (Badings, 1965). Deca-2,4-dienal, derived from the oxidation of linoleic acid, has a typical deep-fat-fried flavour (Kinsella, 1969). Oct-1-en-3-ol, produced by copper/ascorbic acid oxidation of butterfat (Stark & Forss, 1964), is the key compound which converts blue-vein-cheese flavour into Camembert flavour (Moinas et al., 1973). Here a mould enzyme system presumably brings about the specific oxidation. When butterfat from oxidized butter with a fishy flavour was heated for 1 h at 80°C with a small amount (0·5 mg/litre) of hydrogen sulphide, the fishy off-flavour disappeared and changed into the pleasant smell of frying (Badings et al., 1975). Pokorny (1976) also found that fishy butter may become quite attractive after treatment with casein. Similar mechanisms may also be responsible for the fact that oxidized butterfat can produce perfectly satisfactory baked goods.

In spite of these few examples of desirable flavours, the main products of the oxidation of unsaturated fatty acids are highly undesirable. Table 6 shows some of the compounds from milkfat which contribute to oxidation flavours (Badings & Neeter, 1980).

Alkanals and alkan-1-ols are produced simultaneously even in the early stages of oxidation of milkfat, the major components being hexanal and pentan-1-ol (Forss, 1972); pentan-1-ol is always present in fresh milk (Urbach & Milne, 1987).

The alkanals range in flavour from green for hexanal (C_6) to tallowy for undecanal (C_{11}). The alk-2-enals range from green to fatty and the alka-2,4-dienals from oily to deep fried. Hex-*cis*-3-enal is reminiscent of green grass and hept-*cis*-4-enal is creamy tending to putty at higher concentrations. The nonadienals smell distinctly of cucumber and nona-*trans*-2,*cis*-6-enal is actually found in cut cucumbers (Forss *et al.*, 1962). Deca-2,4,7-trienal is fishy and beany and oct-1-en-3-one and oct-1-*cis*-5-dien-3-one are metallic. The precursors of the metallic, fishy taint in butterfat are eicosapenta-5,8,11,14,17-enoic acid and docosapenta-7,10,13,16,19-enoic acid (Swoboda & Peers, 1976). Forss (1972) summarized the flavour thresholds of various saturated and unsaturated aldehydes (C_{2-14}) in water, milk, vegetable oil and paraffin oil and a comprehensive list of flavour properties is given by Badings (1970).

Not all aldehydes in butterfat originate from the oxidation of unsaturated fatty acids. Parks *et al.* (1961) found the C_{9-18} *n*-alkanals, several branched aldehydes, the $C_{12,18,20}$ alk-2-enals and traces of three dienals in butter oil as ether derivatives.

Although the $C_{2\&3}$ alkan-1-ols and alkan-2-ol are relatively flavourless,

TABLE 6
Compounds contributing to typical oxidation flavours (Badings & Neeter, 1980) (with permission from the *Netherlands Milk & Dairy Journal*)

Compounds	Flavours
Alkanals C_{6-11}	Green to tallowy
Alk-2-enals C_{6-10}	Green to fatty
Alka-2,4-dienals C_{7-10}	Oily to deep fried
Hex-*cis*-3-enal	Green
Hept-*cis*-4-enal	Cream/putty
Nona-2,4- and 3,6-dienals	Cucumber
Deca-2,4,7-trienal	Fishy, sliced beans
Oct-1-en-3-one	Metallic
Octa-1,*cis*-5-dien-3-one	Metallic
Oct-1-en-3-ol	Mushroom

the flavour becomes stronger as the carbon chain becomes longer, and the $C_{7,9,10}$ alkanols have the same order of threshold as the corresponding alkan-2-ones, i.e. approximately 0·2 ppm in water for $C_{9\&10}$ (Siek et al., 1971). Table 7 lists the flavour thresholds of some alkanols in water and in milk, Table 8 lists the odour thresholds of some unsaturated alcohols in cottonseed oil and Table 9 lists the flavour thresholds of pent-1-en-3-ol and oct-1-en-3-ol in various media.

TABLE 7
Flavour thresholds (ppm) of alkanols in water and milk

Compound	Water[a]	Milk[b]
Ethyl alcohol	200	
Propan-1-ol	45	
Butan-1-ol	7·5	0·5
Pentan-1-ol	4·5	0·5
Hexan-1-ol	2·5	0·5
Heptan-1-ol	0·52	
Nonan-1-ol	0·086	
Decan-1-ol	0·18	
Propan-2-ol	190	
Butan-2-ol	5·1	
Pentan-2-ol	8·5	
Hexan-2-ol	6·7	
Heptan-2-ol	0·41	
Nonan-2-ol	0·28	
Decan-2-ol	0·33	

[a] Siek et al. (1971).
[b] Honkanen et al. (1964).

TABLE 8
Odour thresholds of some alk-1-en-3-ols in cottonseed oil

Compound	Odour threshold	Flavour (2 ppm) description
Hex-1-en-3-ol	0·5	Rubbery, rancid, hydrocarbon
Hept-1-en-3-ol	3·0	
Oct-1-en-3-ol	0·9	Musty, foreign
Non-1-en-3-ol	1·3	

Vinyl ketones have strong unpleasant flavours and are often found among the oxidation products of unsaturated fatty acids. Oct-1-en-3-one is the compound responsible for the metallic flavour in oxidized butter oil, washed cream and safflower oil (Stark & Forss, 1962) and it, as well as pent-1-en-3-one, has been found in autoxidized soybean oil (Hill & Hammond, 1965). Pent-1-en-3-one also contributes to the flavour of copper/ascorbic acid-oxidized buttermilk (Stark et al., 1967).

H_2S has been implicated in cream off-flavours (*vide infra*). Unsaturated compounds often react with H_2S and methane thiol (CH_3SH) to give highly odorous products, e.g. the reaction product of the acetone dimers, mesityl oxide (2-methyl-pent-1 or 2-en-4-one) with H_2S gives the intensely odorous

TABLE 9
Flavour thresholds of pent- and oct-1-en-3-ols in various media

Compound	Water	Skim milk	Butter	Butter oil	Paraffin oil
Pent-1-en-3-ol	3^a	3^a	10^a	10^a	$4\cdot2^b$
Oct-1-en-3-ol	$0\cdot001^c$	$0\cdot01^c$		$0\cdot1^c$	$0\cdot0075^b$

[a] Stark et al. (1967).
[b] Badings (1970).
[c] Stark & Forss (1964).

TABLE 10
Sensory evaluation of reaction products of some unsaturated carbonyl compounds and H_2S or methane thiol (Badings et al., 1975) (with permission from PUDOC—Centre for Agricultural Publishing and Documentation)

Reagents	Flavour
But-2-enal + H_2S	Onion, leak
Hex-2-enal + H_2S	Floral (lantana), rhubarb
Non-2-enal + H_2S	Bast-like, floral
But-2-enal + CH_3SH	Cheese-like (brie)
Hex-2-enal + CH_3SH	Cabbage, rubbery
Hept-2-enal + CH_3SH	Unripe tomato
Non-2-enal + CH_3SH	Bast-like, slightly floral
Oct-1-en-3-one + CH_3SH	Radish-like

2-methyl-2-thio-pentan-4-one which has an odour of cat's urine at 1 part in 10^{10} (Badings, 1967). But-2-enal reacts with H_2S to give a series of compounds whose odour ranges from omelette to onion to asparagus (Badings et al., 1976a). Table 10 gives the results of sensory evaluation of the reaction products of some unsaturated carbonyl compounds and H_2S or methane thiol (Badings et al., 1975).

Esters (methyl, ethyl, propyl and butyl formates; ethyl and propyl acetates and propionates) were identified in copper-oxidized butterfat (Forss et al., 1967b) and (methyl, ethyl, vinyl, propyl, butyl and isoamyl formates; methyl acetate and caproate) in gamma-irradiated butterfat (Merritt et al., 1967). The flavour thresholds of esters range from 0·012 ppm for ethyl hexanoate to 6·6 ppm for ethyl formate (Forss, 1972). Here also the aroma properties of optical isomers differ (Mosandl & Deger, 1987).

The factors which cause the well-known phenomenon of spontaneous oxidation of some milks are not well understood (Bruhn et al., 1976). Neimann-Sorensen et al. (1973) suggested that the tendency of milk to spontaneously oxidize could be reduced by appropriate breeding (Kristensen & Jensen, 1975).

The oxidative stability of butter varies throughout the year. Petrukhina & Piskarev (1975) found that winter butter had low oxidative stability and was unsuitable for prolonged storage at $-18°C$. This lack of oxidative stability is likely to be due to change of diet. When cows are fed safflower oil protected against biohydrogenation in the rumen, the resulting butter is much more susceptible to oxidation than conventional butter (Goering et al., 1976) and the flavour is more bland (Badings et al., 1976b). However, when cream and sunflower or soyabean oil were churned to contain one of these oils as 20% of the fat phase, both oils produced blends of satisfactory flavour stability if the initial peroxide value of the oil was less than 1·0 (Black, 1976). Supplementation of cows' diet with α-tocopherol acetate or the direct addition of α-tocopherol to the milk effectively prevented development of oxidized off-flavours (Goering et al., 1976) in the butterfat from cows fed on protected polyunsaturated oil.

Light induces oxidation of butter oil under conditions of time (more than 2 weeks) and temperature (5–10°C) that would occur in retail stores (Harwalkar et al., 1974).

It has been claimed that the addition of 1–6% cyclodextrin masks the bad flavour of butter or butter oil (Lotte KK, 1982).

OFF-FLAVOURS FROM IRRADIATED FATS

Ionizing radiation is recognized as the most effective method for eliminating the bacteria that cause most kinds of food poisoning (Wills, 1986).

However, in some foods it causes off-flavours, e.g. gamma-ray pasteurization of milk gives the milk a burnt taste (Carrard, 1983). Gamma-irradiation of milkfat under reduced pressure produced monocarbonyls at a rate proportional to the irradiation dose. The flavour of the irradiated fat had three components: hydrolytic rancidity, oxidized and candle-like; the candle-like flavour was due to long-chain aldehydes (Day & Papaioannou, 1963). When anhydrous butterfat was irradiated at 6 Mrad, carbon dioxide was produced in the greatest amount. Of the remaining compounds, aliphatic hydrocarbons were predominant both in number and amount (C_{1-13} n-alkanes, C_{4-9} 2-methylalkanes, C_{2-14} alk-1-enes, C_{2-9} alk-1-ynes) (Merritt et al., 1967).

Ivanov & Stamatov (1975) investigated the effects of γ-irradiation with 10^4, 10^6 and 10^7 rad at 18°C on butter, sunflower oil and lard, and the after-effects of this irradiation during 16 and 32 days' storage in the dark at 25°C. Low level irradiation had only a slight effect on oxidation, medium and high level irradiation increased all types of change, especially oxidation. All medium and high level irradiated fats and low level irradiated stored fats were unpalatable. Butter was unpalatable at a relatively lower level of oxidation than the other fats, possibly due to the production of low molecular weight fatty acids.

EFFECT OF ANTI-OXIDANTS

Wyatt & Day (1965) found that various anti-oxidants were effective in stabilizing butter oil stored for 12 months at 30°C, providing the oil was first steam deodorized (200°C at 15 mmHg for 3 h) (Patton, 1964b). If the oil was not deodorized a flavour defect, described as stale or lactone, developed but chemical indices showed that oxidation was still controlled. In the deodorized butter oil, non-fat milk solids proved to be the most effective for retaining flavour stability.

Studies on the effect of incorporating Maillard reaction products from casein and glucose on the oxidative stability of full-cream milk powders have indicated that Maillard reaction products can retard fat oxidation and increase shelf life of these powders (McGookin, 1991; McGookin and Augustin, 1991).

The stability of ghee can probably be attributed to the effect of heated non-fat milk solids.

FLAVOURS NOT FORMED DIRECTLY FROM THE FAT

Milkfat and other oils and fats also contain many flavours by virtue of their solvent properties. Flavours may have been transferred from the feed to the milk; they can be the result of the cows' metabolism, or they can be the result of minor conversions of milk constituents by chemical (oxidative, thermal, etc.), microbial or enzymatic reactions. However, it must be remembered that the cow, being a ruminant, modifies her feed more than a non-ruminant and is therefore less directly affected by diet.

The tendency of undisturbed milk to absorb odour has been reported (Hess & Lutz, 1972) as being directly proportional to the fat content of the milk, the concentration of odour in the air above the milk, the air temperature and the time of contact. Stirring increased the rate of odour uptake 50–90-fold. Removal of odour from the tainted milk by degassing was practically impossible.

In Australia, butter from King Island in Bass Staight is highly prized for its flavour. King Island has lush pastures practically all the year round and this is reflected in the butter by a high content of phytol and phytanol (Urbach, unpublished data; Flanagan et al., 1975).

Consumers tend to regard the butter which they normally buy as 'good' and any variant as 'off'. In Australia and New Zealand butter tends to have a strong yellow colour because of the high concentration of carotene, the result of pasture feeding. In the Middle East such butter is suspect as being oxidized. The Japanese dislike what they term 'grassy, animally' notes in Australian and New Zealand butter and also object to the yellow colour (Keen, 1986).

In a monumental work, Badings isolated 114 compounds with flavour significance in low-temperature-pasteurized milk (Badings & Neeter, 1980). These are listed in Table 11. However, when milk is subjected to ultra-high-temperature treatment (142°C for 4·6 s), 40 of these compounds are lost and 39 new flavour compounds are formed. Most of these compounds will be more soluble in the fat than in the aqueous phase; therefore the extent of heat treatment which the milk has had before the milkfat is separated will have a great influence on the flavour of the milkfat.

Although most of the flavours and flavour precursors of milkfat originate via the biosynthetic processes of the cow, the diet of the cow can have a strong effect on flavour. Thus Dumont & Adda (1978) found sesquiterpenes in mountain cheese from summer milk from cows grazing on high altitude pastures. Butterfat from cows fed on clover had a stronger flavour than that from cows fed on rye grass (Forss, 1979). Wilson (1989) isolated six terpenes from New Zealand milkfat but not from Finnish

TABLE 11
Compounds isolated from milk (Badings & Neeter, 1980) (with permission from the *Netherlands Milk & Dairy Journal*)

Acetaldehyde	1,2-Dimethylnaphthalene	δ-Tridecalactone
Diacetyl	1,3-Dimethylnaphthalene	Dimethyl sulphide
Isobutanol	δ-Decalactone	Ethyl acetate
2-Methylbutanal	Ionol	But-2-enal
Pentanal	C_3-naphthalene	Butan-1-ol
Pentan-3-one	δ-Dodecalactone	Pyrazine
Methyl butyrate	δ-Tetradecalactone	N-Methylpyrrole
2-Methylbut-2-enal	Methane thiol	4-Methylpentan-2-one
Dimethyl disulphide	Butan-2-ol	2-Methylbutan-1-ol
Pyridine	3-Methylbutanal	Toluene
Pent-4-enonitrile	Pentan-2-one	Hexan-2-one
3-Methylpentan-3-ol	i-Butyl mercaptan	Pentan-1-ol
Methyl thiophene	Pentan-3-ol	Hexan-2-ol
Pyrrole	Methyl isothiocyanate	Hex-2^t-enal
3-Methylbut-2-en-1-ol	1,1-Diethoxyethane	Cyclohexanone
Hexanal	Hexan-3-one	2,4-Dithiapentane
2-Methylpyrazine	Ethyl isothiocyanate	Styrene
2-Methylpyrrole	Ethyl butyrate	Furylmethylketone
Allyl isothiocyanate	Furfural	Ethyl benzoate
Hexan-1-ol	Ethylbenzene	2,3,4-Trithiapentane
Hept-4^c-enal	Heptan-2-one	Oct-1-en-3-one
Methyl hexanoate	Heptanal	Ethyl hexanoate
But-3-enyl isothiocyanate	Benzaldehyde	Indene
Cyanobenzene	Heptan-1-ol	Acetophenone
Trimethylbenzene	Octanal	Nonan-2-one
p-Cymene	Phenol	Nonanal
2-Ethylhexan-1-ol	Benzyl alcohol	Benzyl cyanide
Octan-1-ol	δ-Hexalactone	Nona-2^t,6^c-dienal
Methyl benzoate	α-Terpinolene	Nonan-1-ol
p-Cresol	m-Cresol	Naphthalene
Propyl pyridine	Non-2^t-enal	Decanal
3-Methylindene	Methylindene	Iso-quinoline
Dimethylethylbenzene	Decan-2-one	δ-Octalactone
Ethyl octanoate	Nona-2^t,4^t-dienal	Indole
Tetraethylurea	γ-Octalactone	1-Methylnaphthalene
Benzthiazole	Undecan-2-one	δ-Nonalactone
2,3,5-Trimethylanisole	γ-Nonalactone	Ethyl decanoate
Decan-1-ol	Skatole	Tetraethylurea
2-Methylnaphthalene	Ethylnaphthalene	γ-Decalactone
Methyl decanoate	Tridecan-2-one	Acenaphthene
Biphenyl	Dimethylbiphenyl	γ-Dodecalactone
Dodecan-2-one	δ-Undecalactone	

milkfat and showed that 1 ppm of D-limonene was responsible for the green/grassy flavour present in New Zealand milkfat at certain times of the year.

Butterfat contains a series of hydrocarbons, the most abundant in the C-20 region being phyt-1-ene, phyt-2-ene, neophytadiene and some of its isomers (Urbach & Stark, 1975). Although it is often assumed that aliphatic hydrocarbons, particularly relatively long-chain ones, are odourless, the solution containing the isolated hydrocarbons had quite a strong odour and it is possible that hydrocarbons, particularly branched-chain unsaturated hydrocarbons, do make a contribution to butter flavour.

Unfortunately, many of the flavours originating from feed are undesirable. Thus, benzyl nitrile and benzyl isothiocyanate are caused by Cruciferae (Walker & Gray, 1970), and peppercress (*Lepidium*) interferes with the tryptophan metabolism and causes an excessive build up of indole and skatole which impart a faecal odour (Conochie, 1953), although in lower quantities they contribute to the desirable flavour of butter (Urbach *et al.*, 1972). Honkanen *et al.* (1970) found that when cows were fed on a purely synthetic diet (zero milk) the milk contained less indole and skatole than the milk from conventionally fed cows. Land cress causes a scorched flavour which is intensified by heat. Compounds such as benzyl isothiocyanate, benzyl methyl sulphide and benzyl cyanide occur in the milk of cows which have ingested land cress and of these compounds benzyl methyl sulphide and possibly benzyl mercaptan impart a typical burnt, land cress flavour to milk (Park *et al.*, 1969).

The phenolic compounds in butterfat are also desirable near threshold level but cause phenolic off-flavours at higher concentrations. They are thought to be the breakdown products of lignin (Stark *et al.*, 1976*b*) although *para*substituted phenols could arise from the breakdown of tyrosine. Ha and Lindsay (1991) found that Pyrenees sheep's milk cheese contained significant amounts of methyl- and ethyl-substituted phenols, which contributed characterizing sheep-like flavour notes to this cheese variety. Diacetyl, which is desirable in small quantities even in sweet-cream butter, results from the action of lactic bacteria (Marth, 1974).

Dimethyl sulphide, which has been isolated from both sweet-cream and cultured-cream butter, has an optimum flavour level of 40 ppb (40 in 10^9) in butter oil and is claimed, at that level, to smooth the flavour of cultured-cream butter. At 200 ppb it is definitely feedy (Day *et al.*, 1964).

COOKED (H_2S) FLAVOUR

High-temperature pasteurization of cream is useful in producing butter with a reduced susceptibility to autoxidation and improved keeping quality during storage. However, if too high a temperature is used during cream pasteurization, excessive amounts of H_2S are formed; this causes a cooked flavour in the butter. The conditions of pasteurization must therefore be carefully balanced between excess H_2S and lack of stability in the butter. Badings & van der Pol (1967) developed a simple method for measuring H_2S in cream which makes use of the length of a black zone on a Draeger tube (Drager–Rohrchen H_2S 1/b, Fabrikant Dragerwerk, Lubeck). The length of the zone also gives a measure of the SH-group content. Badings (1971) found that several factors besides the actual temperature of pasteurization also affected the level of H_2S in the butter. Winter butter had less H_2S than summer butter; early lactation produced less H_2S than late lactation; if the milk was kept cold before cream separation, this also reduced the level of H_2S; if the pasteurized cream contained less than 30% fat, the butter contained less H_2S than if the level of fat in the cream was higher; culturing the cream lowered the level of H_2S and some starters were more effective than others. Badings (1971) concluded that the H_2S originated from the protein in the fat globule membrane. In Australia the temperature of cream pasteurization can range from 80°C to over 100°C (under pressure) depending on the equipment in the particular factory. Off-flavours due to H_2S are avoided by vacreation.

NEUTRAL VOLATILES

Siek & Lindsay (1968) vacuum steam-distilled butter oil, fresh raw cream, fresh pasteurized cream and pasteurized stored cream (7 days at 3°C). Distillation at 210°C of different butter oils yielded over 120 volatile compounds of which more than 100 were identified. Many of the volatile compounds were obviously heat-produced as they were either absent from or present in small quantities in fresh raw cream compared to heated cream and butter oil. Aromatic compounds and aliphatic hydrocarbons were found in fresh raw cream but not in controls of laboratory contaminants and distillation artefacts. All substrates studies contained the odd C-numbered methyl ketones from C_3 to C_{15}, n-hexanal, butanone, ethanol, ethyl formate, ethyl acetate, chloroform, δ-$C_{8,10,12}$ lactones, acetaldehyde, benzene and toluene. Low levels of $C_{5,7,8,9,11}$-alkanals as well as hex-2-enal and hept-2-enal were present in some fractions. Storage allowed such

TABLE 12
Concentration of volatiles found in fresh butter (ppm) (Siek & Lindsay, 1970) (with permission from the *Journal of Dairy Science*)

Compound	Found	Compound	Found
Acetone	0·13	Butanone	0·16
Pentan-2-one	0·14	Hexan-2-one	0·004
Heptan-2-one	0·35	Octan-2-one	0·013
Nonan-2-one	0·21	Decan-2-one	0·08
Undecan-2-one	0·24	Tridecan-2-one	0·35
Acetaldehyde	0·45	n-Propanal	0·07
n-Butanal	0·06	n-Pentanal	0·04
n-Hexanal	0·20	n-Heptanal	0·03
n-Decanal	0·01	n-Undecanal	0·08
Diacetyl	0·10	δ-Octalactone	0·40
δ-Decalactone	2·0	δ-Undecalactone	0·10
δ-Dodecalactone	5·0	δ-Tetradecalactone	4·0
γ-Undecalactone	0·05	γ-Decalactone	0·10
γ-Dodecalactone	0·3	Ethyl formate	0·99
Ethyl acetate	2·0	Ethyl butyrate	0·01
Ethyl hexanoate	0·02	Ethyl octanoate	0·03
Methyl acetate	0·01	Ethanol	0·53
Chloroform	0·02	n-Octane	0·006
n-Nonane	0·006	n-Decane	0·006
Benzene	0·0005	Toluene	0·011
Ethylbenzene	0·007	p-Xylene	0·002
m-Xylene	0·004	o-Xylene	0·003
p-Cymene	0·0006	Methylethylbenzene	0·005
Benzaldehyde	0·002	p-Dichlorobenzene	0·005
o-Dichlorobenzene	0·05		

organisms as *Pseudomonas* to reduce aldehydes to the corresponding alcohols (Keenan *et al.*, 1967) producing propan-1-ol, butan-1-ol and hexan-1-ol in pasteurized sweet cream which was stored for 7 days at 3°C. Odd C-number methyl ketones, lactones, dimethyl sulphide, butanone and hydrogen sulphide increased during pasteurization of cream. Heating butter oil produced $C_{6,8,10,12}$-alkan-2-ones. Ethyl esters of butanoate, hexanoate, octanoate, decanoate and dodecanoate were more prominent in heated milk or in heated butter oil which contained some butter serum. 2-Furfural and 5-methyl-2-furfural were detected when cream or butter oil containing some butter serum was heated above 140°C. Diethyl acetals were present either in substrates which were aged or in butter oil substrates distilled in the presence of butter serum, but not in fresh butter oil or fresh cream. Ethanol increased drastically in aged or decomposed cream. The results of a quantitative analysis of volatiles from fresh sweet-cream butter are given in Table 12 (Siek & Lindsay, 1970).

FLAVOURS OF BUTTERFAT FRACTIONS

When milkfat was crystallized from acetone at $-13°C$, the low-melting fraction (about 30% of the milkfat) had a lactone potential (i.e. total lactones produced by heating the milkfat for 16 h at 120°C) 1·5–2 times greater than the original milkfat (Walker & Keen, 1976; Walker et al., 1977), i.e. the low-melting fraction would be expected to be of the order of twice as flavoursome as the original milkfat. During the commercial preparation of anhydrous milkfat, up to 10% of the lactone potential and 20% of the methyl ketone potential of freshly-secreted milkfat was lost. When milkfat (softening point 33°C) was divided into high-melting (softening point 45°C) and low-melting (softening point 25°C) fractions, the concentrations of lactone and methyl ketone precursors in the low-melting fractions were somewhat higher than those in the anhydrous milkfats. High-melting fractions contained only 50–70% of the lactone and methyl ketone potential of the low-melting fractions and had reduced flavour intensity. Aldehydes derived from oxidation were somewhat concentrated in the low-melting fraction and hence only good quality milkfat with no detectable off-flavour should be regarded as suitable for fractionation. When Cupakova et al. (1981) separated milkfat into fractions which were either solid or liquid at 15°C, the greatest quantity of carbonyls (C_{1-5}) was in the fraction liquid at 15°C. This fraction had the strongest odour. The high-melting fraction of anhydrous milkfat has a possible use in the preparation of pastries but, where the full flavour potential of milkfat is required, the low-melting fraction is more suitable (Walker, 1972). The high-melting fraction is of no advantage in chocolate because it forms a eutectic with cocoafat; this eutectic has a lower melting point than either of the component fats (Timms, 1980; Timms & Parekh, 1980). A concentrated butterfat aroma product obtained from butterfat fractions has been marketed (Andres, 1983; Petersen, 1986).

SUPERCRITICAL EXTRACTION OF BUTTER FLAVOUR

Schaap (Private communication) suggested that the supercritical extraction of milkfat with carbon dioxide could be used to extract 5% of the flavour, yielding a valuable product without removing enough flavour from the milkfat to make any significant difference to its flavour. Schaap *et al.* (1986) found that free fatty acids can be extracted from milkfat with supercritical carbon dioxide and that components reacting with thiobarbituric acid but not peroxides are concentrated in the extract. However, the storage stability of the residue was no better than that of fresh milkfat.

BUTTER POWDER

A dry powder containing 80% butterfat may be manufactured by spray drying the butterfat with a suitable mixture of milk constituents and other ingredients. The product has use in the food industry wherever butterfat must be blended with other dry ingredients. Its major use is as a shortening in baked goods (Hansen, 1963a, b; Snow et al., 1967a, b; Townsend et al., 1968). The flavour of butter powder as distinct from butter does not appear to have been investigated.

GHEE

Ghee is the usual Indian name for clarified butterfat usually from cow or buffalo milk. Middle-Eastern countries refer to it either as *maslee* or by some variant of the Arabic term *samn*. In these areas milk from goats, sheep and camels is also used, sheep milk being the most common source in Iran where it is called *roghan*. Until recently ghee was exclusively prepared by fermenting whole milk to curd, churning the curd to butter and boiling down the latter to give ghee. The degree of heating used depends on the local taste. Where a slightly burnt flavour is desired, the final temperature may be as high as 150°C; if a raw, buttery flavour is to be retained, the butter is heated just to boiling; but, in general, 115°C is regarded as the optimum. A considerable part of the aroma, flavour, colour and texture of ghee is developed during clarification. For maximum stability development on heating, it is necessary for phospholipid, lactose, casein and albumin to be present all together. Ghee must be made from soured cream, otherwise it is regarded as flat and tasteless. A pleasant, nutty, lightly-cooked or caramelized aroma is generally prized. From its method of manufacture, it is to be expected that ghee contains all the compounds discussed above which result from the action of heat on butterfat as well as the products of the Maillard reaction between the butterfat and the constituents of the butter serum. In practice, home-made ghee is also likely to contain products from the oxidation of unsaturated fatty acids as well as microbial metabolites (Rangappa & Achaya, 1973).

VOLOGDA BUTTER

Vologda butter is an unsalted butter made in the former USSR from sweet cream which is pasteurized at 95–98°C for not less than 15 min, thus producing a butter with a stronger flavour than butter used in Western Europe (Ramonas & Berzhinskas, 1970). In an investigation into the origins of the characteristic pasteurized flavour of Vologda butter, Vyshemirskii & Vasilisin (1972) found that free amino acids, lactose, inorganic salts, carbonyl compounds and volatile fatty acids all participated to some extent in the formation of the flavour. The most intensive flavour was achieved in the presence of maximum quantities of SH-groups. Kotova et al. (1973) found that butter from Ayrshire cows was paler and had an insufficiently pronounced taste to be accepted as Vologda butter when compared with butter from other breeds (Russian Black Pied, Kholmogor, Yaroslavl).

HEATED BUTTER

In an unpublished experiment, Urbach heated butter under vacuum to remove the water. The residual dry butter was then heated at 140–160°C for 0·75 h producing a delicious odour of butter biscuits. The odour could be removed from the heated butter with water and was shown to contain acids and esters. These polar carbonyl compounds were not identified but it is likely that they contribute to the flavour of both ghee and Vologda butter.

Sulser & Büchi (1974) produced the flavour of browned butter by heating dried buttermilk, containing 3–5% lactic acid, to 145–155°C and workers at Unilever NV (1968) found that, if 5–10% milk with a cooked flavour (produced by heating the milk above 116°C) was added to margarine, it produced, without the addition of further flavour compounds, a particularly mild creamy flavour which was in no way reminiscent of the cooked flavour of the milk.

EFFECT OF STORAGE ON BUTTER FLAVOUR

Danish workers showed that reducing the storage temperature of butter to $-30°C$, but not beyond, improved the keeping quality of butter. Salted sweet-cream butter was best suited for long storage whereas ripened-cream butter, especially if salted, was less suitable for storage. Milder ripening (to pH 5·2) did not impair the special consumer appeal of Danish butter but improved its keeping quality (Pedersen, 1977). Petrukhina & Piskarev (1976) also found that at $-30°C$ butter retained its flavour constituents and there was a lower accumulation of secondary oxidation products than at higher temperatures.

PART 2: FLAVOUR OF FATS OTHER THAN MILKFAT
(M. H. Gordon)

INTRODUCTION

In contrast to milkfat, most other fats are not valued for their flavour, and manufacturers include a deodorization stage during refining in order to convert most fats to bland tasting products. The challenge for the manufacturer is to maintain the product free of off-flavours during the shelf-life of the product. However, a few fats are prized for their flavour and are frequently sold without refining. The main fats in which the flavour is an important quality attribute are cocoa butter, olive oil and lard.

COCOA BUTTER

Although the flavour of chocolate is mainly dependent on the cocoa powder, sugar and dairy ingredients, cocoa butter contributes an attractive cocoa flavour to white chocolate. Many of the flavour components present in cocoa powder are fat soluble and the cocoa butter therefore has a similar, although weaker and less bitter, flavour to cocoa powder.

Cocoa butter is derived from the cocoa bean by a number of processing steps. The beans are fermented, dried and roasted and these steps are critical in the generation of flavour compounds (Rohan, 1969). After roasting the beans are dehulled and the nib is ground to give cocoa liquor (also known as cocoa mass). Cocoa liquor contains about 55% cocoa butter and 45% cocoa powder, and the bulk of the cocoa butter can be separated from the cocoa powder by pressing with only 10–12% fat remaining in the cocoa powder.

Fermentation results in the production of many flavour precursors including peptides, amino acids, reducing sugars and others. The amino acids and peptides formed during the anaerobic phase of the fermentation take part in non-oxidative carbonyl–amine condensation reactions which occur both during fermentation and later heating phases such as drying, roasting and grinding (Hoskin & Dimick, 1988). The second phase of fermentation is aerobic and the oxidative reactions are important for reducing the astringency and bitterness of the cocoa mass (Zak & Keeney, 1976).

Roasting at approx. 140°C for about 30 min converts the flavour of the beans from an astringent, bitter, acidic, unclean, nutty flavour to the typical intense flavour of cocoa (Maniere & Dimick, 1979). Undesirable low-boiling compounds are driven off during the early stages of roasting, and subsequently water and astringent organic volatile components including short chain fatty acids are evaporated. Flavour precursors including amino acids, sugars, sugar breakdown products and polyphenols react by a variety of mechanisms to produce the typical cocoa flavour. The main reactions leading to flavour formation are described as Maillard or non-enzymic browning reactions, because they also lead to colour formation.

Nursten (1980) classified volatile Maillard reaction products into three groups, namely:

1. 'Simple' sugar dehydration/fragmentation products: furans, pyrones, cyclopentenes, carbonyl compounds, acids.
2. 'Simple' amino acid degradation products: aldehydes, sulphur compounds.
3. Volatiles produced by further reaction: pyrroles, pyridines, imidazoles, pyrazines, oxazoles, thiazoles and compounds from aldol condensations.

TABLE 13
Number of aroma compounds in cocoa classified by structure (Ziegleder, 1983)

	Aliphatic	Aromatic	Heterocyclic
Hydrocarbons	15	32	10 Pyrrole
Alcohols	23	5	8 Pyridine
Aldehydes	18	6	1 Quinoline
Ketones	25	5	74 Pyrazine
Esters	44	12	3 Quinoxaline
Ethers	8	3	3 Thiazole
Nitrogen compounds	9	4	4 Oxazole
Sulphur compounds	12	2	19 Furan
Acids	22	15	4 Pyrone
Phenols	—	7	6 γ-Lactone

Ziegleder (1983) reported 399 compounds in cocoa flavour which were categorized in 20 classes as shown in Table 13. Further volatile components were detected by Gill et al. (1984), who identified 13 new compounds, and Ney (1985), who identified 14 keto-acids. Thus the volatile constituents of cocoa liquor number well over 400 and it is expected that most of these components are present in the fat phase.

OLIVE OIL

Virgin olive oil has a very attractive flavour and is widely used in Europe in salad dressings and other applications despite its high price compared with other liquid oils. Virgin olive oil is defined by the International Olive Oil Council as the oil obtained from the fruit of the olive tree solely by mechanical or other physical means under conditions, particularly thermal conditions, that do not lead to alterations in the oil, and which has not undergone any treatment other than washing, decantation, centrifugation and filtration. Thus all fat-soluble flavour components are retained during isolation of the oil.

The International Olive Oil Council uses three groups of descriptors in panel assessments of virgin olive oil. Group I comprises only olive fruity which is the most desired flavour characteristic of virgin olive oil. Group II includes descriptors that define a variety of sensory notes which when encountered are considered positive in the overall flavour, particularly if they are detected along with the basic olive-fruity note. These descriptors comprise apple, green bitter, pungent, sweet and other ripe fruit. Group III comprises attributes that are tolerable only at low levels and attributes that are undesirable at all levels. These include rough, metallic, winy, 'atrojado' (fusty) and rancid.

The organoleptic characteristics of olive oil depend both on the phenolic and the volatile constituents (Vasquez Roncero, 1978). The non-volatile components are mainly responsible for the bitter flavour note, although some non-volatile components also contribute a hot flavour (Gutierrez et al., 1989). The individual phenolic compounds present in virgin olive oil have been studied by several authors and are shown in Table 14. The total phenolic content is typically about 40–160 mg kg^{-1} oil (Montedoro et al., 1978; Gutfinger, 1981).

There have been many studies of the volatile components of olive oil (Gutierrez et al., 1972, 1975; Flath et al., 1973; Olias et al., 1974, 1977, 1978, 1980; Fedeli, et al., 1976; Fedeli, 1977; Montedoro et al., 1978). Olias et al.

TABLE 14
Phenolic compounds in virgin olive oil

(3-Hydroxyphenyl) ethanol	Vanillic acid
B (3,4-Hydroxyphenyl) ethanol	Homovanillic acid
Oleuropein	Gentisic acid
Oleuropeilaglycone	Quinic acid
Coumaric acids	Shikimic acid
Caffeic acid	p-Hydroxybenzoic acid
Hydrocaffeic acid	Syringic acid
Sinapic acid	Apigenin
Protocatechuic acid	Luteolin

Data from Montedoro & Cantarelli, 1969; Ragazzi & Veronese, 1973; Solinas et al., 1975; Vasquez Roncero (1978).

(1980) categorized the volatile components into nine groups, namely saturated hydrocarbons, unsaturated hydrocarbons, aromatic hydrocarbons, alcohols, aldehydes, ketones, esters, ethers and unidentified. The volatile components of olive oil which have been identified are shown in Table 15. The aldehydes are present in the highest concentration (40–80%) followed by the alcohols (7–25%), esters (3–10%) and hydrocarbons (0·1–5·2%) (Montedoro et al., 1978).

Of the compounds identified, only ethylbenzene and hexyl acetate had olive in the odour description given by Montedoro et al. (1978). These compounds were only present at low concentrations and it is clear that the aroma of olive oil is complex and dependent on subtle interactions between the odour of many volatile components. The octane level correlated with the 'atrojado' (stored or fusty) flavour generated as the oil deteriorated during storage (Barrio et al., 1981).

OTHER FATS

The flavour of other fats has not been studied in such detail. However, the flavour of lard has been found to be due to a mixture of aldehydes and nitrogen compounds with a mixture of 2,4-nonadienal; 2,4-decadienal; 2,5-dimethyl-3-ethylpyrazine, tetramethylpyrazine and pyroligneous acid representing a useful flavour concentrate (Burke, 1973).

The flavour of groundnut oil has been shown to depend on the nut roasting temperature used. A roasting temperature of 185°C produced the

TABLE 15
Volatile components of virgin olive oil (Montedoro et al., 1978)

Hydrocarbons	Aldehydes	Furan derivatives
Isopentane	Acetaldehyde	2-Propyl furan
2-Methyl pentane	Propanal	2-*n* Propyl dihydrofuran
Hexane	2-Methyl-butanal	2-*n* Pentyl-3-methyl furan
Octane	3-Methyl-butanal	
Nonane	Butanal	Thiophene derivatives
Naphthalene	Pentanal	2-Isopropenyl thiophene
Ethyl-naphthalene	*cis*-2-Pentenal	2-Ethyl-5-hexylthiophene
Acenaphthene	*trans*-2-Pentenal	2,5-Diethylthiophene
Aromatic hydrocarbon	Hexanal	2-Ethyl-5-hexyldihydrothiophene
Ethyl benzene	*cis*-2-Hexenal	2-Ethyl-5-methyldihydrothiophene
	trans-2-Hexenal	2-Octyl-5-methylthiophene
	Heptanal	
Aliphatic alcohols	2,4-Hexadienal	Esters
Methanol	*cis*-2-Heptenal	Methyl acetate
Ethanol	*trans*-2-Heptenal	Ethyl acetate
Isopropyl alcohol	Benzaldehyde	Ethyl propionate
1-Pentanol	Octanal	Methyl butyrate
3-Methylbutan-1-ol	2,4-Heptadienal (isomer A)	Ethyl-2-methyl propionate
1-Penten-3-ol	*trans*-2-Octenal	2-Methyl-1-propyl acetate
cis-3-Hexen-1-ol	Nonanal	Methyl-3-methyl butyrate
trans-2-Hexen-1-ol	*trans*-2-Nonenal	Methyl-2-methyl butyrate
1-Hexanol	2,4-Nonadienal	Ethyl butyrate
1-Heptanol	*trans*-2-Decenal	Propyl propionate
1-Octanol	2,4-Decadienal (isomer A)	Methyl pentanoate
1-Nonanol	2,4-Decadienal (isomer B)	Ethyl-2-methyl butyrate
2-Phenylethanol	*trans*-2-Undecenal	

TABLE 15 (cont.)

Oxygenated terpenes
  1,8-Cineole
  Linalol
  Terpineol
  Lovandulol

Ethers
  Methoxybenzene
  1,2-Dimethoxybenzene

Ketones
  Acetone
  3-Methylbutan-2-one
  3-Pentanone
  2-Hexanone
  4-Methyl-3-penten-2-one
  2-Methyl-2-hepten-6-one
  2-Octanone
  2-Nonanone
  Acetophenone

Esters (*cont.*)
  Ethyl-3-methyl butyrate
  1-Propyl-2-methyl propionate
  3-Methyl-1-butyl acetate
  2-Methyl-1-butyl acetate
  2-Methyl-1-propyl-2-methyl propionate
  Methyl hexanoate
  Cis-3-hexenyl acetate
  Hexyl acetate
  Methyl heptanoate
  Methyl octanoate

best flavour. Roasting at 190° or 200°C produced a bitter flavour in the oil, whereas roasting at 165°, 175° and 180° produced an inferior flavour (Huang et al., 1988).

OIL DEGRADATION PRODUCTS AND FLAVOUR

Edible oils consist mainly of triglycerides which are usually present at levels in excess of 95% after refining. Freshly-refined edible oils are flavourless but oxidation and hydrolysis of triglycerides may lead to degradation products. As mentioned earlier, fat oxidation or hydrolysis products may produce desirable flavours in some dairy products, but for most edible oils the flavours that are formed are perceived as off-flavours.

AUTOXIDATION MECHANISM

Edible fats that are in contact with air develop off-flavours at a rate that depends mainly on the fatty acid composition, the concentration of pro- and anti-oxidants and the temperature. The rate of deterioration is not constant. An induction period occurs during which very little deterioration of the fat can be detected, but after this period the fat deteriorates rapidly. This behaviour is characteristic of a free radical chain reaction and the mechanism shown in Fig. 6 is now accepted. The reaction consists of primary initiation reactions during which free radicals are formed from

Initiation
$ROOH^* \rightleftharpoons ROO\cdot + H\cdot$
$ROOH \rightleftharpoons RO\cdot + \cdot OH$
$2ROOH \rightleftharpoons RO\cdot + H_2O + ROO\cdot$

Propagation
$R\cdot + O_2 \rightleftharpoons ROO\cdot$
$ROO\cdot + R'H \rightleftharpoons ROOH + R'\cdot$

Termination
$ROO\cdot + R'OO\cdot \rightleftharpoons ROOR' + O_2$
$RO\cdot + R'\cdot \rightleftharpoons ROR'$

* Formed by various pathways including reaction of $^1O_2$ with unsaturated lipids or lipoxygenase-catalysed oxidation of polyunsaturated fatty acids.

Fig. 6. Mechanism of autoxidation.

lipid molecules; propagation reactions in which one lipid radical is converted to a different radical; and termination reactions in which two radicals combine to form stable end products. The hydroperoxides which are formed decompose much more readily than other lipid molecules, and therefore when the hydroperoxide concentration builds up the secondary initiation reactions become a major source of new free radicals. Hence, the reaction becomes autocatalytic and the rate increases with time.

The formation of a free radical from a lipid molecule is fastest for a polyunsaturated fatty acid since the bond dissociation energy of the C—H bond on the intervening methylene group between two carbon–carbon double bonds is lower than that for a C—H bond on a methylene group next to one carbon–carbon double bond. Saturated fatty acids are most stable because the bond dissociation energy for a C—H bond in a methylene group in a saturated fatty acid is very high. The relative rates of oxidation of methyl oleate, methyl linoleate and methyl linolenate were found to be $1:12:25$ (Lea, 1952).

Formation of Volatile Products by Autoxidation

Triglycerides are highly involatile and therefore do not contribute to oil flavour. However, radicals derived from hydroperoxides readily cleave producing volatile components as shown for linoleic acid in Fig. 7. The two hydroperoxides produced from linoleic acid initially are the 9- and 13-hydroperoxides in which the carbon–carbon double bonds are conjugated. The alkoxy radicals derived from these hydroperoxides cleave either side of the C—O bond leading to low molecular weight fragments which have been detached from the glycerol backbone. Diperoxides and other more highly oxidized peroxides are formed in the later stages of lipid oxidation and these compounds contribute to the complex mixture of volatiles that are formed.

In the case of oleic acid, the hydroperoxides formed are the 8-, 9-, 10- and 11-hydroperoxides while for linolenic acid eight isomers of *cis*, *trans*- and *trans*-, *trans*-conjugated trienes are formed with the hydroperoxy group at the 9, 12, 13 and 16 positions. Some of the volatiles formed from these fatty acids are shown in Table 16. Aldehydes, ketones, hydrocarbons, oxo esters, alcohols, esters and furans are common lipid oxidation product classes. The threshold values of a large number of volatile compounds formed by lipid oxidation have been determined (Forss, 1973). Some of these have been reported earlier in this chapter. A wide variety of flavour descriptions have been reported for these compounds. Some compounds with particularly low flavour thresholds are shown in Table 17.

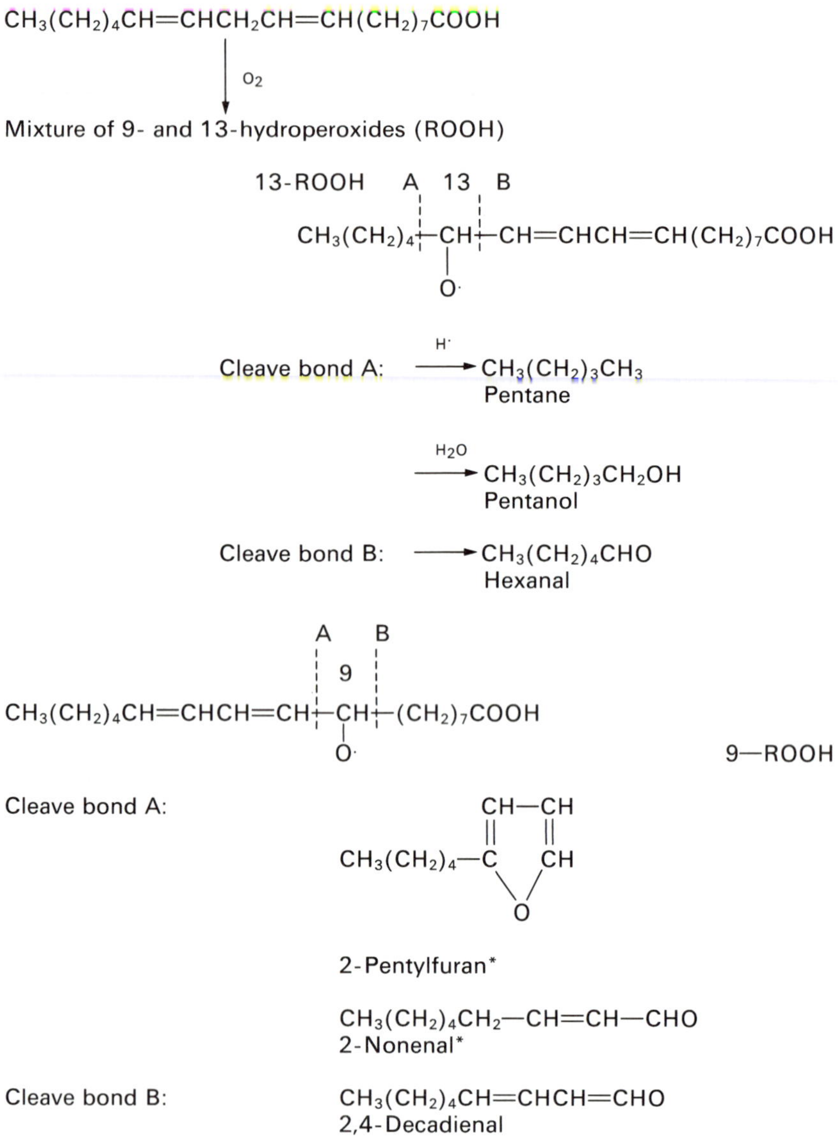

$CH_3(CH_2)_4CH=CHCH_2CH=CH(CH_2)_7COOH$

$\downarrow O_2$

Mixture of 9- and 13-hydroperoxides (ROOH)

13-ROOH   A   13   B

$CH_3(CH_2)_4\!\!-\!\!CH\!\!-\!\!CH=CHCH=CH(CH_2)_7COOH$
$\quad\quad\quad\quad\quad\quad\quad\;\;|$
$\quad\quad\quad\quad\quad\quad\quad\;\;O\cdot$

Cleave bond A: $\xrightarrow{H\cdot}$ $CH_3(CH_2)_3CH_3$
Pentane

$\xrightarrow{H_2O}$ $CH_3(CH_2)_3CH_2OH$
Pentanol

Cleave bond B: $\longrightarrow$ $CH_3(CH_2)_4CHO$
Hexanal

$\quad\quad\quad\quad\quad\quad\quad\;\;A\;\;\;B$
$\quad\quad\quad\quad\quad\quad\quad\;\;|\;9\;|$
$CH_3(CH_2)_4CH=CHCH=CH\!\!-\!\!CH\!\!-\!\!(CH_2)_7COOH$
$\quad\quad\quad\quad\quad\quad\quad\quad\quad\quad\;\;|$
$\quad\quad\quad\quad\quad\quad\quad\quad\quad\quad\;O\cdot$ 9—ROOH

Cleave bond A:

$\quad\quad\quad\quad CH_3(CH_2)_4\!-\!\!\underset{\underset{O}{\diagdown\;\;\diagup}}{\overset{CH-CH}{\overset{||\quad\;\;||}{C\quad\quad CH}}}$

2-Pentylfuran*

$CH_3(CH_2)_4CH_2\!-\!CH=CH\!-\!CHO$
2-Nonenal*

Cleave bond B: $\quad CH_3(CH_2)_4CH=CHCH=CHO$
2,4-Decadienal

* Formed via $CH_3(CH_2)_4CH=CHCH=CHOOH$

Fig. 7. Formation of some volatile oxidation products from linoleic acid.

TABLE 16
Volatile oxidation products from triolein, trilinolein and trilinolenin

Precursor	Triolein[a]	Trilinolein[b]	Trilinolenin[c]
Products	Decanal	3-Nonenal	2,4,7-Decatrienal
	2-Undecenal	2,4-Decadienal	3-Hexenal
	Nonanal	Hexanal	2,4-Heptadienal
	2-Decenal	Acrolein	Propanal
	Octanal	Propanal	Acrolein
	Pentanal	Butanal	Acetaldehyde
	Hexanal	Heptanal	2-Butenal
	Heptanal	Hexenal	2-Pentenal
	2-Heptenal	2-Heptenal	4,5-Epoxy-2-heptenal
	2-Octenal	2-Octenal	Ethanol/furan
	2-Nonenal	4,5-Epoxy-2-decenal	2-Butylfuran
	Heptane	Pentane	
	Octane	1-Pentanol	
	1-Heptanol	Oct-1-en-3-ol	
		2-Pentylfuran	

[a] Selke et al. (1977); [b] Selke et al. (1980); [c] Selke & Rohwedder (1982).

TABLE 17
Threshold values of some autoxidation products

Compounds	Threshold value (ppm)	Precursor
Hexanal	0·08–0·6[b]	Linoleic acid
Pentane	340[b]	Linoleic acid
Pentanol	45[a]	Linoleic acid
2-Pentylfuran		Linoleic acid
trans-2-Nonenal	0·1[b]	Linoleic acid
trans-2,cis-4-Decadienal	0·020[b]	Linoleic acid
trans-2,trans-4-Decadienal	0·1	Linoleic acid
3,6-Nonadienal	0·0015	αLinolenic acid
2,6-Nonadienal	0·002	αLinolenic acid
2-Pentenal	0·046	αLinolenic acid
trans-2, cis-Heptadienal	0·055	αLinolenic acid
cis-3-Hexenal	0·09	αLinolenic acid
2,4,7-Decatrienal	0·15	αLinolenic acid
cis-6-Nonenal	0·002	Isolinoleic acid
cis-7-Nonenal	0·0003	Isolinoleic acid
2,7-Decadienal	0·02	Isolinoleic acid

[a] Tested from water; [b] Tested from paraffin oil.

FACTORS AFFECTING THE RATE OF OFF-FLAVOUR DEVELOPMENT

As mentioned earlier, several factors affect the rate of oxidative deterioration of fats. Polyunsaturated fatty acids oxidize much more rapidly than mono-unsaturated fatty acids. In addition an increase in temperature causes a marked acceleration in autoxidation. An increase of 10°C is sufficient to reduce the induction period by a factor of about 2. This was demonstrated by Kochhar & Rossell (1989) who found that the induction periods of hard soya 45, brazilian cocoa butter and beef oleo were 178 h, 87·3 h and 110·4 h at 100°C but fell to 70 h, 39·6 h and 45·8 h at 110°C. Above 150°C, hydroperoxides decompose spontaneously and the radical concentration increases markedly leading to much greater polymer formation.

Other factors affecting the rate of off-flavour development include the effect of light, the presence of pro-oxidants or anti-oxidants and the sensitivity of the palate to oxidation products. Light in the ultraviolet (> 280 nm) or visible regions catalyses oxidation of oils in the presence of small quantities of pigments, such as chlorophyll, and other substances. Therefore, packaging which excludes light is preferred for the storage of fatty foods.

Metal ions such as copper and iron, which possess two valency states are very strong pro-oxidants. These ions catalyse the decomposition of hydroperoxides by rapid electron donation, when present in their lower valence state according to eqn (1)

$$ROOH + M^{n+} \rightarrow RO\cdot + {}^-OH + M^{(n+1)+} \qquad (1)$$

However, the oxidized metal ion is reduced back to its original valence state by a further slow reaction with hydroperoxides according to eqn (2)

$$ROOH + M^{(n+1)+} \rightarrow ROO\cdot + H^+ + M^{n+} \qquad (2)$$

Since metal ions are continuously regenerated in their lower valence state by these reactions, they are highly effective pro-oxidants even at very low levels. The concentration of metals required to reduce the keeping time of lard by 50% at 98°C was found to be 0·05 ppm for copper, 0·6 ppm for iron and 2·2 ppm for nickel (Ohlson, 1973).

There are several types of anti-oxidants used in foods to extend the induction period of an edible fat. Primary or chain-breaking anti-oxidants include the tocopherols, and also common synthetic anti-oxidants such as butylated hydroxyanisole (BHA) and butylated hydroxytoluene (BHT). These compounds act by reacting rapidly with lipid radicals involved in propagating the autoxidation chain reaction (ROO· OR RO·) according to eqn (3) (AH_2 = anti-oxidant).

$$AH_2 + ROO· \rightarrow ·AH + ROOH \qquad (3)$$

The radical derived from the anti-oxidant is relatively stable due to the aromatic character of the molecule. In addition for some anti-oxidants such as tocopherol, two radicals can react to produce two stable molecules of which one is oxidized to a quinone and the other is the original anti-oxidant.

$$·AH + ·AH \rightarrow AH_2 + A \qquad (4)$$

Other components may retard the oxidation of fats by processes other than that of interrupting the autoxidation chain by reacting with free radicals. These are termed secondary anti-oxidants, and a combination of a primary and a secondary anti-oxidant is often highly effective in retarding oxidative deterioration. Secondary anti-oxidants include sequestering agents such as citric acid which is added to fats to form complexes with metal ions, and hence extend the induction period by reducing or eliminating the pro-oxidant effect of the metal ions. Thus, the induction period of coconut oil at 120°C in the Rancimat test was extended from 1·9 h to 13·5 h by the addition of 100 ppm citric acid (Rahman & Gordon, unpublished data).

Another type of secondary anti-oxidant comprises oxygen scavengers such as ascorbic acid or ascorbyl palmitate. Ascorbic acid can be used in complex foods containing an aqueous phase such as emulsions, but, because of its low solubility in vegetable oils, ascorbyl palmitate is usually used in oils. Cort (1974) reported that 0·01% ascorbyl palmitate is more effective in stabilizing vegetable oils than either BHA or BHT. Ascorbyl palmitate scavenges oxygen and is oxidized to dehydroascorbyl palmitate. It is also very effective in combination with α-tocopherol by regenerating the primary anti-oxidant from the tocopheroxyl radical or the tocoquinone (Packer *et al.*, 1979).

Phospholipids also act as secondary anti-oxidants. Phosphatidyl ethanolamine (PE) is the most common phospholipid with strong anti-oxidant properties. The induction period of degummed coconut oil was extended from 11·6 h to 99·7 h by the addition of PE (Rahman & Gordon, unpublished data). PE, phosphatidyl inositol, phosphatidyl serine and phosphatidic acid were effective at protecting α-tocopherol during the autoxidation of methyl linoleate, whereas phosphatidyl choline was not effective (Ishikawa *et al.*, 1984). PE may act by releasing protons and bringing about the rapid decomposition of hydroperoxides without forming free radicals (Tai *et al.*, 1974; Pokorny *et al.*, 1981) or by regenerating primary anti oxidants (Brandt *et al.*, 1973).

Reversion Flavours
Fats such as soybean oil or fish oil, which contain significant concentrations of linolenic (18:3) or more highly unsaturated acids, develop reversion off-flavours even when the peroxide value is quite low, whereas fats which contain linoleic acid (18:2) as the only polyunsaturated fatty acid do not produce off-flavours until higher peroxide values are reached (Frankel, 1980). Reversion off-flavours are detected at an early stage of oxidation because unsaturated aldehydes with a double bond at the n-3 (also called ω-3) position, e.g. 3,6-nonadienal and 2,6-nonadienal, have very low threshold values (Table 17). Hydrogenated soybean oil also produces volatiles of particularly low threshold values because of the formation of isolinoleic acid during hydrogenation.

Ketonic Rancidity
Ketonic rancidity is a problem that can be encountered with some products such as desiccated coconut which contain short-chain saturated fatty acids. Kellard *et al.* (1985) showed that moulds such as *Eurotium amstelodami* degrade the triglycerides in the presence of limited amounts of air and water. Free fatty acids are liberated initially and these subsequently suffer β-oxidation with the formation of methyl ketones and aliphatic alcohols. A musty, stale note in the product is characteristic of ketonic rancidity.

Hydrolytic Rancidity
Free fatty acids are removed from vegetable oils during refining, but they can re-form in foods exposed to microorganisms because of the presence of lipolytic enzymes in these organisms. The problem of off-flavours due to free fatty acids is mainly encountered in products containing lauric fats such as coconut or palm kernel oil because of the low threshold value for shorter-chain fatty acids (see Table 5). Thus soapy off-flavours may develop in chocolate-flavoured coatings prepared from palm kernel stearines if poor quality cocoa powder is used.

CONCLUSIONS

The flavour of butter is affected by feed, season, stage of lactation, breed of cow (or other animal) and method of manufacture of the butter. The action of enzymes on butterfat can produce a series of commercially useful flavours.

The special appeal of butter for cooking is due to the production of flavour compounds in the cooking process. Compounds produced include lactones and ketones from the fat and reaction products from the serum. In contrast, it is clear that only a few non-milk fats have attractive flavours, and for many fats problems of off-flavour development may occur. Off-flavours develop most commonly by autoxidation, and this problem, which is most severe for polyunsaturated fats, may be avoided by low temperature storage and the use of anti-oxidants.

REFERENCES

ANDRES, C. (1983) Concentrated natural butter flavor. *Food Process. (Chicago)*, **44** (12), 76.

ANET, E. F. L. J. (1975) Precursors of C-12 lactones in animal products. *CSIRO Div. Food Res. Rep. Res., 1974–75*, pp. 41–2.

ASTRUP, H. N. (1980) Relationship between lipase activation and rancid flavour in milk. *Meieriposten*, **69**, 122–4.

BADINGS, H. T. (1965) The flavour of fresh butter and of butter with cold-storage defects in relation to the presence of 4-*cis*-heptenal. *Neth. Milk Dairy J.*, **19**, 69–72.

BADINGS, H. T. (1967) Causes of ribes flavor in cheese. *J. Dairy Sci.*, **50**, 1347–51.

BADINGS, H. T. (1970) Cold-storage defects in butter and their relation to the autoxidation of unsaturated fatty acids. *Neth. Milk Dairy J.*, **24**, 145–256.

BADINGS, H. T. (1971) (Ed.) Investigations into the H_2S-off-flavour in butter. *Verslagenreeks, K. Ned. Zuivelbond*, **(19)**, 64 pp.

BADINGS, H. T. & NEETER, R. (1980) Recent advances in the study of aroma compounds of milk and dairy products. *Neth. Milk Dairy J.*, **34**, 9–30.

BADINGS, H. T. & VAN DER POL, J. J. G. (1967) A fast and simple method of determining the H_2S-content of pasteurized cream. *Misset's Zuivel*, **73**, 10–13.

BADINGS, H. T., MAARSE, H., KLEIPOOL, R. J. C., TAS, A. C., NEETER, R. & TEN NOEVER DE BRAUW, M. C. (1975) Formation of odorous compounds from hydrogen sulphide and methanethiol, and unsaturated carbonyls. In: *Aroma Research, Proc. Int. Symp. Aroma Res.*, Maarse, H. & Groenen, P. J. (Eds), Centre for Agricultural Publishing and Documentation (PUDOC), Wageningen, pp. 63–73.

BADINGS, H. T., MAARSE, H., KLEIPOOL, R. J. C., TAS, A. C., NEETER, R. & TEN NOEVER DE BRAUW, M. C. (1976*a*) Formation of odorous compounds from hydrogen sulphide and 2-butenal. *Z. Lebensm. Unters.-Forsch.*, **161**, 53–9.

BADINGS, H.T., TAMMINGA, S. & SCHAAP, J. E. (1976*b*) Production of milk with a high content of polyunsaturated fatty acids. II. Fatty acid composition of milk in relation to the quality of pasteurized milk, butter and cheese. *Neth. Milk Dairy J.*, **30**, 118–31.

BARRIO, P.-C. A., GUTIERREZ, F. & GUTIERREZ, R. (1981) Application of GLC headspace technique to the olive oil atrojado problem I. *Grasas y Aceites*, **32** (3), 155–61.
BLACK, R. G. (1976) Oxidative flavour stability of milkfat–vegetable oil blends. *Aust. J. Dairy Technol.*, **31**, 22–6.
BOLDINGH, J. & TAYLOR, R. J. (1962) Trace constituents of butterfat. *Nature (London)*, **194**, 909–13.
BRANDT, P., HOLLSTEIN, E. & FRANZKE, C. (1973) Pro- and antioxidative effects of phosphatides. *Lebensmitt. Ind.*, **20**, 31–3.
BRENNAND, C. P., HA, J. K. & LINDSAY, R. C. (1989) Aroma properties and thresholds of some branched-chain and other volatile fatty acids occurring in milkfat and meat lipids. *J. Sensory Studies*, **4**, 105–20.
BREWINGTON, C. R., PARKS, O. W. & SCHWARTZ, D. P. (1973) Conjugated compounds in cow's milk. *J. Agric. Food Chem.*, **21**, 38–9.
BREWINGTON, C. R., PARKS, O. W. & SCHWARTZ, D. P. (1974) Conjugated compounds in cow's milk. II. *J. Agric. Food Chem.*, **22**, 293–4.
BRUHN, J. C., FRANKE, A. A. & GOBLE, G. S. (1976) Factors relating to development of spontaneous oxidized flavor in raw milk. *J. Dairy Sci.*, **59**, 828–33.
BURKE, C. E. (1973) Lard flavour concentrate. US Patent 3, 767, 429.
CARRARD, G. (1983) New method of preserving food. *AAEC Nucl. News* (15), 1–4.
CHANG, S. S. & MAY, W. A. (1973) Deep fat fried flavor. US Patent 3 767 427.
CONOCHIE, J. (1953) Indole and skatole in the milk of the ruminant feeding on *Lepidium* spp. *Aust. J. Exp. Biol. Med. Sci.*, **31**, 373–84.
COOK, L. J. & SCOTT, T. W. (1975) *In vivo* biogenesis in sheep fed protected lipids. *Proc. Aust. Biochem. Soc.*, **8**, 46.
CORT, W. M. (1974) Antioxidant activity of tocopherols, ascorbyl palmitate and ascorbic acid and their mode of action. *J. Am. Oil Chem. Soc.*, **51** (7), 321–5.
CUPAKOVA, M., PALO, V. & GÖRNER, F. (1981) Study of carbonyl compounds in milk fat and its fractions. In: *Zbornik Prednasok, V Celostatneho Sympozia o Aromatickych Latkach v Pozivztinach, Bratislava, 1981*, pp. 68–73.
DAY, E. A. & PAPAIOANNOU, S. E. (1963) Irradiation-induced changes in milk fat. *J. Dairy Sci.*, **46**, 1201–6.
DAY, E. A., LINDSAY, R. C. & FORSS, D. A. (1964) Dimethyl sulfide and the flavor of butter. *J. Dairy Sci.*, **47**, 197–9.
DIMICK, P. S. & HARNER, J. L. (1968) Effect of environmental factors on lactone potential in bovine milk fat. *J. Dairy Sci.*, **51**, 22–7.
DIMICK, P. S. & WALKER, H. M. (1968) Effect of environmental factors on monocarbonyl potential in fresh bovine milk fat. *J. Dairy Sci.*, **51**, 478–82.
DIMICK, P. S., WALKER, N. J. & KINSELLA, J. E. (1966) Aliphatic delta-lactones: determination in bovine milk from animals on normal and fat-depressing diets. *Cereal Sci. Today*, **11**, 479–80, 502.
DUMONT, J.-P. & ADDA, J. (1978) Occurrence of sesquiterpenes in mountain cheese volatiles. *J. Agric. Food Chem.*, **26**, 364–7.
EFTHYMIOU, C. C. & MATTICK, J. F. (1964) Development of domestic Feta cheese. *J. Dairy Sci.*, **47**, 593–8.
FEDELI, E. (1977) Volatile components of olive oil *Riv. Ital. Sost. Gras.*, **54**, 202–5.
FEDELI, E., CAMURATI, F., CORTESI, N., FAVINI, G., CIRIO, V. & VITO, G. (1976) *Lipids*, Vol. 2, Raven Press, New York, p. 385.

FERON, R. & GOVIGNON, M. (1961) The relation between free acidity and taste of edible oils. *Ann. Falsif. Expert. Chim.*, **54**, 308–14.

FLANAGAN, V. P., FERRETTI, A., SCHWARTZ, D. P. & RUTH, J. M. (1975) Characterization of two steroidal ketones and two isoprenoid alcohols in dairy products. *J. Lipid Res.*, **16**, 97–101.

FLATH, A. R., FORREY, R. R. & GUADAGNI, D. G. (1973) Aroma components of olive oil. *J. Agric. Food Chem.*, **21**, 948–52.

FODA, E. A., HAMMOND, E. G., REINBOLD, G. W. & HOTCHKISS, D. K. (1974) Role of fat in flavor of Cheddar cheese. *J. Dairy Sci.*, **57**, 1137.

FORSS, D. A. (1972) Odor and flavor compounds from lipids. *Prog. Chem. Fats Other Lipids*, **13**, 177–258.

FORSS, D. A. (1979) Mechanisms of formation of aroma compounds in milk and milk products. *J. Dairy Res.*, **46**, 691–706.

FORSS, D. A. & PATTON, S. (1966) Flavor of Cheddar cheese. *J. Dairy Sci.*, **49**, 89–91.

FORSS, D. A., DUNSTONE, E. A., RAMSHAW, E. H. & STARK, W. (1962) Flavour of cucumbers. *J. Food Sci.*, **27**, 90–3.

FORSS, D. A., STARK, W. & URBACH, G. (1967a) Volatile compounds in butter oil. I. Lower boiling compounds. *J. Dairy Res.*, **34**, 131–6.

FORSS, D. A., ANGELINI, P., BAZINET, M. L. & MERRITT, C. (1967b) Volatile compounds produced by copper-catalyzed oxidation of butterfat. *J. Am. Oil Chem. Soc.*, **44**, 141–3.

FRANKEL, E. N. (1980) Lipid oxidation. *Prog. Lipid Res.*, **19**, 1–22.

GILL, M. S., MACLEOD, A. J. & MOREAU, M. (1984) Volatile components of cocoa with particular reference to glucosinolate products. *Phytochemistry*, **23**, 1937–42.

GOERING, H. K., GORDON, C. H., WRENN, T. R., BITMAN, J., KING, R. L. & DOUGLAS, F. W. (1976) Effect of feeding protected safflower oil on yield, composition, flavour, and oxidative stability of milk. *J. Dairy Sci.*, **59**, 416–25.

GUTFINGER, T. (1981) Polyphenols in olive oils. *J. Am. Oil Chem. Soc.*, **58**, 966–8.

GUTIERREZ, R., NOSTI VEGA, M., COLAKOGLU, M. & CABRERA, J. (1972) Profile evolution of the aromagrams during the oxidation of olive oils. *Grasas y Aceites*, **23**, 351–8.

GUTIERREZ, R., OLIAS, J. M., GUTIERREZ, F., CABRERA, J. & BARRIO, P. (1975) The chromatographic and organoleptic methods in the evaluation of the aromatic characteristics in virgin olive oil. *Grasas y aceites*, **26**, 21–32.

GUTIERREZ, F., ALBI, M. A., PALMA, R., RIOS, J. J. & OLIAS, J. M. (1989) Bitter taste of virgin oil: Correlation of sensory evaluation and instrumental HPLC analysis. *J. Food Sci.*, **54** (1), 68–70.

HA, J. K. & LINDSAY, R. C. (1990) Method for the quantitative analysis of volatile free and total branched-chain fatty acids in cheese and milk fat. *J. Dairy Sci.*, **73**, 1988–99.

HA, J. K. & LINDSAY, R. C. (1991) Contribution of cow, sheep, and goat milks to characterizing branched-chain fatty acid and phenolic flavors in varietal cheeses. *J. Dairy Sci.*, **74**, 3267–74.

HANSEN, P. M. T. (1963a) Manufacture of butter powder. *Aust. J. Dairy Technol.*, **18**, 79–86.

HANSEN, P. M. T. (1963b) The baking performance of butter powder. *Aust. J. Dairy Technol.*, **18**, 86–91.

HARPER, W. J. (1959) Chemisty of cheese flavors. *J. Dairy Sci.*, **42**, 207–13.
HARWALKAR, V. R., EMMONS, D. B. & GILCHRIST, M. R. (1974) Light-induced oxidation of butter oil. *Br. Commun. XIX Int. Dairy Congr*, 1E, 236.
HAVERKAMP BEGEMANN, P. & KOSTER, J. C. (1964) 4-*cis*-Heptenal: a cream-flavoured component of butter. *Nature (London)*, **202**, 552.
HESS, E. & LUTZ, H. (1972) The affinity of milk for odours. *Arch. Lebensmittelhyg.*, **23**, 224–5.
HILDITCH, T. P. & WILLIAMS, P. N. (1964) *The Chemical Constitution of Natural Fats*, 4th edn, Chapman & Hall, London.
HILL, F. D. & HAMMOND, E. G. (1965) Studies on the flavor of autoxidized soybean oil. *J. Am. Oil Chem. Soc.*, **42**, 1148–50.
HILL, R. D. (1979) Oxidative enzymes and oxidative processes in milk. *CSIRO Food Res. Q.*, **39**, 33–7.
HONKANEN, E., KARVONEN, P. & VIRTANEN, A. I. (1964) Studies on the transfer of some flavour compounds to milk. *Acta Chem. Scand.*, **18**, 612–18.
HONKANEN, E., MOISIO, T., KARVONEN, P., VIRTANEN, A. J. & PAASIVIRTA, J. (1968) On the occurrence of a new lactone compound, *trans*-4-methyl-5-hydroxy-hexanoic acid lactone in milk. *Acta Chem. Scand.*, **22**, 2041–3.
HONKANEN, E., MOISIO, T., KARVONEN, P. & VIRTANEN, A. I. (1970) Comparative studies on the flavour compounds of milk produced with urea and normal feeding. *Suom. Kemistil. B*, **43**, 1–3.
HOSKIN, J. C. & DIMICK, P. S. (1988) Chemistry of flavour development. In: *Industrial Chocolate Manufacture and Use*, Beckett, S. T. (Ed.), Blackie, Glasgow.
HUANG, H. T. & DOOLEY, J. G. (1976) Enhancement of cheese flavors with microbial esterases. *Biotechnol. Bioengng*, **18**, 909–19.
HUANG, J. J., SHYU, S. L. & CHANG, R. L. (1988) Effect of roasting on the quality of groundnut oil. *J. Chin. Agric. Chem. Soc.*, **26** (4), 466–75.
ISHIKAWA, Y., SUGIYAMA, K. & NAKABAYASHI, K. (1984) Stabilisation of tocopherol by three components synergism involving tocopherol, phospholipid and amino compound. *J. Am. Oil Chem. Soc.*, **61** (5), 950–4.
IVANOV, S. A. & STAMATOV, S. D. (1975) The γ-irradiation-induced degree of oxidation, hydrolysis and polymerization of sunflower oil, lard and butter. *Seifen, Oele, Fette, Wachse*, **101**, 589–92.
JURRIENS, G. & OELE, J. M. (1965) Determination of hydroxy-acid triglycerides and lactones in butter. *J. Am. Oil Chem. Soc.*, **42**, 857–61.
KAMI, T. (1983) Composition of the essential oil of alfalfa. *J. Agric. Food Chem.*, **31**, 38–41.
KANEKO, H. & MITA, M. (1969) Isolation from cigar tobacco leaves of 2,3-dimethyl-4-hydroxy-2-nonenoic acid lactone. *Agric. Biol. Chem.*, **33**, 1525–6.
KAWADA, T., KRISHNAMURTHY, R. G., MOOKHERJEE, B. D. & CHANG, S. S. (1967) Chemical reactions involved in the deep fat frying of foods. II. Identification of acidic volatile decomposition products of corn oil. *J. Am. Oil Chem. Soc.*, **44**, 131–5.
KEEN, A. R. (1986) Future developments for profitable dairy farming: the colour and flavour of dairy products. *Dairyfarming Ann., 1986*, 36–9.
KEENAN, T. W., BILLS, D. D. & LINDSAY, R. C. (1967) Dehydrogenase activity of *Pseudomonas* species. *Appl. Microbiol.*, **15**, 1216–18.

KELLARD, B., BUSFIELD, D. M. & KINDERLERER, J. L. (1985) Volatile off-flavour compounds in desiccated coconut. *J. Sci. Food Agric.*, **36**, 415.
KIKOMAN SYOYU Co. Ltd (1971) Butter-like taste enrichment—using 2-en-5-ol-1-carboxylic-acid lactone. Jpn Patent 71041183.
KILARA, A. (1985) Enzyme-modified lipid food ingredients. *Process Biochem.*, **20**, 35–44.
KINDELERER, J. L. & KELLARD, B. (1984) Ketonic rancidity in coconut due to xerophilic fungi. *Phytochemistry*, **23** (12), 2847–9.
KINSELLA, J. E. (1969) The flavour chemistry of milk lipids. *Chem. Ind. (London)*, (2) 36–42.
KINSELLA, J. E. & HWANG, D. H. (1976) Enzymes of *Penicillium roqueforti* involved in the biosynthesis of cheese flavor. *CRC Crit. Rev. Food Sci. Nutr.*, **8**, 191–228.
KINTNER, J. A. & DAY, E. A. (1965) Major free fatty acids in milk. *J. Dairy Sci.*, **48**, 1575–81.
KOCHHAR, S. P. & ROSSELL, J. B. (1989) Leatherhead Food Research Association Report No. 412.
KOTOVA, O. G., ZHILOV, V. N., ZHUKOV, P. I., MAKAROVA, V. T., TIKHOVA, N. T. & KABANOV, N. YA. (1973) Characteristics and keeping quality of butters made from milk of different breeds. In: *Sovershenstvovanie tekhnologicheskikh protsessov v molochnoi promyshlennosti. Tom I Chast' I.* Leningradskii Tekhnologicheskii Institut Kholodil'noi Promyshlennosti, pp. 94–6.
KRISTENSEN, J. M. B. & JENSEN, S. G. (1975) Investigations into the effect of hereditary factors on the occurrence of oxidized flavour in milk from different breeds of cow. *Maelkeritidende*, **88**, 168–71.
KURTZ, F. E. (1974) The lipids of milk: composition and properties. In: *Fundamentals of Dairy Chemistry*, Webb, B. H., Johnson, A. H. & Alford, J. A. (Eds), 2nd edn, AVI Publ. Co., Westport, CN, pp. 125–219.
LABUSCHAGNE, J. H. (1975) Identification and determination of off-flavour compounds in butter. *S. Afn. J. Dairy Technol.*, **7**, 229–33.
LAWRENCE, R. C. (1967) The possible role of milk fat in the formation of Cheddar cheese flavour. *N. Z. J. Dairy Technol.*, **2**, 55–7.
LEA, C. H. (1952) Methods for determining peroxides in lipids. *J. Sci. Food Agric.*, **3**, 586–94.
LINDSAY, R. C., DAY, E. A. & SATHER, L. A. (1967) Preparation and evaluation of butter culture flavor concentrates. *J. Dairy Sci.*, **50**, 25–31.
LOTTE, K. K. (1982) Production of ice cream having improved taste—with addition of cyclodextrin to butter and butter oil. Jpn Patent 82046348.
MAGA, J. A. (1976) Lactones in foods. *CRC Crit. Rev. Food Sci. Nutr.*, **8**, 1–56.
MANIERE, H. Y. & DIMICK, P. S. (1979) Effects of conching on the flavour and volatile components of dark semi-sweet chocolate. *Lebensm. Wiss. Technol.*, **12**, 102–7.
MARTH, E. H. (1974) Fermentations. In: *Fundamentals of Dairy Chemistry*, Webb, B. H., Johnson, A. H. & Alford, J. A. (Eds), 2nd edn, AVI Publ. Co., Westport, CN, p. 819.
MAY, W. A. (1972) Identification of the decomposition products produced by triolein under simulated deep fat frying conditions and a preliminary study of sensory characteristics and physiological effects of unsaturated lactones. Thesis, Rutgers University. *Diss. Abstr. Int. B.*, **32**, 5240.

MAY, W. A., PETERSON, R. J. & CHANG, S. S. (1978) Synthesis of some unsaturated lactones and their relationship to deep-fat fried flavor. *J. Food Sci.*, **43**, 1248–52.

MCDANIEL, M. R., SATHER, L. A. & LINDSAY, R. C. (1969) Influence of free fatty acid on sweet cream butter flavor. *J. Food Sci.*, **34**, 251–4.

MCGOOKIN, B. J. (1991) Casein–sugar reaction products as anti-oxidants. *CSIRO Food Research Quarterly*, **51**, 55–9.

MCGOOKIN, B. J. & AUGUSTIN, M.-A. (1991) Antioxidant activity of casein and Maillard reaction products from casein–sugar mixtures. *J. Dairy Res.*, **58**, 313–20.

MCNEILL, G. P., O'DONOGHUE, A. & CONNOLLY, J. F. (1986) Quantification and identification of flavour components leading to lipolytic rancidity in stored butter. *Ir. J. Food Sci. Technol.*, **10**, 1–10.

MERRITT, C., FORSS, D. A., ANGELINI, P. & BAZINET, M. L. (1967) Volatile compounds produced by irradiation of butterfat. *J. Am. Oil Chem. Soc.*, **44**, 144–6.

MICK, S., MICK, W. & SCHREIER, P. (1982) The composition of neutral volatile constituents of sour cream butter. *Milchwissenschaft*, **37**, 661–5.

MOINAS, M., GROUX, M. & HORMAN, I. (1973) Flavour of cheese. I. New method for isolation of volatile constituents. Application to Roquefort and Camembert. *Lait*, **53**, 601–9.

MONTEDORO, G. & CANTARELLI, C. (1969) Phenolic constituents of olive oil. *Riv. Ital. Sost. Gras.*, **46**, 115–20.

MONTEDORO, G., BERTUCCIOLI, M. & ANICHINI, F. (1978) Aroma analysis of virgin olive oil by head space (volatiles) and extraction (polyphenols) techniques. In: *Flavour of Foods and Beverages*, Charalambous, G. & Inglett, G. E. (Eds), Academic Press, New York, pp. 247–81.

MOSANDL, A. & DEGER, W. (1987) Stereoisomeric flavour compounds XVII. Chiral carboxylic esters—synthesis and properties. *Z. Lebensm. Unters. Forsch.*, **185**, 379–82.

MOSANDL, A. & GESSNER, M. (1988) Stereoisomeric flavour substances. XXIII. δ-Lactone flavour compounds—structure and properties of enantiomers. *Z. Lebensm. Unters. Forsch.*, **187**, 401–4.

NEDERLANDS INSTITUUT VOOR ZUIVELONDERZOEK (1975) Method for preparing a food product having the flavour of butter made from microbiologically ripened cream. Netherlands Patent Application 7 311 820.

NEIMANN-SORENSEN, A., NIELSEN, E. O., POULSEN, P. R. & JENSEN, G. K. (1973) Investigations concerning the influence of heritability on the development of oxidized flavour in cow's milk. *Beret. Statens Forsoegsmejeri* (199), 42 pp.

NELSON, J. H. (1972) Enzymatically produced flavors for fatty systems. *J. Am. Oil Chem. Soc.*, **49**, 559–62.

NEY, K. H. (1985) Cocoa aroma—new α-keto acids in cocoa aroma. *Gordian*, 88–92.

NOBUHARA, A. (1968) Synthesis of unsaturated lactones. I. Some lactones of 5-substituted-5-hydroxy-2-enoic acids as synthetic butter or butter cake flavor. *Agric. Biol. Chem.*, **32**, 1016–20.

NOBUHARA, A. (1969a) Synthesis of unsaturated lactones. II. Flavorous nature of some 4- and 5-substituted-5-hydroxy-2-enoic acid lactones. *Agric. Biol. Chem.*, **33**, 225–9.

Nobuhara, A. (1969b) Synthesis of unsaturated lactones. III. Flavorous nature of some δ-lactones having the double bond at various sites. *Agric. Biol. Chem.*, **33**, 1264–9.

Nobuhara, A. (1970) Synthesis of unsaturated lactones. IV. Flavour of some aliphatic γ-lactones. *Agric. Biol. Chem.*, **34**, 1745–7.

Nursten, H. E. (1980) Recent developments in studies of the Maillard reaction. *Food Chem.*, **6**, 263–77.

Ohlson, R. (1973) Antioxidant activity of phenols as related to effects of substituent groups. In: *Proceedings of the 3rd International Symposium on Metal-Catalysed Lipid Oxidation*, Institut des Corps Gras, Paris, pp. 184–92.

Olias, J. M., Cabrera, J. & Gutierrez, R. (1974) Relation between GLC and sensory properties of the aroma of packaged olive oil. *Grasas y Aceites*, **25**, 34–41.

Olias, J. M., Del Barrio, A. & Gutierrez, R. (1977) Volatile components in the aroma of virgin olive oil. *Grasas y Aceites*, **28**, 107–12.

Olias, J. M., Dobarganes, C., Gutierrez, F. & Gutierrez, R. (1978) Volatile components in the aroma of virgin olive oil II. Identification and sensorial analysis of the chromatographic eluents. *Grasas y Aceites*, **29** (3), 211–18.

Olias, J. M., Gutierrez, F., Dobarganes, M. C. & Gutierrez, R. (1980) Volatile components in the aroma of olive oil IV. Their evolution and influence in the aroma during the fruit ripening process in the Picual and Hojiblanca varieties. *Grasas y Aceites*, **31** (6), 391–402.

Packer, J. E., Slater, T. F. & Willson, R. L. (1979) Direct observation of a free radical interaction between vitamin E and vitamin C. *Nature*, **278**, 737.

Palm, U., Askari, C., Hener, U., Jakob, E., Mandler, C., Gessner, M., Mosandl, A., König, W. A., Evers, P. & Krebber, R. (1991) Stereoisomeric aroma compounds. XLVII. Direct chirospecific high resolution GC analysis of natural δ-lactones. *Z. Lebensm. Unters. Forsch.*, **192**, 209–13.

Park, R. J., Armitt, J. D. & Stark, W. (1969) Weed taints in dairy produce. II. *Coronopus* or land cress taint in milk. *J. Dairy Res.*, **36**, 37–46.

Park, R. J., Murray, K. E. & Stanley, G. (1974) 4-Hydroxydodec-*cis*-6-enoic acid lactone: an important component of lamb flavour from animals fed a lipid-protected dietary supplement. *Chem. Ind. (London)*, 380–2.

Park, R. J., Ford, A. L. & Ratcliff, D. (1976) The influence of two kinds of protected lipid supplement on the flavour of lamb. *J. Food Sci.*, **41**, 633–5.

Parks, O. W., Keeney, M. & Schwartz, D. P. (1961) Bound aldehydes in butteroil. *J. Dairy Sci.*, **44**, 1940–3.

Patton, S. (1964a) Flavor thresholds of volatile fatty acids. *J. Food Sci.*, **29**, 679–80.

Patton, S. (1964b) Flavor stabilization of milk fat. US Patent 3 127 275.

Paulet, G., Mestres, G. & Cronenberger, L. (1974) Soapy taste in foods: effect of the lipase of white pepper. *Rev. Fr. Corps Gras*, **21**, 611–16.

Pedersen, A. H. (1977) Major challenges to dairying. State Research Dairy involved in the controversy over packaging and re-use. Butter and related products: practical methods of improving butter consistency. *Maelkeritidende*, **90**, 295–6, 298–9, 354–9.

Petersen, J. (1986) Improvement of product technology using modified butterfat. *Dtsch. Milchwirtsch.*, **37**, 1156.

PETRUKHINA, E. P. & PISKAREV, A. I. (1975) The effect of seasonal conditions in butter manufacture on variations in its quality under cold storage. *Kholod. Tekh.* (5), 34–7.

PETRUKHINA, E. P. & PISKAREV, A. I. (1976) Study of changes in butter quality at storing in various negative temperatures. *Kholod. Tekh.*, (3), 41–4.

POKORNY, J. (1976) Effect of nonlipidic substances on rancid off flavor of lipids. In: *Lipids, 2, Technology*, Paoletti, R., Jacini, G. & Porcellati, R. (Eds), Raven Press, New York, pp. 475–80.

POKORNY, J., POSKOCILOVA, H. & DAVIDEK, J. (1981) Effect of phospholipids on the decomposition of hydroperoxides. *Nahrung*, **25**, K29–31.

RAGAZZI, E. & VERONESE, G. (1973) Phenolic constituents of olive oil. *Riv. Ital. Sost. Gras.*, **50**, 443–8.

RAMONAS, R. & BERZHINSKAS, G. (1970) Improving butter quality in continuous butter making. *Trudy. Fil. Vses. Nauchno-Issled. Inst. Maslodel. Syrodel'n. Prom*, **5**, 71–4. *Dairy Science Abs.* (1971) **33**, Abs. No. 2751.

RANGAPPA, K. S. & ACHAYA, K. T. (1973) *Indian Dairy Products*, Asia Publishing House, Bombay.

ROHAN, T. A. (1969) The flavour of chocolate, its precursors and a study of their reaction. *Gordian*, 443–7, 500–1, 542–4, 587–90.

SCHAAP, J. E., STRAATSMA, J., ESCHER, J. T. M. & BADINGS, H. T. (1986) Extraction of milk fat with supercritical carbon dioxide. In: *Milk the Vital Force: Posters Presented at the XXII International Dairy Congress, The Hague, 1986*, Reidel, Dordrecht, p. 26.

SCHOGT, J. C. M., HAVEKAMP BEGEMANN, P., DE JONG, K. & RADEMAKER-KOSTER, J. (1967) Butter flavour in food products. British Patent 1 068 712.

SCOTT, T. W., BREADY, P. J., ROYAL, A. J. & COOK, L. J. (1972) Oil seed supplements for the production of polyunsaturated ruminant milk fat. *Search*, **3**, 170–1.

SELKE, E. & ROHWEDDER, W. K. (1982) Paper presented at AOCS meeting, Toronto, Canada.

SELKE, E., ROHWEDDER, W. K. & DUTTON, H. J. (1977) Volatile components from triolein heated in air. *J. Am. Oil Chem. Soc.*, **54**, 62–7.

SELKE, E., ROHWEDDER, W. K. & DUTTON, H. J. (1980) Volatile components from trilinolein heated in air. *J. Am. Oil Chem Soc.*, **57**, 25–30.

SEVENANTS, M. R. (1983, 1986) Cream flavor composition for use with buttery flavored food products. US Patent 4 411 924; European Patent 0 074 140 B1.

SIEK, T. J. & LINDSAY, R. C. (1968) Volatile components of milk fat steam distillates identified by gas chromatography and mass spectrometry. *J. Dairy Sci.*, **51**, 1887–96.

SIEK, T. J. & LINDSAY, R. C. (1970) Semiquantitative analysis of fresh sweet-cream butter volatiles. *J. Dairy Sci.*, **53**, 700–3.

SIEK, T. J., ALBIN, I. A., SATHER, L. A. & LINDSAY, R. C. (1969) Taste thresholds of butter volatiles in deodorized butteroil medium. *J. Food Sci.*, **34**, 265–7.

SIEK, T. J., ALBIN, I. A., SATHER, L. A. & LINDSAY, R. C. (1971) Comparison of flavor thresholds of aliphatic lactones with those of fatty acids, esters, aldehydes, alcohols, and ketones. *J. Dairy Sci.*, **54**, 1–4.

SMITH, P. W., PARKS, O. W. & SCHWARTZ, D. P. (1984) Characterization of male goat odors: 6-*trans* nonenal. *J. Dairy Sci.*, **67**, 794–801.

Snow, N. S., Buchanan, R. A., Freeman, N. H. & Bready, P. J. (1967a) Manufacturing conditions for butter powder. 1. Powder removal from the drier and fluidized-bed cooler. *Aust. J. Dairy Technol.*, **22**, 122–5.

Snow, N. S., Townsend, F. R., Bready, P. J. & Shimmin, P. D. (1967b) Manufacturing conditions for butter powder. 2. The effect of manufacturing conditions on baking performance. *Aust. J. Dairy Technol.*, **22**, 125–34.

Solinas, M., Di Giovacchino, L. & Cucurachi, A. (1975) Phenolic constituents of olive oil. *Annali 1st Elaiotec.*, 5.

Stark, W. & Forss, D. A. (1962) A compound responsible for metallic flavour in dairy products: I. Isolation and identification. *J. Dairy Res.*, **29**, 173–80.

Stark, W. & Forss, D. A. (1964) A compound responsible for mushroom flavour in dairy products. *J. Dairy Res.*, **31**, 253–9.

Stark, W. & Urbach, G. (1974) The level of saturated and unsaturated γ-dodecalactones in the butter fat from cows on various rations. *Chem. Ind. (London)*, 413–14.

Stark, W. & Urbach, G. (1976) The effect of storage of butter at $-10°C$ on the level of free delta-lactones and free fatty acids in the butterfat. *Aust. J. Dairy Technol.*, **31**, 80–2.

Stark, W. & Urbach, G. (1979) *CSIRO Div. Food Res. Rep. Res., 1978–1979*, 113.

Stark, W., Smith, J. F. & Forss, D. A. (1967) n-Pent-1-en-3-ol and n-pent-1-en-3-one in oxidized dairy products. *J. Dairy Res.*, **34**, 123–9.

Stark, W., Urbach, G. & Hamilton, J. S. (1976a) Volatile compounds in butter oil. IV. Quantitative estimation of free fatty acids and free δ-lactones in butter oil by cold-finger molecular distillation. *J. Dairy Res.*, **43**, 469–77.

Stark, W., Urbach, G. & Hamilton, J. S. (1976b) Volatile compounds in butter oil. V. The quantitative estimation of phenol, o-methoxypenol, m- and p-cresols, indole and skatole by cold-finger molecular distillation. *J. Dairy Res.*, **43**, 479–89.

Stark, W., Urbach, G., Cook, L. J. & Ashes, J. R. (1978) The effect of diet on the γ- and δ-lactone and methyl ketone potentials of caprine butterfat. *J. Dairy Res.*, **45**, 209–21.

Sulser, H. & Büchi, W. (1974) Method for producing a roasted-type flavouring. Swiss Patent 555 143.

Suzuki, J. & Bailey, M. E. (1985) Direct sampling capillary GLC analysis of flavor volatiles from ovine fat. *J. Agric. Food Chem.*, **33**, 343–7.

Swern, D. (1964) Composition and characteristics of individual fats and oils. In: *Bailey's Industrial Oil and Fat Products*, Swern, D. (Ed.), 3rd edn, Interscience, New York, pp. 165–247.

Swoboda, P. A. T. & Peers, K. E. (1976) (n-3) Pentaenoic fatty acids in butterfat as precursors of metallic, fishy taint. *Chem. Ind. (London)*, 160–1.

Tai, P. T., Pokorny, J. & Janicek, G. (1974) Non-enzymic browning X. Kinetics of the oxidative browning of phosphatidylethanolamine. *Z. Lebensm. Unters. Forsch.*, **156** (5) 257–62.

Tharp, B. W. & Patton, S. (1960) Coconut-like flavor defect of milk fat. IV. Demonstration of δ-dodecalactone in the steam distillate from milk fat. *J. Dairy Sci.*, **43**, 475–9.

Thompson, J. A., May, W. A., Paulose, M. M., Peterson, R. J. & Chang, S. S. (1978) Chemical reactions involved in the deep-fat frying of foods. VII.

Identification of volatile decomposition products of trilinolein. *J. Am. Oil Chem. Soc.*, **55**, 897–901.

TIMMS, R. E. (1980) The phase behaviour of mixtures of cocoa butter and milk fat. *Lebensm.-Wiss. Technol.*, **13**, 61–5.

TIMMS, R. E. & PAREKH, J. V. (1980) The possibilities for using hydrogenated, fractionated or interesterified milk fat in chocolate. *Lebensm.-Wiss. Technol.*, **13**, 177–81.

TOWNSEND, F. R., SNOW, N. S., BREADY, P. J. & THOMPSON, H. (1968) Manufacturing conditions for butter powder. 3. Butter powder for domestic cake baking. *Aust. J. Dairy Technol.*, **23**, 85–9.

TUYNENBURG MUYS, G., VAN DER VEN, B. & DE JONGE, A. P. (1962) Synthesis of optically active γ- and δ-lactones by microbiological reduction. *Nature (London)*, **194**, 995.

UNILEVER NV (1968) Margarine with a particularly butter-like flavour. Dtsch. Off. Schr. 1 692 540.

URBACH, G. (1979) The flavour of milk fat. In: *Proceedings of Milk Fat Symposium*, held at Dairy Research Laboratory, Division of Food Research, CSIRO, 10th October, 1979, Australian Society of Dairy Technology, Melbourne, pp. 18–27.

URBACH, G. (1982) The effect of different feeds on the lactone and methyl ketone precursors of milk fat. *Lebensm.-Wiss. Technol.*, **15**, 62–7.

URBACH, G. (1990) The effect of feed on flavor in dairy foods. *J. Dairy Sci.*, **73**, 3639–50.

URBACH, G. & MILNE, T. (1987) The concentration of volatiles in pasteurized milk as a function of storage time and storage temperature—a possible indicator of keeping quality. *Aust. J. Dairy Technol.*, **42**, 53–8.

URBACH, G. & STARK, W. (1975) The C-20 hydrocarbons of butterfat. *J. Agric. Food Chem.*, **23**, 20–4.

URBACH, G. & STARK, W. (1978a) Unsaturated tetradecalactones in butterfat. *Br. Commun. 20th Int. Dairy Congr.*, E, 887.

URBACH, G. & STARK, W. (1978b) The effect of diet on the γ- and δ-lactone and methyl ketone potentials of bovine butterfat. *J. Dairy Res.*, **45**, 223–9.

URBACH, G., STARK, W. & FORSS, D. A. (1972) Volatile compounds in butter oil. II. Flavour and flavour thresholds of lactones, fatty acids, phenols, indole and skatole in deodorized synthetic butter. *J. Dairy Res.*, **39**, 35–47.

VAN DER VEN, B. (1964) Detection of γ- and δ-keto acids in butterfat. *Rech. Trav. Chim. Pays-Bas*, **83**, 976–82.

VAN DER VEN, B., HAVERKAMP BEGEMANN, P. & SCHOGT, J. C. M. (1963) Precursors of methyl ketones in butter. *J. Lipid Res.*, **4**, 91–5.

VAN DER ZIJDEN, A. S. M., DE JONG, K., SLOOT, D., CLIFFORD, J. & TAYLOR, R.-J. (1966) Components of butter fat occurring in traces. IV. Isolation and identification of unsaturated aliphatic lactones. *Rev. Fr. Corps Gras*, **13**, 731–5.

VAN NIEL, C. B., KLUYVER, A. J. & DERX, H. G. (1929) The butter aroma. *Biochem. Z.*, **210**, 234–51.

VASQUEZ RONCERO, A. (1978) The polyphenols of olive oil and their effect on oil characteristics. *Rev. Franc. des Corps Gras*, **25**, 21–6.

VERINGA, H. A., VAN DEN BERG, G. & STADHOUDERS, J. (1976) An alternative method for the production of cultured butter. *Milchwissenschaft*, **31**, 658–62.

VYSHEMIRSKII, F. A. & VASILISIN, S. V. (1972) Effect of heat treatment on composition of cream and intensity of pasteurized flavour. *Tr., Vses. Nauchno-Issled. Inst. Maslodel. Syrodeln. Promsti* (9), 77–102.

WALKER, N. J. (1972) Distribution of flavour precursors in fractionated milkfat. *N.Z. J. Dairy Sci. Technol.*, **7**, 135–9.

WALKER, N. J. & GRAY, I. K. (1970) The glucosinolate of land cress (*Coronopus didymus*) and its enzymic degradation products as precursors of off-flavor in milk—a review. *J. Agric. Food Chem.*, **18**, 346–52.

WALKER, N. J. & KEEN, A. R. (1976) Lactones in solvent-fractionated milkfat. *Ann. Rep., N.Z. Dairy Res. Inst.*, 1975–76, pp. 36–7.

WALKER, N. J., PATTON, S. & DIMICK, P. S. (1968) Incorporation of [1-^{14}C]acetate into the aliphatic δ-lactones of ruminant milk. *Biochim. Biophys. Acta*, **152**, 445–53.

WALKER, N. J., CANT, P. A. E. & KEEN, A. R. (1977) Lactones in fractionated milkfat and spreadable butter. *N.Z. J. Dairy Sci. Technol.*, **12**, 94–100.

WIGAN, F. (1951) *Judging Milk Products*, P. V. Turk Dairy Publications, Sydney.

WILKINSON, R. A. (1964) Theories of the mechanisms of oxidized flavour development in dairy products. *CSIRO Div. Dairy Res. Internal Rep.*, No. 4, 39 pp.

WILLS, P. A. (1986) Radiation treatment of food. *Nucl. Spectrum*, **2** (2), 5–10.

WILSON, R. D. (1989) Flavour volatiles from New Zealand milkfat. Paper presented to Fats For the Future II, Auckland, NZ, February 13–17, 1989. Int. Conf. Fats, R. Soc. N.Z., Wellington, NZ.

WINTER, M., STOLL, M., WARNHOFF, E. W., GREUTER, F. & BÜCHI, G. (1963) Volatile carbonyl constituents of dairy butter. *J. Food. Sci.*, **28**, 554–61.

WYATT, C. J. & DAY, E. A. (1965) Evaluation of antioxidants in deodorized and nondeodorized butteroil stored at 30°C. *J. Dairy Sci.*, **48**, 682–6.

YAMAGUCHI, S. (1979) The umami taste. In: *Food Taste Chemistry*, Boudreau, J. C. (Ed.), American Chemical Society, Washington, DC (ACS Symp. Ser. no. 115), pp. 33–51.

ZAK, D. L. & KEENEY, P. G. (1976) Changes in cocoa protein during ripening of fruit. Fermentation and further processing of cocoa beans. *J. Agric. Food Chem.*, **24**, 483–8.

ZIEGLEDER, G. (1983) New knowledge of cocoa aroma fermentation and its modification through technical processes. *Lebensmitt. Gerichtliche Chemie*, **37**, 63–9.

BIBLIOGRAPHY

Reviews

ARNOLD, R. G., SHAHANI, K. M. & DWIVEDI, B. K. (1975) Application of lipolytic enzymes to flavor development in dairy products. *J. Dairy Sci.*, **58**, 1127–43.

CUPAKOVA, M. (1979) Review of aromatic substances in milk fat. *Zbornik Prednasok, IV Celostatnaho Sympozia o Aromatickych Latkach v Pozivatinach, Bratislava, 1979*, pp. 68–73.

DAY, E. A. (1966) Role of milk lipids in flavors of dairy products. In: *Flavor Chemistry*, Hornstein, I. (Ed.), American Chemical Society, Washington, DC (Adv. Chem. Ser. No. 56), pp. 94–120.

DEETH, H. C. & FITZ-GERALD, C. H. (1983) Lipolytic enzymes and hydrolytic rancidity in milk and milk products. In: *Developments in Dairy Chemistry, —2. Lipids*, Fox, P. F. (Ed.), Applied Science Publishers, London, pp. 195–239.

DIMICK, P. S., WALKER, N. J. & PATTON, S. (1969) Occurrence and biochemical origin of aliphatic lactones in milk fat—A review. *J. Agric. Food Chem.*, **17**, 649–55.

FORSS, D. A. (1969) Flavors of dairy products: A review of recent advances. *J. Dairy Sci.*, **52**, 832–40.

FORSS, D. A. (1971) The flavors of dairy fats—A review. *J. Am. Oil Chem. Soc.*, **48**, 702–10.

IWAI, M. (1978) Enzymic production of flavour of milk products. II. Production by lipase. *Kagaku To Kogyo (Osaka)*, **52**, 93–9.

JACOBS, M. B. (1946) Diketone components of butter flavors. *Am. Perfumer*, **48** (11), 59, 61, 63.

KARG, J. E. (1983) Specific flavour substances for margarine; their occurrence in milk and milk products. *Seifen, Oele, Fette, Wachse*, **109**, 327–30.

KEPPLER, J. G. (1970) Synthetic flavors for fatty foods. *J. Agric. Food Chem.*, **18**, 988–91.

KEPPLER, J. G. (1977) Twenty-five years of flavor research in a food industry. *J. Am. Oil Chem. Soc.*, **54**, 474–7.

KINSELLA, J. E. (1975) Butter flavor. *Food Technol.*, **29** (5), 82, 84, 86, 88, 90, 92, 96, 98.

KINSELLA, J. E., PATTON, S. & DIMICK, P. S. (1967) The flavor potential of milk fat. A review of its chemical nature and biochemical origin. *J. Am. Oil Chem. Soc.*, **44**, 449–54.

KRUKOVSKY, V. N. (1961) Review of biochemical properties of milk and the lipid deterioration in milk and milk products as influenced by natural varietal factors. *J. Agric. Food Chem.*, **9**, 439–47.

LANG, F. & LANG, A. (1977) New developments in butter and in uses of butterfat. *Milk Ind.*, **79**, (10), 4–5; (9) 19–20.

MANNING, D. J. & NURSTEN, H. E. (1985) Flavour of milk and milk products. In: *Developments in Dairy Chemistry—3. Lactose and Minor Constituents*, Fox, P. F. (Ed.), Elsevier Applied Science, London, pp. 217–38.

MIN, B. D. & SMOUSE, T. H. (Eds) (1985) *Flavor Chemistry of Fats and Oils*, American Oil Chemists' Society, Champaign, IL.

MONCRIEFF, R. W. (1965) The butter flavour—recent work extends knowledge. *Food Process. Mark.*, **33** (401), 51–4.

MOORE, J. H., DOWNEY, W. K., OLIVECRONA, T., COGAN, T. M., JELLEMA, A., FLEMING, M. G., KUZDZAL-SAVOIE, S., CONNOLLY, J. F., MURPHY, J. J., O'CONNOR, C. B. & HEADON, D. R. (1980) Flavour impairment of milk and milk products due to lipolysis. *Bull., Int. Dairy Fed.* No. 118, 76 pp.

MORRISON, W. R. (1970) Milk lipids. In: *Topics in Lipid Chemistry, Vol. 1*, Gunstone, F. D. (Ed.), Logos Press, London, pp. 51–106.

NELSON, J. H., JENSEN, R. G. & PITAS, R. E. (1977) Pregastric esterase and other oral lipases—a review. *J. Dairy Sci.*, **60**, 327–62.

Ramshaw, E. H. (1974) Volatile components of butter and their relevance to its desirable flavour. *Aust. J. Dairy Technol.*, **29**, 110–15.

Richardson, T. & Korycka-Dahl, M. (1983) Lipid oxidation. In: *Developments in Dairy Chemistry—2. Lipids*, Fox, P. F. (Ed.), Applied Science Publishers, London, pp. 241–363.

Shipe, W. F., Lee, E. C. & Senyk, G. F. (1975) Enzymatic modification of milk flavor. *J. Dairy Sci.*, **58**, 1123–6.

Supran, M. K. (Ed.) (1978) *Lipids as a Source of Flavor*, American Chemical Society, Washington DC, (ACS Symp. Ser. No. 75).

Patents Dealing with Cultured-Butter Flavour

Epstein, A. K. & Harris, B. R. (1934) Margarine. US Patent 1,945,347.

Merker, D. R. (1956) Method of imparting a butter-like flavouring to fat-containing food products and the resulting product. US Patent 2,773,772.

Nederlands Instituut voor Zuivelonderzoek (1960) Method for the flavouring of butter and margarine. Netherlands Patent 93,517.

Poppe, M. (1899) German Patent 128,729.

Patents Dealing with Sweet-Cream-Butter Flavour

Boldingh, J., Begemann, P. H., Lardelli, G., Taylor, R. J. & Weller, W. T. (1956) Improvements in or relating to new chemical compounds. British Patent 748,661.

Firmenich & Co. (1963) French Patent 1,319,516.

Folliet, P. (1963) Process for the preparation of dioxane derivatives. Australian Patent Application 27065/63.

Givaudan & Cie (1975) Novel dioxolane derivatives and their use as flavouring agents. British Patent 1352092.

Hatori, T. & Shinoda, A. (1967) Manufacture of flavouring compounds having the aroma of butter. Japan Patent 22 193/67.

Lamparsky, D. (1975) Method for producing novel flavouring substances. Swiss Patent 557 814.

Naarden Int. NV (1979) Preparation of sweet, buttery flavouring for food—by heating fructose and proline in polar solvent. Netherlands Patent 7712745.

Nakel, G. M. (1967) Flavored fat or oil. US Patent 3,336,138.

Soda Koryo Kk (1983) Perfume composition containing straight chain 5-alkenic acid—giving the product an odour like butter, cheese or milk, used in cosmetics and foodstuffs. Japan Patent 58096014.

Taylor, R. J. & Weller, W. T. (1956) Improvements in or relating to new chemical compounds. British Patent 748801.

Unilever, NV (1965) Potential flavoring compounds. Netherlands Patent Application 6 503 576.

Wode, N. G. & Holm, U. (1959) Process for improving the taste and flavor of margarine and other foods and edible substances. US Patent 2,903,364.

Yamamoto, K. (1976) Method for giving foodstuffs a flavour resembling that of a dairy product—8-nonen-2-one. British Patent 1 423 004.

Sweet-Cream Butter Flavour

Kameoka, H., Maseki, M. & Hirao, N. (1974) The constituents of natural butter flavour and synthesis of δ-alkylvalerolactones. *Yukagaku*, **23**, 400–4.

KAWANISHI, G. & SAITO, K. (1965) Studies on the volatile compounds of butter. I. Fresh sweet-cream butter. *Jpn J. Zootech. Sci.*, **36**, 436–42.

KAWASHIRO, I., TANABE, H. & ISHII, A. (1960) Application of gas chromatography for food analysis. I. Fatty acids in butter and cheese. *Shokuhin Eiseigaku Zasshi*, **1**, 78–83.

KUZDZAL-SAVOIE, S. (1968) Unsaponifiable matter of butter. I. Methods. *Ann. Technol. Agric.*, **17**, 115–50.

KUZDZAL-SAVOIE, S., LANGLOIS, D., TROTIER, D. & DZIK, B. (1975) Comparative study of some hydrocarbons in butter and modified butters. *Ann. Falsif. Expert. Chim.*, **68**, 577–617.

LEESMENT, H. (1960) Estimation of malt flavour in milk, starters and butter. *Sven. Mejeritidn.*, **52**, 181–4.

Cultured-Cream Butter Flavour

GRINENE, E. K. (1982) Influence of technological factors on the formation of aromatic properties in cultured butter. *Br. Commun. 21st Int. Dairy Congr.*, Moscow, Vol 1 (bk. 1), p. 328.

KAWANISHI, G. & SAITO, K. (1965) Studies on the volatile compounds of butter. II. Fresh ripened-cream butter. *Jpn J. Zootech. Sci.*, **36**, 443–50.

KAWANISHI, G. & SAITO, K. (1966) Free amino acids and related compounds in sweet- and ripened-cream butters. *Nippon Chikusan Gakkai Ho*, **37**, 430–5.

KIERMEIER, F. & RENNER, E. (1962) Effect of silage on the quality of butter. *Milchwissenschaft*, **17**, 495–8.

Action of Enzymes on Milkfat

KANISAWA, T. (1983) Production of flavours by biochemical methods. II. Production of ethyl ester mixture from butterfat by *Candida cylindracea* lipase. *Nippon Shokuhin Kogyo Gakkaishi*, **30**, 572–8.

KANISAWA, T., YAMAGUCHI, Y. & HATTORI, S. (1982) Production of flavours by biochemical methods. I. Production of dairy flavours by microbial lipase. *Nippon Shokuhin Kogyo Gakkaishi*, **29**, 693–9.

KIHARA, K. & SHIROMOTO, T. (1982) Method for making a butter flavour. Japan Patent 57 41 898 83.

SHIMAMOTO, S. & KUROKI (1974) Butter flavours. Japan Patent 4 922 696.

TANABE SEIYAKU CO. LTD (1970) Butter flavour. Japan Patent 3 187/70.

Ghee

ABD EL-SALAM, M. H., OSMAN, Y. M., FAHMI, A. H. & SHARARA, H. A. (1973) The flavour compounds in samn. II. Effect of processing on the monocarbonyl contents of samn. *Milchwissenschaft*, **28**, 338–40.

FAHMI, A. H., SHARARA, H. A., OSMAN, Y. M. & ABD EL-SALAM, M. H. (1973) The flavour compounds in samn. I. Isolation and characterization. *Milchwissenschaft*, **28**, 223–5.

GABA, K. L. & JAIN, M. K. (1973) A note on the flavour changes in ghee on storage: their sensory and chemical assessment. *Indian J. Anim. Sci.*, **43**, 67–70.

GABA, K. L. & JAIN, M. K. (1975) Organoleptic and chemical evaluation of flavour changes during storage of ghee prepared from fresh and ripened desi butters. *Indian J. Dairy Sci.*, **28**, 278–88.

GABA, K. L. & JAIN, M. K. (1976) A comparative appraisal of the total carbonyls in fresh and stored desi ghee. *Indian J. Dairy Sci.*, **27**, 81–9.

GABA, K. L. & JAIN, M. K. (1976) Head-space carbonyls in fresh and stored desi ghee. *Indian J. Dairy Sci.*, **29**, 1–6.

GABA, K. L. & JAIN, M. K. (1976) Carbonylic flavour profiles of butter oil and ghee and attempts to simulate ghee flavour in butter oil. *Milchwissenschaft*, **31**, 32–4.

LATIF, A., SALIB, A. & KHALIL, F. (1973) Some chemical studies on the odour of samn. II. Physical and chemical characterization of the odour of samn. *Egypt. J. Food Sci.*, **1**, 119–36.

WADHWA, B. K. & JAIN, M. K. (1985) Simulation of ghee flavour in butter oil with synthetic flavouring compounds. *J. Food Sci. Technol.*, **22**, 24–7.

WADHWA, B., BINDAL, M. P. & JAIN, M. K. (1977) Simulation of ghee flavour in butter oil. *Indian J. Dairy Sci.*, **30**, 314–18.

WADHWA, B., BINDAL, M. P. & JAIN, M. K. (1978) Lactone flavour components of ghee. *Br. Commun., XX Int. Dairy Congr.*, **E**, 890–1.

WADHWA, B., BINDAL, M. P. & JAIN, M. K. (1979) A comparative study of the keeping quality of butter oil, flavour induced butter oil and ghee. *Indian J. Dairy Sci.*, **32**, 227–30.

WADHWA, B., BINDAL, M. P. & JAIN, M. K. (1979) Isolation, fractionation and characterization of lactonic components of cow ghee. *Milchwissenschaft*, **34**, 481–3.

WADHWA, B., BINDAL, M. P. & JAIN, M. K. (1980) Variations in lactone profiles of ghee prepared from milks of different species. *Milchwissenschaft*, **35**, 355.

Index

acidolysis reactions 320
acoustic cavitation 51
agglutination 41
air/water interface 34
alcoholysis reactions 320
aldehydes 365, 375
Alfa Laval Centrifixator 122, 127
Alfa Laval Clarifixator 125–7
alkanals 365
alkanols 360, 365, 366
Alkonix system 89
amylose 231
anhydrous butteroil 112
anhydrous milkfat (AMF) 111–54, 245, 261, 284, 286, 289
  addition of hard fraction 146
  alternatives to 113
  blended products 113
  consumer packaging 139–41
  definition 112
  dehydration 131–4
  direct-from-cream process 114–17
  dissolved oxygen in 137–8
  FFA levels 135, 136
  flavour deterioration 136
  handling 137–41
  high-melting fraction 375
  hydrogenation 310
  IDF specification 112
  in recombined products 149
  in spreads 145–6
  manufacture from butter 123–5
  manufacturing processes 114–25
  melting 143
  moisture content 135, 136
  packaging 137–41
  phase inversion 117–23, 125–7
  plant and equipment for manufacture of 125–31
  plant performance monitoring 134–5
  post-dehydration handling 137
  post-pasteurization processing 116–17
  pre-treatment of cream 114–16
  quality assessment 135–7
  raw materials 114
  recombining techniques 141–9
  separation devices 127–31
  uses 112–13
  vegetable oil blends 146–7
  world production 113–14
anionic emulsifiers 231
anisidine value (AV) 136
antioxidants 174, 369
aroma 348
ascorbic acid 389
ascorbyl palmitate 389
Aspergillus 336
Aspergillus niger 339, 340
autoxidation
  formation of volatile products 385–6
  mechanisms of 384–5
  threshold values of products 387

Bacillus cereus 55
Bacillus coli 99
bakery coatings 305
bakery fats 220–9
  functions of 221
backer products 213–53, 286–90
  emulsifiers in 230–1
Barnicoat softening point 10
batter preparation 223
beef tallow blends 330
benzyl mercaptan 372
benzyl methyl sulphide 372
binary diagram 8
boiling point elevation 75
bovolide 356
boxed fats 242
bread improver 240
bread staling 231
buffer storage 91–2
butter 156
  AMF manufacture from 123–5

407

408 Index

butter *contd*
 and allied products 69–109
 body 102
 buggy flavour 99
 bulk packing 92–5
 cholesterol level 205
 colour 101–2
 consistency of 184
 cream cooling and holding 76–7
 cream handling 69–70
 cream treatment 70–77
 crumbly or brittle body 102
 cultured 348–9
 cultured-cream 372
 effect of storage on flavour 377–8
 fishy flavour 100
 flat flavour 99
 flavour 98–101, 348
 flavour defects caused by micro-organisms 99
 hardness 104, 169
 harsh flavour 99
 heated 377
 interesterified 332
 laminations 95
 leaky body 102
 modified 180
 moisture measurement and control 88–9
 mouldy flavour 99
 neutralized or soda flavour 98
 oxidative flavours 100
 oxidative stability 368
 packing and handling 91–6
 palletization 96
 patting 96
 porous or ice-cream texture 102–3
 primrose colour 102
 production 214, 272
 puff pastry 228
 putrefactive taint 99
 rancid flavour 99
 recombined 142, 284
 reduced fat 195
 rheological properties 220–1
 salt incorporation into 91
 scorched or cooked flavour 98
 softening methods 104–6
 soured cream 172
 spotted colour 101–2
 spreadability 183, 184, 205
 spreadable 291
 sticky body 102
 storage flavour 100
 streaky colour 101
 supercritical extraction of flavour 375–6
 sweet-cream 348, 349, 372
 synthetic 360–1
 transport 91
 types 103–6
 unsalted 375
 winter 368
 yield stress 183
butter fractions 252
butter powders 290, 376
 methods of manufacture 239–40
buttercreams 289–90
butterfat 351, 352, 356–60, 371–2
 interesterification 332
 modified 360–1
butterfat fractions 199
 flavours of 375
buttermaking 77–82
 auger angle 86
 beater speed 83–4
 buffer storage 91–2
 buttermilk draining 80
 churning section 80
 computer control 181
 continuous 118, 173
 cooling regime 87
 cream acidity 88
 cream fat content 86–7
 cream feed pump 78
 cream flow rate 85
 cream temperature 87
 cream variables 82, 86–8
 dosing 81
 draining cylinder 85
 fat globule size distribution 87–8
 Fritz-type continuous machines 78–81
 goose neck height 86
 machine variables 82–6
 multiple machines 81
 outlet cone 81
 processing faults 98–9
 processing variables 82–8
 recombined 81–2, 142
 reworking 97
 salt addition rate 86
 salting 81, 89–91
 seasonal factors 88
 setting 96–7
 shock cooling 87

theories of churning 77
two-stage fractionation 106
vacuum working 81
worker speeds 84–5
working 80
buttermilk 23, 80, 130, 141, 197, 331
butteroil 112, 113
butterolein 199

cake mixes 241
cake recipes 223–7
high ratio 227
cakes 287–8
Candida cylindracea 339, 340
candies 312–13
caramelization 274
carbon number 329, 330
carotene 284
β-carotene 101
carotenoids 101, 284
i-carrageenan 47
k-carrageenan 47
casein 32, 44–6
β-casein 33, 35
k-casein 46, 47
casein protein 62
cheese manufacture 149
chemical interesterification *see* interesterification
chicken fat 305
chiller units 176
chocolate 256, 259–74, 292–3, 307–9, 313
flavour of 378
softening by milkfat 261
cholesterol contents 252
churning theories 77
citric acid treatment 304
clumping 41
coalescence 41
cocoa, aroma compounds in 379
cocoa butter 259–74, 292–3, 378–80
solid fat content 260
cocoa butter equivalents (CBE) 322
cocoa butter/milkfat phase diagram 15
cocoa butter replacers (CBR) 274, 322
cocoa butter substitutes (CBS) 307–8, 313, 322, 327
cocoa mass 273
coconut oil 309
coffee cream 55–6
coffee whiteners 241
compound formation 160

condensed milk 148
confectionery industry 255, 259–75
confectionery products 307–9
confocal microscopy 185
contact angle 42
continuous phase in oil/water emulsions 44–50
continuous vacuum dehydrators 132
cooked flavour 373
cooking oils 246–7
cooking products 213–53, 304–5
copper catalysts 303
cottonseed oil 309
cream 29–67, 285
biological membrane 31–2
composite interface 35–7
composition of 54
crystallization 39–44
emulsifiers 37–9
free fatty acid development 53
high fat 118
interfaces 30–9
liqueur 56–7
low-fat 54–7
molecular arrangement at interface 34–5, 39
pasteurization of 116
polysaccharide stabilisers 45
processing variables 50–3
product considerations 53–63
recombined 147–9
ripening 172–3
see also anhydrous milkfat (AMF): butter; buttermaking
croissants 228–9
crystal growth 20–1
crystal network 161–2
crystal separation 280
crystallization 17–22, 39–44, 157–9, 216–19, 280, 292–3
crystallizer units 174, 176, 178

Danish butter concentrate (DBE) 260–1
Danish cookies 288
Danish pastry 228, 288
dark chocolate 313
δ-decalactones 353, 357
decanter 131
deep fat frying 248–51
diagram of oil changes 249
deep-fried foods 362
dehydration 131–4
deodorization 72

desludging 130
detergent fractionation 279
dewaxing 279
dialkyl glyceryl ethers 194
dielectric constant 181
  measurement of 89
differential scanning calorimetry (DSC) 11, 182
dilatometry 182–3
dimethyl sulphide 372
directed interesterification 319, 323, 328, 329
δ-dodecalactones 353
γ-dodecalactones 354
γ-dodec-cis-6-enolactone 353, 354, 355
dropping point 10
dry fractionation 279–80

Economajor 72
edible beef tallow 326
edible fats, off-flavours 384
electrical conductance 185
electron microscopy 184–5
emulsifiers 37–9, 143, 173, 191, 198, 202
  function of 230
  in bakery products 227, 230–1
  lipoprotein-type 198
  low-molecular-weight 37–9, 57, 61
  non-protein 37–9
emulsion sizing 185
emulsion stability 23–4
  and functionality 40–4
emulsions 305–7
  reduced fat 252
  see also oil-in-water emulsions; water-in-oil emulsions
enzymic interesterification 319, 333–40
  of milkfat 339–40
  of non-milkfats 333–9
erucic acid esters 198
ester interchange reaction 320
esters 368
ethylbenzene 381
Eurotium amstelodami 90
eutectic formation 160
evaporating milk 173
extrusion tests 184

fat blends in low-fat spreads 202
fat crystals 160–1
fat globules 61, 87–8, 117, 121, 162, 173
fat globules damage 129

fat powders
  applications of 240–6
  fat content 240
  methods of manufacture 238–9
  non-fat dry solids in 241
fat substitutes 252
fats
  basic principles 4–11
  colour 284
  flavour 281–4
  functional properties 281–4
  physical chemistry 1–27
  physical properties 281
  rheology 157–69
  taste 281–4
fatty acid content, of milkfat 257, 258
fatty acid esters 320
fatty acid profiles, hydrogenated fractions 311
fatty acids 327, 328, 335, 339, 349, 360–1
  flavour thresholds 362
fish oil 307
flaked fats, methods of manufacture 236–8
flash pasteurization 75
flash pasteurizers 72
flavours 347–403
  and oil degradation products 384
  elements of 347–8
  milkfat 347–8
  not formed directly from fat 370–2
  of butterfat fractions 375
  of fats other than milkfat 378–91
  originating from feed 370, 371
  precursors 283, 349–68
  see also off-flavours
flocculation 41
flowing fats 175
fractional crystallization 21–2
fractionated fats, applications 284–93
fractionation 277–93
  definition 277
  processes 278–80
free fat 120, 121, 122
free fatty acids (FFA) 115, 135, 136, 259, 360, 390
fryer design 248–51
frying fats and oils 247–51, 284–6
  monitoring quality of 250
  odours 251
  selection 251
  testing 250
frying products 304–5

gelatin 46
gelling agents 167
Geotrichum candidum 335
ghee 125, 242–5, 374–5
   flavour 243–4
   free fatty acid content 243
   keeping quality 244
   methods of manufacture 242–3
   quality control 243–5
   storage 244
   uses 245
globular fat 206, 221
globule disintegration 51
globule distribution 52
gluten 231
glyceride composition 330, 331
groundnut oil, flavour of 381

δ-hexalactone 352
hexyl acetate 381
high melting fractions (HMF) 12–13, 291, 293
   recombination 272
high melting glycerides (HMG) 328, 329
higher melting component (HMC) 16
HLB number 38, 166, 306
hollow gelatin spheres 194
homogenization 50–2, 119, 121
H_2S 367–8
   flavours due to 373
Humicola lanuginosa 336
hydraulic cavitation 51
hydrocolloids 197
hydrogenated fats 274
hydrogenated milkfat products 309–13
hydrogenation 293–313
   catalyst activity 294–5, 297–9
   catalyst re-use 296–7
   definition 277
   food applications 299–309
   mass transfer mechanisms 296
   milkfat 13, 14
   process 294–9
   reaction mechanisms 294
   selective 299, 305, 312
   selectivity 295–6
   technology and equipment for 294
hydrolytic rancidity 390
hydroxyacids 349–57

ice cream 29–67, 285–6
   biological membrane 31–2
   composite interface 35–7
   composition 54, 61
   compositional and processing changes 61
   crystallization 39–44
   destabilizing 62
   emulsifiers 37–9
   interfaces 30–9
   molecular arrangement at interface 34–5, 39
   polysaccharide stabilizers 45
   processing 60–3
   processing variables 50–3
   product considerations 53–63
   recombined 147
   soft scoop 62–3
ice crystals 48–50, 62
ice recrystallization 49
IDF Oven Method 88
infant feeds 290
infra-red determination of fat content 181
instant desserts 241
Instron compression testing 332
interesterification 190, 319–45
   butter 332
   butterfat 332
   catalysts for 324
   definition 319
   directed 319, 328, 329
   enzymic 333–40
   function of 320–2
   milkfat 327–33
   non-milkfat 325–7
   random 319, 322, 329
   raw materials 324–5
   reactions 320
   types of 319, 322
   with vegetable oils 180
interfacial properties 22–4
iodine value 294, 305, 308, 312, 313
ionizing radiation 368–9
irradiated fats, off-flavours from 368–9
ISO-NMR curves 261
isosolid diagrams 14, 16

jojoba oil 307

Karl Fischer method 135
β-ketoacids 357–60
β-ketoacids-glycerides 359
ketonic rancidity 390
kitchen products 213–53

α-lactalbumin 33, 35, 46, 304–5
lactic butter 103
β-lactoglobulin 33, 35, 46
lactones 283, 284, 356, 357
  flavour thresholds and optimum levels 352
  δ-lactones 351, 354, 357
  γ-lactones 351, 352, 354–6
lactose crystallization 49
lard 325–6
lauric oils 306, 326
LDL (low density lipoproteins) 332
lecithin 174
Leuconostoc cremoris 501 103
light microscopy 185
linoleic acid 305, 308, 355, 385, 386, 390
linolenic acid 303
lipases 333, 336, 337, 338, 339, 359
lipids 167–8
  Class I 22
  Class II 22
  Class III 22
  interaction with water 22–3
lipolysis 114–15
lipoprotein-type emulsifiers 198
liquid oils 281
loop-train-tail model 35
low melting fractions (LMF) 12–13, 284, 291
  recombination 272
low-molecular-weight emulsifiers 37–9, 57, 61
lower fatty acids 360–2

Maillard reaction 274, 379
margarine 24, 143, 145, 156, 163, 164, 174, 177, 284
  blended 219
  flavour 356
  for bakery use 220
  hardness 169
  manufacture 306, 326
  oils 326, 327
  plasticity of 225
  processing 219–20
  production 215–16
  puff pastry 228
  quality control 232–4
  rheological properties 220–1
mayonnaise 247
melting curves 20
melting point 9–11
  methods of determination 10

methyl ketones 283, 358–9
methyl polysiloxane 250
MFGM 31–3, 58
  heat treatment 31
microencapsulation 239
microwave heating 175
middle melting fraction (MMF) 12–13
milk
  compounds isolated from 371
  free fatty acids in 259
  recombined 147–9, 149
  surface-active components in 32–3
milk chocolate 256, 274–5, 292, 313
milk composition 256–7
milk crumbs 273
milk powder 272–3, 273
milk protein interface 32–4
milk solids not fat (MSNF) 61, 62, 143, 148
milk/vegetable fat products 205
milkfat 1–2
  and fractions 11–14
  blends with beef tallow 330
  chemical composition 2
  enzymic oxidation 363
  fatty acid composition 3
  fatty acid content 257, 258
  flavour 259, 274, 312
  fractionation 14, 180, 272, 274
  functional properties 257, 283
  hydrogenated 13, 14
  mixtures with other fats 14–15
  modified 273–4
  phase behaviour of 12
  physical properties 3, 11–17
  rapid cooling 12
  recombination 272
  re-esterification of 329
  softening points 328, 329–30
  solubility in liquid oils 16–17
  summer 293
  supercritical extraction of 375–6
  trans unsaturation content in 259
  winter 293
  see also anhydrous milkfat (AMF) *and under specific applications*
milkfat-in-water emulsions 23
moisture content, monitoring 234
moisture measurement and control 88–9
molten emulsions 167
monitoring of AMF plant performance 134–5
monoglycerides 23, 167, 174, 198

Mucor miehei 191, 335, 336, 338–40
multiple fractionations 282

natural fats, phase behaviour of 6–9
neutral volatiles 373–5
neutralization 70
nickel catalysts 297, 301, 303, 304, 312
nickel–copper chromite catalysts 304
NIZO process 114
non-protein emulsifiers 37–9
Nuclear magnetic resonance (NMR)
 techniques 10, 175, 181–3, 233
nucleation 19–20
 fat crystallization 175
 nutritional attitudes 251–2

δ-octalactones 352, 353
off-flavours
 edible fats 384–5
 factors affecting rate of development 388–90
 from irradiated fats 368–9
oil degradation products and flavour 384
oil exudation 184
oil-in-water emulsions 30, 34, 117, 130, 149
 continuous phase in 44–50
 instability in 40–4
 see also water-in-oil emulsions
oil-in-water interface 23
oleic acid 135, 385
olive oil 247, 380–1
 organoleptic characteristics 380
 volatile components of 380–1
optical microscopy 184
oxidation flavours 364, 365

palm kernel oil 303, 308
palm oil 306, 307–8
palm olein 308
partial esters 168
pasteurization 70–1, 331, 364
 of cream 116
 temperature 88
peanut butter 306, 307
penetrometry 183
Penicillium roqueforti 359
phase behaviour 5
 of milkfat 12
 of natural fats 6–9
phase concept 5–6
phase diagrams 6, 9, 219

phase inversion 117–23, 125–7
phenolic compounds 372, 381
phospholipids 23, 389–90
Pickering type mechanism 166
plain chocolate 256, 273–4, 292
plastic deformation 165
plastic fats 281
plasticity 286–7
polymorphism 4–5
polyol fatty acid esters 194
polysaccharides 44, 47–8, 167, 171, 193
polyunsaturated fatty acids 252
powdered fats
 applications of 240–6
 methods of manufacture 236–8
pregastric esterases 360–1
Pricat 9906 298–9
protein/polysaccharide interactions 46–7
proteins 44–6
 in low fat spreads 196–7, 201
Pseudomonas 372
Pseudomonas fluorescens 339, 340
Pseudomonas fragii 337
Pseudomonas putrefaciens 99
puff pastry 227–9, 289
PV test 136

random interesterification 319, 322, 329
rapeseed oil 279, 302, 304, 305, 339
rapid cooling 21, 160
re-esterification of milkfat 329
reversion flavours 390
rework system 174
Rhizopus arrhizus 336, 337
Rhizopus delemar 336
Rhizopus niveus 337
rice bran oil 308–9

salad dressings 247, 284–6, 380
salad oils 246–7, 284–6, 299–304
salt 89–91
 particle size 90–1
 slurries 90
 storage 90
samna balady 125
saturated fats 252
saturation–supersaturation diagram 18
sauces 247, 290
scraped wall heat exchangers 173–5
SDS 42
separation devices 127–31
separation process 52–3

Shainin control chart 89
short pastry 221–3
shortenings 215–16, 219, 284, 287
   fluid 235–6
   for bakery use 220
   liquid 234–5
   plasticity of 225
   processing 219–20
   production 325
   quality control 232–4
   slurry 236
single cream 54–5
skim-milk 130, 141, 143, 197
skim-milk powder (SMP) 148
slip point 10
sodium caseinate 56
sodium methoxide 326
sodium–potassium alloy catalyst 327
solid fat content (SFC) 3, 9–11, 15, 298, 299
solid fat index (SFI) 306
solid solutions 6, 12
solids measurement in spreads 181–3
solvent fractionation 279
soy trisaturate blends 326
soyabean oil 279, 299, 300, 303–5, 326
spontaneous oxidation 368
spray driers 238
spreads 155–211, 290, 305–7
   alterations in properties 179–81
   AMF in 145–6
   appearance of 185–6
   aqueous phase 162–3, 196, 197
   churning of butter type products 171–3
   composition of 157
   consumption patterns 157
   double emulsion low fat 196
   double emulsion products 191
   effect of fats on product texture 186–7
   effects of emulsifiers 191
   fat blends 199
   fat continuous 169–71, 194–200, 204
   fat solids indices 179
   fat substitutes 192–4
   functions of 187, 189
   granular fat structure 181
   health aspects 191–2
   high fat 172, 180, 189–90
   low calorie 193
   low fat 163, 164, 167, 171, 181, 194–202
   major categories 157
   microbiological stability 199–200
   miscellaneous developments 191
   multifunctional 205
   natural 206
   novel textured 206
   nutritionally adapted 204–5
   palate behaviour 187–9
   performance of 181–9
   phase behaviour 159–60
   post-manufacture plasticizing 180
   processing 169–81, 206
   processing and product structure relationship 175–9
   research and development 20–7
   rheology 157–69
   solids measurement in 181–3
   structure 163–5, 168–9, 189
   structure tests 183–5
   very low fat 202–4
   viscosity enhancement in aqueous phase 203
   water continuous 171, 200–2, 204
   yellow fat 291–2
stabilizers 197, 201–2
   polysaccharide 44
stirred crystallisers 173–5
Streptococcus cremoris 270 103
Streptococcus diacetylactis DRC_1 103
sucrose fatty acid polyesters 193
sugar esters 193
supercritical extraction of milkfat 375–6
supersaturation 17–19
surface-active components 32–3, 167–8
surface-active lipids 168
sweet whey 197

taint removal 71–5
tainting substances 71–2
taste 347–8
δ-tetradecalactone 353
Thermomyces ibandanensis 336
thiobarbituric acid test (TBA) 136
Tirtiaux process, fractionation 280
toffee 274–5
total fat 120, 121
Totox value 136
trans unsaturation content in milkfat 259
triacylglycerols 291, 293
triglycerides 12, 159, 168, 190, 193, 216–19, 320, 324, 329, 330, 335, 336, 339, 349, 384, 385

classification 13
high melting 331
trilinolein 356, 387
trilinolenin 387
triolein 387
trisaturated glycerides 329
trisaturated triglycerides 330
tristearin 216

UFT milk 41, 148
UHT cream 55
UHT whipping cream 59, 60
Ultitem 72
ultrasonic techniques 182
unsaturated fatty acids 362–8

vacreation 75, 88
Vacreator 72, 75
vacuum pasteurization 71, 75
valve homogenizer 50–2
vanaspati 245–6
 formulation 246
 manufacture 246
 oxidative stability 246
 texture requirements 246
vegetable fats 298
vegetable gums 202
vegetable oil 245, 304, 390
 blends 146–7
 churning with 181
 interesterification with 180

vegetable oil based low-fat spread 201
vinyl ketones 367
volatile components of olive oil 380, 381
volatile compounds 283, 349, 373–5
Vologda butter 377
votator A-units 174

water-in-oil emulsions 24, 166, 198
 see also oil-in-water emulsions
whey 141
whey proteins 44–6, 62
whipped toppings 241
whipping cream 57–60, 285
white chocolate 378
whole-milk powder (WMP) 148
Wiley melting point 10
winterization 279
work softening 169, 224

xanthan gum 202
xanthine oxidase/lactoperoxidase system 363–4
X-ray diffraction 11, 159
X-ray powder spectroscopy 158

yellow fat spreads 291–2
yoghurts 149